ORGANIC CHEMISTRY SERIES
Series Editor: J E Baldwin, FRS

Volume 1

Stereoelectronic Effects in Organic Chemistry

Related Pergamon Titles of Interest

BOOKS

Organic Chemistry Series

Volume 2
DAVIES: Organotransition Metal Chemistry: Applications to Organic Synthesis

Volume 3
HANESSIAN: Synthetic Design with Chiral Templates

Major Works

BARTON & OLLIS: Comprehensive Organic Chemistry
KATRITZKY & REES: Comprehensive Heterocyclic Chemistry
WILKINSON et al: Comprehensive Organometallic Chemistry

Also of Interest

BARTON et al: R B Woodward Remembered
COETZEE: Recommended Methods for Purification of Solvents and Tests for Impurities
FREIDLINA & SKOROVA: Organic Sulfur Chemistry
HERAS & VEGA: Medicinal Chemistry Advances
MIYAMOTO & KEARNEY: Pesticide Chemistry: Human Welfare and the Environment
NOZAKI: Current Trends in Organic Synthesis
PERRIN et al: Purification of Laboratory Chemicals, 2nd Edition
RIGAUDY & KLESNEY: Nomenclature of Organic Chemistry, "The Blue Book"
TROST & HUTCHINSON: Organic Synthesis: Today and Tomorrow

JOURNALS

Chemistry International (IUPAC news magazine for all chemists)
Polyhedron (primary research and communication journal for inorganic and
    organometallic chemists)
Pure and Applied Chemistry (official IUPAC research journal for all chemists)
Tetrahedron (primary research journal for organic chemists)
Tetrahedron Letters (rapid publication preliminary communication journal for organic
    chemists)

*Full details of all Pergamon publications/free specimen copy of any Pergamon journal available on reque
from your nearest Pergamon office.*

Stereoelectronic Effects in Organic Chemistry

PIERRE DESLONGCHAMPS

Département de Chimie, Université de Sherbrooke, Québec, Canada

PERGAMON PRESS

OXFORD · NEW YORK · TORONTO · SYDNEY · PARIS · FRANKFURT

U.K.	Pergamon Press Ltd., Headington Hill Hall, Oxford OX3 0BW, England
U.S.A.	Pergamon Press Inc., Maxwell House, Fairview Park, Elmsford, New York 10523, U.S.A.
CANADA	Pergamon Press Canada Ltd., Suite 104, 150 Consumers Road, Willowdale, Ontario M2J 1P9, Canada
AUSTRALIA	Pergamon Press (Aust.) Pty. Ltd., P.O. Box 544, Potts Point, N.S.W. 2011, Australia
FRANCE	Pergamon Press SARL, 24 rue des Ecoles, 75240 Paris, Cedex 05, France
FEDERAL REPUBLIC OF GERMANY	Pergamon Press GmbH, Hammerweg 6, D-6242 Kronberg-Taunus, Federal Republic of Germany

First edition 1983

British Library Cataloguing in Publication Data
Deslongchamps, Pierre
Stereoelectronic effects in organic chemistry.—
(Organic chemistry series; vol. 1)
1. Stereochemistry 2. Chemistry, Physical
organic
I. Title II. Series
547.1'223 QD481

ISBN 0-08-026184-1 (Hardcover)
ISBN 0-08-029248-8 (Flexicover)

In order to make this volume available as economically and as rapidly as possible the author's typescript has been reproduced in its original form. This method unfortunately has its typographical limitations but it is hoped that they in no way distract the reader.

Printed in Great Britain by A. Wheaton & Co. Ltd., Exeter

*To those who contributed to the development
of the concept of stereoelectronic effects*

"The first condition to be fulfilled by men of science, applying them-
selves to the investigation of natural phenomena, is to maintain absolute freedom of mind, based on philosophical doubt. Yet we must
not be in the least sceptical; we must believe in science, *i.e.*, in
determinism; we must believe in a complete and necessary relation
between things, among the phenomena proper to living things as
well as in others; but at the same time we must be thoroughly con-
vinced that we know this relation only in a more or less approximate
way, and that the theories we hold are far from embodying change-
less truths. When we propound a general theory in our sciences we
are sure only that, literally speaking, all such theories are false.
They are only partial and provisional truths which are necessary
to us, as steps on which we rest, so as to go on with investigation;
they embody only the present state of our knowledge, and consequently they must change with the growth of science, and all the more
often when sciences are less advanced in their evolution."

(from **Claude Bernard,** *1865)*

FOREWORD

The development, during the past thirty years, of the electronic theory of organic chemistry combining the original Robinson theory with the conformational developments of Barton and others, has led to the stereoelectronic theory of organic chemistry. This approach to the understanding of reactivity of organic compounds is used daily by organic chemists, particularly those who have to deal with structurally complex molecules. Unfortunately much of the underlying experimental basis for this general approach is spread widely through the vast literature of the recent period.

In this book Pierre Deslongchamps, an investigator who has made seminal contributions to the stereoelectronic theory, brings together the experimental data and his conclusions derived from them. It is particularly timely that this be done since, by their very nature, stereoelectronic effects are quite subtle and to see the thread of their influences a large body of experimental data is essential. A similar circumstance applied during the formative years of Robinson's theory.

This book will be particularly valuable to all investigators working with complex organic molecules, whether they be synthetic, medicinal or bioorganic chemists, since it will provide a timely view of stereoelectronic effects and how they may be applied, both to rationalise and to predict organic chemical reactivity.

Professor J. E. Baldwin, FRS

University of Oxford
Dyson Perrins Laboratory

ACKNOWLEDGEMENTS

It is with pleasure that I thank graduate students: C. Bayly, N. Beaulieu, G. Bérubé, M. Caron, L. Deschênes, D. Guay, Y. Nadeau and B. Roy, and post-doctorate collaborators, Drs P.M. Bishop and D.A. Schwartz for their constructive comments. Similar words of appreciation belong to my research associate Drs L. Ruest and P. Soucy and to my colleagues at the Chemistry Department: Professors J. Lessard, R.J. Taillefer, and J.K. Saunders. I also express my sincere thanks to Dr S.E. Thomas for the numerous suggestions she has made throughout the writing of this book.

I should like also to thank Professors B. Belleau (McGill), J.M. Cook (Wisconsin-Milwaukee), M.E. Kuehne (Vermont), S.G. Levine (North Carolina), M.W. Makinen (Chicago), and J.D. Wuest (Montreal) who made several suggestions. Particular words of thanks are due to Professors D. Gravel (Montreal), P. Lavallée (Sherbrooke), and Z. Valenta (New Brunswick) who read and corrected the entire draft manuscript.

The drawing of the chemical structures, the typing, and the preparation of the camera-ready manuscript was the master-work of Mrs M.-M. Leroux, to whom I wish to express my greatest gratitude.

CONTENTS

CHAPTER 1

INTRODUCTION

The organic chemist made an important step in the understanding of chemical
reactivity when he realized the importance of electronic stabilization caused by the delocalization of electron pairs (bonded and non-bonded) in organic molecules. Indeed, this concept led to the development of the resonance
theory for conjugated molecules and has provided a rational for the understanding of chemical reactivity (1, 2, 3). The use of "curved arrows" developed 50 years ago is still a very convenient way to express either the
electronic delocalization in resonance structures or the electronic "displacement" occurring in a particular reaction mechanism. This is shown
by the following examples.

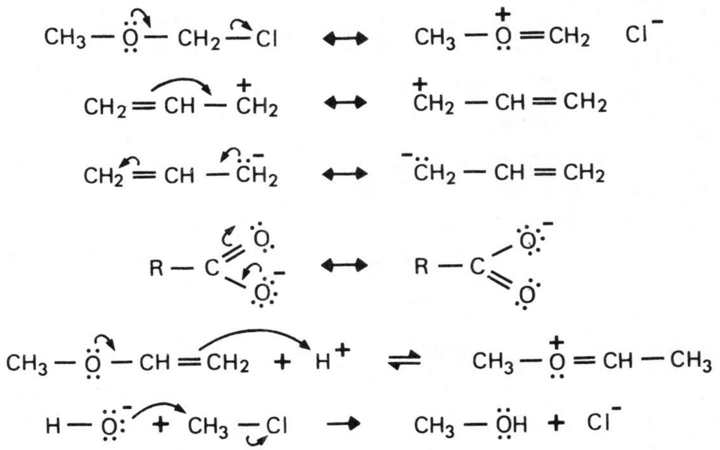

$$CH_3 - \ddot{O} - CH_2 - Cl \longleftrightarrow CH_3 - \overset{+}{\underset{..}{O}} = CH_2 \quad Cl^-$$

$$CH_2 = CH - \overset{+}{C}H_2 \longleftrightarrow \overset{+}{C}H_2 - CH = CH_2$$

$$CH_2 = CH - \ddot{C}H_2^- \longleftrightarrow \overset{-}{\ddot{C}}H_2 - CH = CH_2$$

$$R - C \overset{\ddot{O}}{\underset{\ddot{O}:^-}{\lessgtr}} \longleftrightarrow R - C \overset{\ddot{O}:^-}{\underset{\ddot{O}}{\lessgtr}}$$

$$CH_3 - \ddot{O} - CH = CH_2 + H^+ \rightleftharpoons CH_3 - \overset{+}{O} = CH - CH_3$$

$$H - \ddot{O}:^- + CH_3 - Cl \longrightarrow CH_3 - \ddot{O}H + Cl^-$$

In recent years, experimental evidence has been accumulated which show that this kind of electronic interaction takes place only when the electron pairs are properly oriented in space. Indeed, many results indicate that the reactivity of most types of organic molecules depends upon relative stereochemistry of particular electron pairs, bonded or non-bonded. As a result, the reactivity and the conformational analysis of molecules especially those containing heteroatoms are better understood. It is also through the consideration of the concept of stereoelectronic effects that it becomes possible to acquire the knowledge of the stereochemistry of the transition states of most organic reactions.

This monograph is an attempt to put together all the work which has provided experimental support for the concept of stereoelectronic effects. Reference to theoretical calculation upholding this concept has also been included. The hope of the work is to convince organic chemists that stereoelectronic effects are indeed important and that they should be routinely taken into account when analyzing either the conformation of a particular compound or the course of an organic reaction.

Hydrolysis processes which are key reactions in the biological system are described in the next three Chapters; Chapter 2 covering acetals, Chapter 3 esters, and Chapter 4 amides and related functions. The following three Chapters cover reactions often used by the synthetic organic chemist. Chapter 5 describes reactions which take place at saturated carbon atoms, Chapter 6 deals with reactions on sp_2-type unsaturated systems and Chapter 7 discusses reactions on sp-type unsaturated systems. Chapter 8 is a potpourri of various synthetic organic reactions which could not be appropriately described in the preceding three Chapters. This Chapter describes also how to use stereoelectronic effects to design organic molecules having unusual reactivities and to develop new strategies in organic synthesis. The last Chapter entitled Biological Processes, i.e., Chapter 9, points out the importance of stereoelectronic effects in enzyme-catalyzed reactions and in the understanding of the relationship between chemical structure and biological activity.

It is suggested to read Chapters 2, 3, and 4 first because I believe that the importance of stereoelectronic effects described in the following Chapter will be better appreciated.

REFERENCES

(1) Ingold, C.K. "Structure and Mechanism in Organic Chemistry"; 2nd Ed.;
    Cornell University Press: Ithaca, N.Y., 1969, Chapter 2.
(2) Robinson, Sir R. "Memoirs of a Minor Prophet"; Elsevier: Amsterdam,
    1976, Chapter XI.
(3) Pauling, L. "The Nature of the Chemical Bond"; 3rd Ed.; Cornell Univer-
    sity Press: Ithaca, N.Y., 1960.

CHAPTER 2

ACETALS AND RELATED FUNCTIONS

Conformation of acetals

An acetal function can adopt any of the nine gauche conformations described in Fig. 1. Conformers A, B, and D are the mirror images of conformers A',

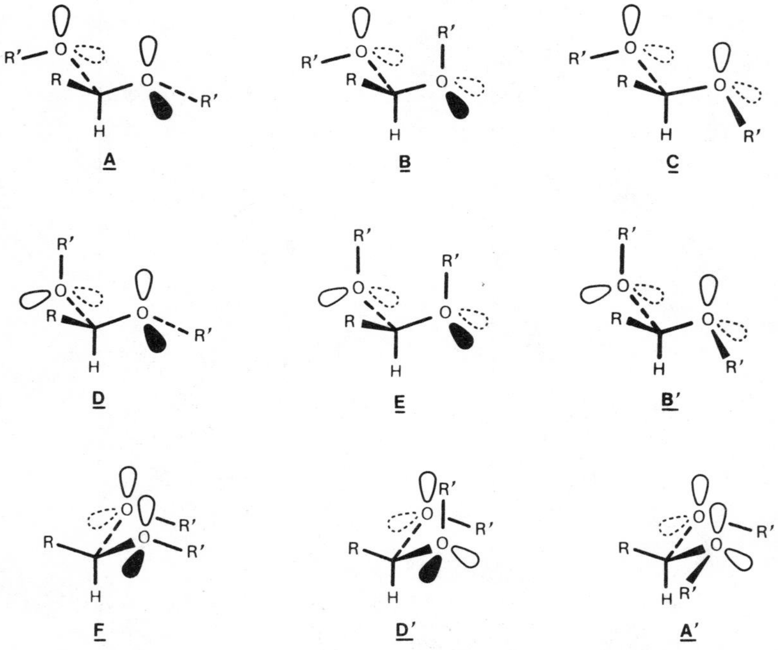

Fig.1

B', and D' respectively. The remaining conformers C, E, and F possess a plane of symmetry. Consequently, the acetal function can in principle exist in the six different conformations A-F. Experimental evidence will be described which shows that the relative stabilities of these various conformers depend on stereoelectronic effects and the standard steric interactions.

Stereoelectronic effects have long been recognized to influence the configuration and the conformation of acetals, particularly in carbohydrates where these effects were first discovered and discussed in terms of the anomeric and the exo-anomeric effects (1-3). The term anomeric effect introduced by Lemieux in 1958 (4) refers to the tendency of an alkoxy group at C-1 of a pyranose ring to assume the axial rather than the equatorial orientation despite unfavorable steric interactions, whereas the term exo-anomeric effect also introduced by the same author (5) concerns the preferred orientation of the O—R bond of the alkoxy group at the anomeric center.

There have been two schools of thought concerning the origin of the anomeric effect (1-3). The first (6-8), considers this electronic effect as destabilizing, due to repulsion by dipole-dipole or electron pair - electron pair (rabbit ear effect) interactions which can be represented by the double headed arrow in structure 1. The second school considers the anomeric effect as a stabilizing electronic effect (9-13) which occurs when an electron pair of an oxygen atom is oriented antiperiplanar to a polar C—X (X= OR, NR_2 or halogen) bond. Stabilization would then be gained by the partial transfer of an electron pair of one heteroatom to another electronegative atom; this electronic transfer being illustrated by the arrows in structure 2.

In practice, it makes little difference whether one believes that the anomeric effect is destabilizing or stabilizing since one can arrive at the same conclusion concerning the relative stability of conformers or isomers. For instance, for a molecule which can take the two conformations 1 and

2, if the anomeric effect is stabilizing, one concludes that 2 is more stable than 1 because 2 has a stabilizing anomeric effect. If one accepts that
the anomeric effect is destabilizing, one concludes that 2 is more stable
than 1 because 1 has a destabilizing anomeric effect.

It is possible and probably very likely that both types of electronic effects are occurring in the acetal function. In other words, 2 could be more
stable than 1 because 2 is stabilized relative to 1 by a partial electron
transfer and 1 is destabilized relative to 2 by electronic repulsions. There
is presently no experimental technique to differentiate between the two
effects. At the present time, many chemists, including myself, prefer to
consider the anomeric effect as a stabilizing rather than a destabilizing
effect. The main reason is that the concept of stabilization of a system
through electronic delocalization is a well established principle in organic chemistry. The resonance theory is indeed based on this principle. I
believe that this concept rather than the dipole - dipole or electron pair
- electron pair repulsions allows the organic chemist to rationalize his
results better.

The anomeric effect in terms of a stabilizing effect can be illustrated
by the concept of "double-bond - no-bond resonance" (14, 15) shown by the
resonance structures 4 and 3 or by the equivalent modern view (16, 17) that
this electronic delocalization is due to the overlap of an electron pair
orbital of an oxygen atom with the antibonding orbital of a C — OR sigma
bond (12).

It has also been suggested (18, 19) that a distinction should be made between the two lone pairs of a given ether-oxygen atom, one being more 2p-
like and a better donor than the other lone pair which is more sp^2-like
and a poorer donor. As a consequence of this, the 2p-like lone pair of the
oxygen atom in a tetrahydropyran ring, for instance, cannot be perfectly
antiperiplanar to an axial alkoxy group (12, 20). I prefer to consider that

the non-bonded electron pairs are equivalent and that the ether oxygen is tetrahedral and thus similar to the tertiary amine nitrogen and the tetra-substituted carbon atom (cf. valence-shell electron pair repulsion theory (21)). However, in the case of the acetal function, when an oxygen atom has one of its electron pairs oriented antiperiplanar to a C — OR bond as in 3, the other resonance structure 4 can be written and the actual situation should correspond to the hybrid of these two resonance structures. The oxygen atom which is donating electrons forms a partial double-bond with the carbon atom, and is therefore partially sp^2 hybridized. Normally, the organic chemist does not write all the resonance structures; he writes one single structure, and applies the resonance theory in his mind. I consider this good practice. I also prefer to use the term "non-bonded electron pair" in place of "lone pair orbital" as it expresses the basis for chemical reactivity (vide infra) in a more precise manner.

The first precise evaluation of the anomeric effect was realized by Descotes and co-workers in 1968 (22). These authors have studied the acid catalyzed isomerization of the cis and trans bicyclic acetals 5 and 6 and found that, at equilibrium, the mixture contains 57% cis and 43% trans at 80°C. The cis isomer is therefore more stable than the trans by 0.17 kcal/mol. The cis isomer 5 has one (stabilizing) anomeric effect whereas the trans isomer 6 has none. Steric interactions in cis acetal 5 were estimated to be 1.65 kcal/mol (one gauche form of n-butane, 0.85 kcal/mol and an OR group axial to cyclohexane, 0.8 kcal/mol). By subtracting an entropy factor (0.42 kcal/mol at 80°C) caused by the fact that the cis acetal 5 exists as a mixture of two conformations (cis decalin system), they arrived at a value of 1.4 kcal/mol for the anomeric effect.

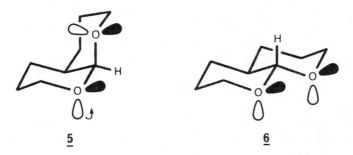

<u>5</u>   <u>6</u>

More recently, the equilibration of the conformationally rigid cis and trans tricyclic acetals 7 and 8 was carried out (23). The cis acetal 7 is less stable (45% at equilibrium) than the trans acetal 8 by 0.14 kcal/mol. Taking into account a value of 1.65 kcal/mol for the steric interactions in **7**, the anomeric effect becomes 1.5 kcal/mol which confirms the value of Descotes.

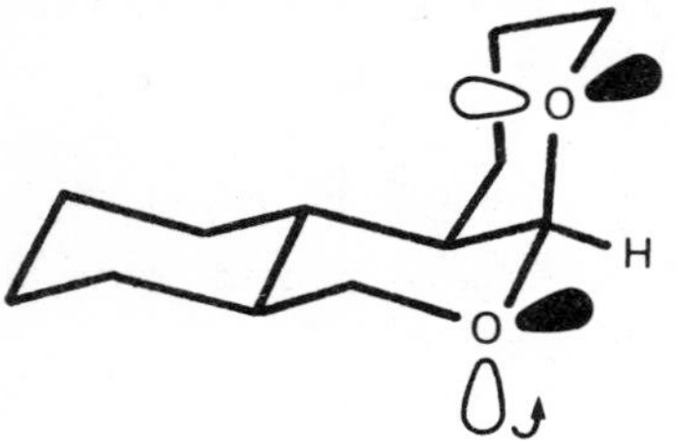

7 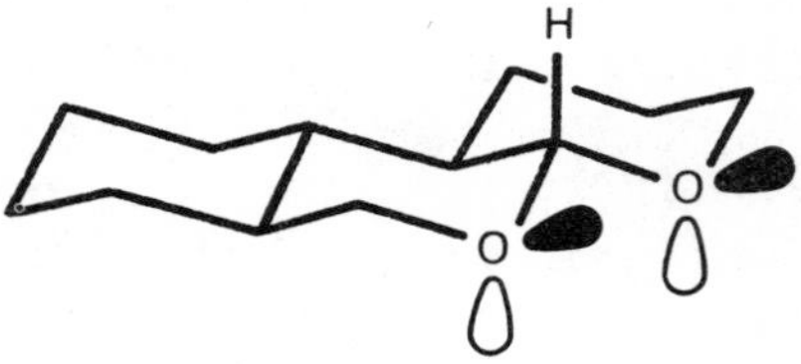   **8**

The first precise evaluation (24, 25) of both the anomeric and the exo-anomeric effects was obtained by studying 1,7-dioxaspiro[5.5]undecane (**9**) (Fig. 2). With this system, conformational analysis by low temperature nmr spectroscopy was possible because each conformational change involves a chair inversion which has a relatively high energy barrier. The steric effect could also be easily evaluated, and by adding appropriate alkyl substituents, it was theoretically possible to isolate isomeric compounds which would exist in different conformations.

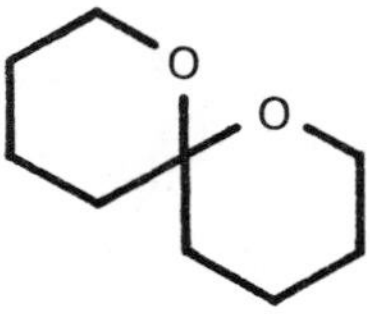

9

9A   ⇌   **9**B   ⇌   **9**C

Fig. 2

Compound **9** can exist in the three conformations **9A**, **9B**, and **9C**. Two oxygens in conformer **9A** have an electron pair antiperiplanar to a C−O bond, while there is only one such oxygen in conformer **9B** and none in conformer **9C**. Thus, **9A** has two anomeric effects, **9B** has one and **9C** has none. The assumption was made that, when there are two anomeric effects in the same conformation, they are additive. Accepting the value of Descotes of 1.4 kcal/mol for one anomeric effect, and considering only electronic effects, then **9A** and **9B** should be more stable than **9C** by 2.8 and 1.4 kcal/mol respectively.

The steric effects were then evaluated in the three conformations. In conformer **9A** the two oxygens are oriented axially. In conformer **9C**, there are two methylene groups oriented axially and in conformer **9B**, there is one oxygen and one methylene group axially oriented. When a methylene group is axially oriented it corresponds to two gauche forms of butane evaluated at 0.9 kcal/mol each. When an oxygen is axially oriented, it corresponds to two gauche forms of $\underline{n}$-propyl ether $(O - CH_2 - CH_2 - CH_2)$ which were evaluated at 0.4 kcal/mol each. Using these values, the steric effects for conformers **9A**, **9B**, and **9C** are evaluated at 1.6, 2.6, and 3.6 kcal/mol respectively. By taking the stabilizing anomeric and the destabilizing steric effects together, conformer **9A** should be more stable than conformers **9B** and **9C** by 2.4 and 4.8 kcal/mol respectively.* Thus, this analysis led to the prediction that the spiro compound must exist essentially in the conformation **9A** only. This prediction was confirmed experimentally by a ^{13}C nmr study at low temperature which clearly showed that compound **9** exists in the conformation **9A** only (24, 25).

The methyl substituted spiro system **10** (Fig. 3) was also studied (24, 25). With this system, two isomers **11** and **12** are possible and molecular models show that they can each exist in four different conformations. Evaluation of the anomeric and the steric** effects of each conformation leads to the prediction that isomer **11** exists in the conformation **11A** only (0 kcal/mol relative to the other conformations) whereas isomer **12** is a mixture of a major (**12A**, 2.4 kcal/mol) and a minor (**12B**, 2.9 kcal/mol) conformer. However, since isomers **11** and **12** are interconvertible (**11** can be converted into the mirror image of **12** by opening and reclosure of the acetal function) and

---

*The relative stabilities are described in terms of relative energy.

**Values of 4.0 and 3.0 kcal/mol were used for the steric interactions of a methyl group 1,3-diaxial to methyl (or methylene) and to oxygen respectively (24).

since **11A** is more stable than **12A** by 2.4 kcal/mol, only isomer **11** should be formed under thermodynamically controlled conditions, and it should exist in the conformation **11A** only. This prediction was completely verified experimentally.

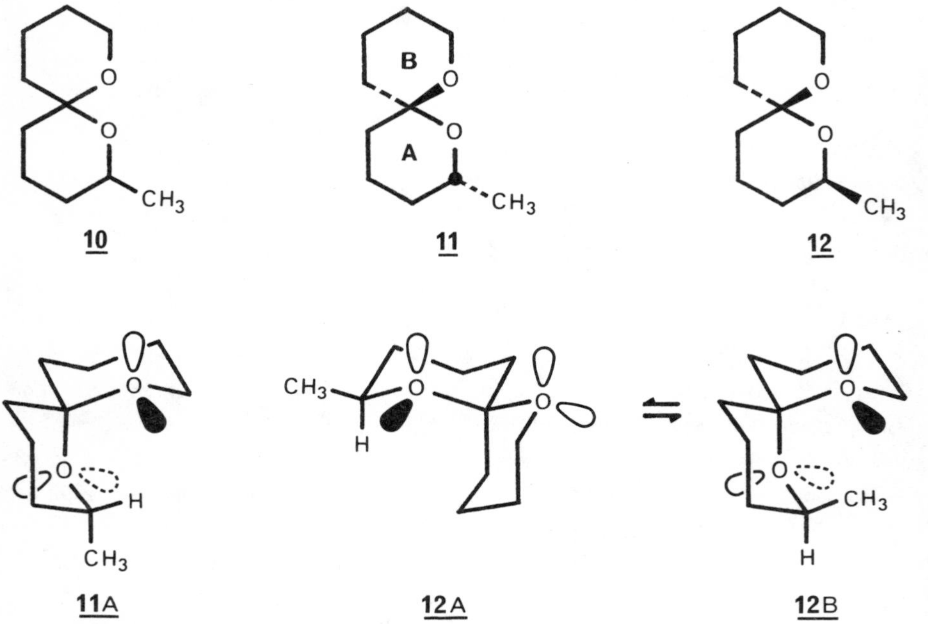

Fig. 3

The spiro compound having two methyl groups was also studied (Fig. 4) (24, 25). With this system there are three possible isomers; two of them (**15** and **16**) are formed from the cyclization of the d*l* dihydroxyketone **13** whereas the third isomer (**17**) comes from the cyclization of the isomeric meso dihydroxyketone **14**. Under acidic conditions, **15** and **16** should be readily interconvertible and if these acidic conditions are strong enough to allow epimerization of the two dihydroxyketones **13** and **14**, the three isomeric compounds **15**, **16**, and **17** should be interconvertible.

Molecular models show that isomers **15** and **16** can exist in three different conformations each whereas four different conformations are possible for isomer **17**. Analysis of the steric and stereoelectronic effects of these different conformations indicated that isomer **15** should exist in only one conformation (**15A**, 0 kcal/mol), isomer **16** should also exist in only one

conformation (**16A**, 1.8 kcal/mol) whereas isomer **17** should exist as a mixture
of a major (**17A**, 3.1 kcal/mol) and a minor (**17B**, 3.7 kcal/mol) conformer.

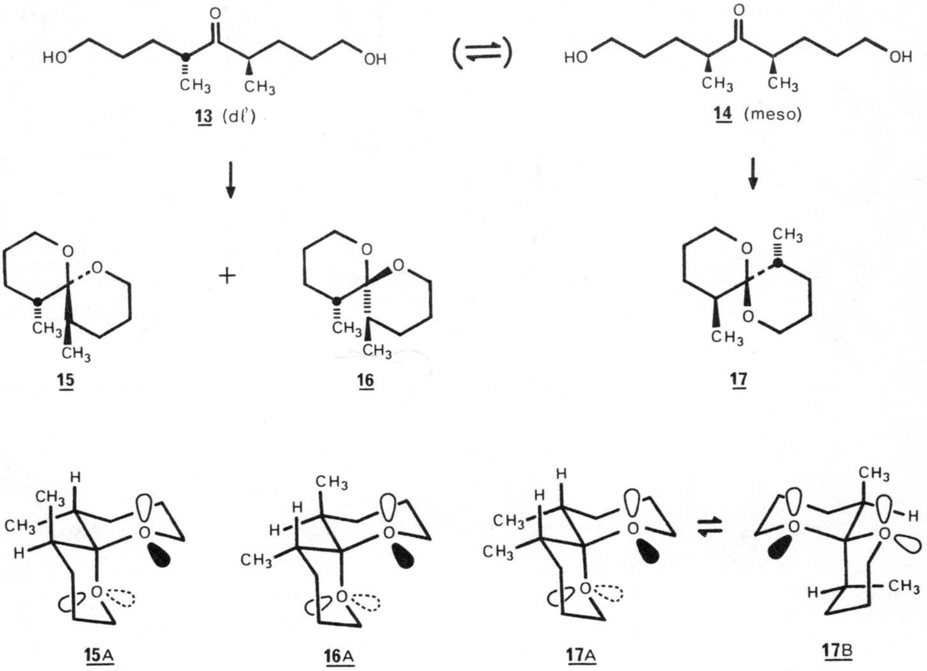

Fig. 4

Cyclization of a mixture of _dℓ_ and _meso_ dihydroxyketones **13** and **14** under
mild acid conditions gave a mixture of the three isomeric spiroketals **15**,
16, and **17**. Low temperature ^{13}C nmr analysis confirmed that isomers **15** and
16 are conformationally rigid and that they exist in the conformations **15A**
and **16A** respectively. Using the same technique, isomer **17** was shown to exist as a mixture of conformers **17A** and **17B** as predicted. Furthermore, acid
equilibration of **15** (or **16**) gave a ≈97:3 mixture of isomers **15** and **16**, and
when isomer **17** was treated under the same conditions it was converted into
a 97:3 mixture of **15** and **16**. These results are completely consistent with
the analysis made above.

With the tricyclic system **18** (Fig. 5), two isomers **20** and **21** are possible
from the cyclization of the dihydroxyketone **19**, and each can exist in two
conformations. It was, however, predicted that each isomer would be conformationally rigid, existing in the conformation **20A** (0 kcal/mol) and **21A**
(2.4 kcal/mol) respectively. Furthermore, since isomers **20** and **21** can be

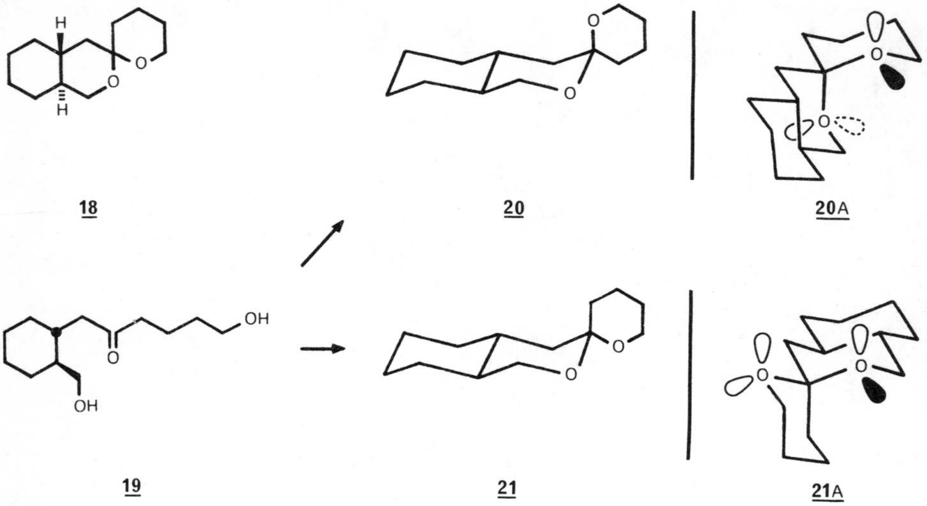

Fig. 5

interconverted by equilibration under acidic conditions, only isomer **20** should be observed experimentally under thermodynamically controlled conditions. Cyclization of dihydroxyketone **19** under very mild acidic conditions gave a mixture of **20** and **21**. Compound **20** was shown to exist in the conformation **20A** and when **21** was equilibrated under acidic conditions, it was completely converted into the more stable isomer **20** (24, 25).

The tricyclic spiro system having one methyl group was also examined (Fig. 6) (24, 25). This system can give the four isomers **23**, **24**, **26**, and **27**. The isomers **23** and **24** come from the cyclization of dihydroxyketone **22** whereas the isomers **26** and **27** come from the cyclization of the isomeric dihydroxyketone **25**. In this case, **22** and **25** are not interconvertible under acidic conditions. Each spiro isomer can exist in two different conformations. The theoretical analysis, however, predicted that isomer **23** exists as a mixture of conformers **23A** (0.5 kcal/mol) and **23B** (0 kcal/mol) whereas isomer **24** exists in the conformation **24A** (0 kcal/mol) (Fig. 6). Approximately, a 1:1 mixture of isomers **23** and **24** should therefore be isolated from the cyclization of dihydroxyketone **22**.

A similar analysis predicted that isomer **26** should exist in the conformation **26A** (0 kcal/mol) whereas isomer **27** should exist as conformer **27A** (4.8 kcal/mol). However, since isomers **26** and **27** are interconvertible under acidic conditions, only isomer **26** existing in the conformation **26A** should be isolated.

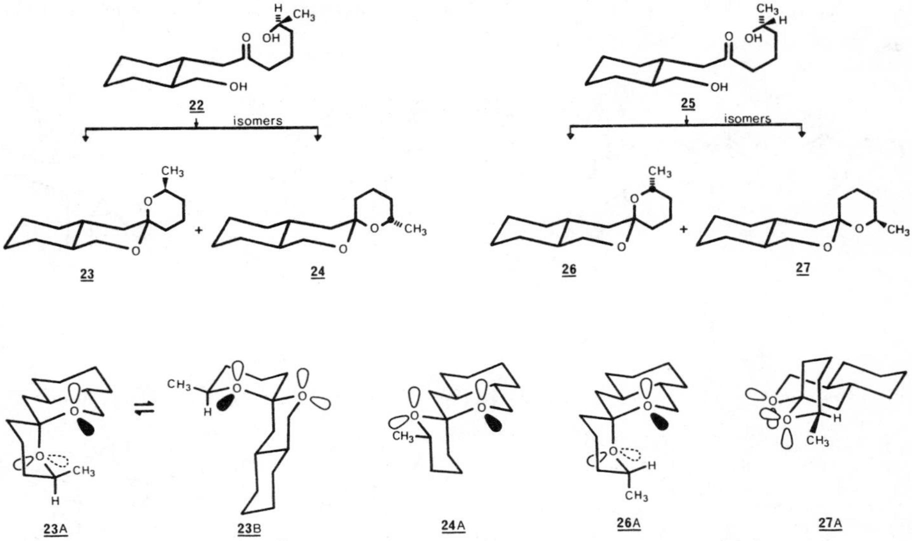

Fig. 6

Cyclization of a mixture of **22** and **25** under acidic conditions gave, as predicted, a mixture of **23**, **24**, and **26**. Furthermore, under acidic conditions, isomer **23** (or **24**) was converted into a ≃1:1 mixture of **23** and **24**, whereas isomer **26** did not equilibrate under acidic conditions.

The masked tetrahydroxyketone **28** (Fig. 7) which can theoretically give isomers **29** and **30**, was found to yield isomer **29** exclusively (26). The structure of **29** was proven by X-ray analysis. Similarly, dibromodihydroxyketone **31** can give either isomer **32** or **33**. Upon cyclization, isomer **32** was the product formed (27) and its structure was also established by X-ray (28). The recently reported total synthesis of ionophore A-23187 (29), a polyether antibiotic whih possesses the 1,7-dioxaspiro[5.5]undecane skeleton having a conformation equivalent to **29** and **32** confirms these results.

This series of experiments establishes that **9A** (Fig. 2) is the most stable spiroketal conformation; it also demonstrates the importance of two electronic effects in the same function. The value of 1.4 kcal/mol for an anomeric effect must, however, be considered as a minimum value, because similar conclusions would have been reached with a value as high as 1.7 kcal/mol.

Also, the steric interactions might be greater (30) in a substituted tetrahydropyran ring than those of a simple cyclohexane system. For those reasons,
the value of 1.4 kcal/mol for an anomeric effect must be considered only
as a reasonable minimum value.

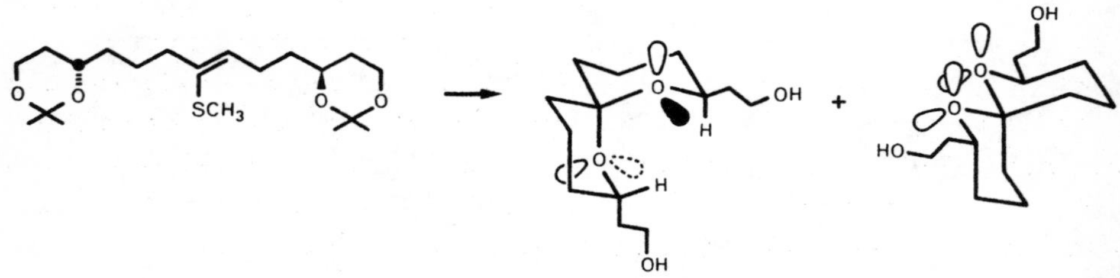

Fig. 7

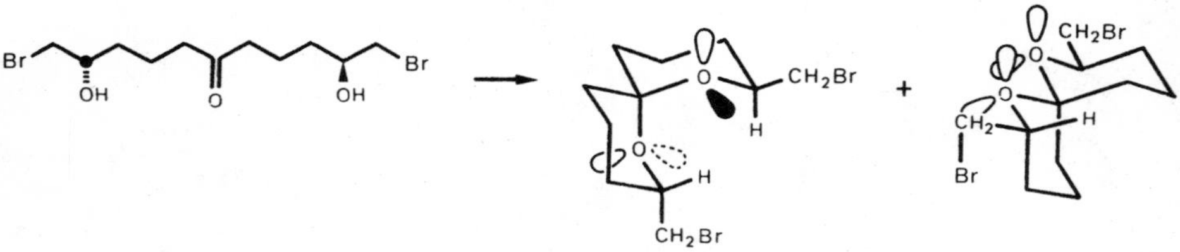

In the synthesis of the pheromone of the common wasp (31, 32), the diastereoisomer **34** (R=CH$_3$) was formed preferentially over the isomer **35** which has
an oxygen atom in the equatorial orientation. Similar results were also
obtained by Kishi and collaborators (33) in the course of the synthesis
of monensin, an antibiotic which incorporates the 1,6-dioxaspiro[4.5]decane
in its structure. Interestingly, several ionophores possess the 1,6-dioxa-
spiro[4.5]undecane unit and X-ray analysis (34) revealed that they all exist
in a relative configuration equivalent to **34**. Descotes and co-workers (35)
have prepared several derivatives of 1,6-dioxaspiro[4.5]decane and have
shown that **34** (R=H) with the oxygen in the axial orientation is the preferred arrangement. They further report that compound **36** exists in the conformation shown. Interestingly, each ether oxygen in **36** has an electron
pair antiperiplanar to a polar C—O bond. This is certainly a direct mani-
festation of the stereoelectronic effects which cause the tetrahydropyran
ring to exist in a twist-boat conformation.

34   **35**

36

The six conformations $\underline{A}$-$\underline{F}$ of 1,1-dimethoxyethane (R=R'=CH_3 in Fig. 1) can now be analyzed. Conformers $\underline{D}$, $\underline{E}$, and $\underline{F}$ have two anomeric effects, $\underline{A}$ and $\underline{B}$ have one and $\underline{C}$ has none. $\underline{E}$ and $\underline{F}$ can be eliminated on the basis of the rather strong steric interactions, and by taking a value of 0.9 kcal/mol for the gauche form of butane and 0.4 for that of $\underline{n}$-propyl ether ($CH_2CH_2CH_2O$), the relative stabilities of the remaining conformers become $\underline{D}$ (0 kcal/mol) > $\underline{A}$ (1.0) > $\underline{B}$ (1.9) > $\underline{C}$ (2.9).

Thus, 1,1-dimethoxyethane should exist as a mixture of a major ($\underline{D}$) and two minor conformers ($\underline{A}$ and $\underline{B}$). It is also conceivable that the anomeric effect for the oxygen of an alkoxy group is larger than that of a tetrahydropyran ring because the oxygen atom of an alkoxy group should more easily become trigonal permitting a larger electronic delocalization (36). In that case, the population of conformer $\underline{D}$ would be even greater. In fact, dipole moment studies revealed that dialkoxyacetals exist in the preferred conformation $\underline{D}$ (37).

The most simple acetal, $\underline{i.e.}$ dimethoxymethane, can adopt only four conformations which correspond to those of $\underline{A}$, $\underline{C}$, $\underline{D}$, and $\underline{E}$ in Fig. 1 (R=H). Conformer $\underline{E}$ can be readily eliminated and the relative stabilities of the others can be evaluated to be $\underline{D}$ (0 kcal/mol) > $\underline{A}$ (1.5) > $\underline{C}$ (2.5). Thus, in dimethoxymethane the preferred conformation is <u>D</u>. In accord with this conclusion, electron diffraction studies of the gas show that <u>D</u> is strongly dominant (38). The same conformation was also found in polyoxymethylene by X-ray analysis (39). The cyclic unstrained oligomers of formaldehyde are also known to adopt this confirmation (40).

As we have seen, the anomeric effect confers a double-bond character to each C−O bond of conformer <u>D</u>; the energy barrier for a C−O bond rotation in acetals must therefore be higher than that observed in simple alkanes. Borgen and Dale (41) may have provided the first evidence for this point by observing that 1,3,7,9-tetraoxacyclododecane (<u>37</u>) has a much higher conformational barrier (11 kcal/mol) than comparable 12-membered rings such as cyclododecane (7.3 kcal/mol (42) or 1,4,7,10-tetraoxacyclododecane (5.5 and 6.8 kcal/mol (43)). It was also shown that the two 1,3-dioxa groupings in <u>37</u> exist in a conformation identical to that of dimethoxymethane, <u>i.e.</u> the conformation <u>D</u>.

Anet and Yavari (44) have studied chloromethyl methyl ether by low temperature proton nmr spectroscopy. Their results show that this compound exists in the gauche conformation <u>38</u> and they observed a barrier of 4.2 kcal/mol for the rotation of the O − CH$_2$Cl bond. This barrier is appreciably higher than that expected on the basis of steric repulsion alone. A rough estimate of the steric barrier is 2 kcal/mol, and they concluded that the anomeric effect increases the barrier to rotation of the O − CH$_2$Cl bond by approximately 2 kcal/mol.

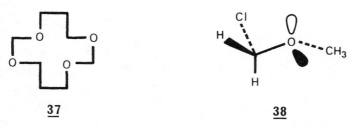

<u>37</u>   <u>38</u>

Recent results by St-Jacques and collaborators (45) have revealed that the ring conformation in 2,4-benzodioxepin which can be the chair <u>39</u> (Fig. 8), the twist boat <u>40</u> or the boat form <u>41</u>, was determined by the nature of substitution. A chair (<u>39</u>) with a minor amount of the twist-boat (<u>40</u>) was observed for the non substituted compound <u>42</u>, whereas the chair form was found for the methyl derivative <u>43</u> and the twist boat-form for the dialkyl derivatives <u>44</u>, <u>45</u>, and <u>46</u>.

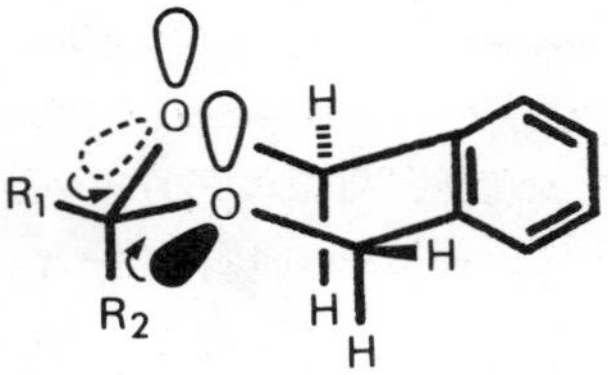

39 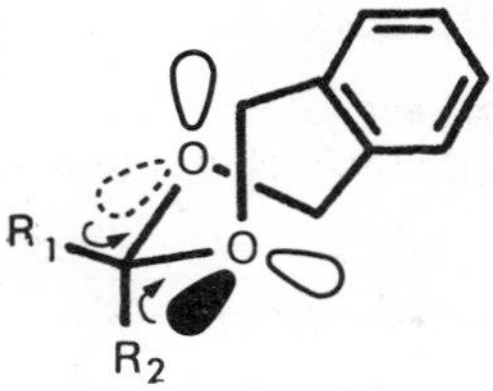   **40** 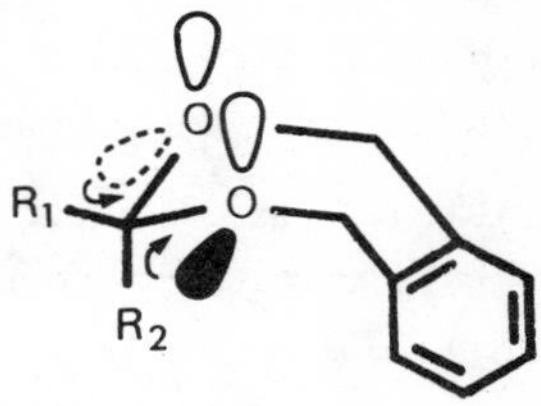   **41**

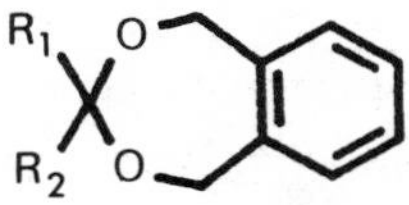

42 R_1 = H, R_2 = H
43 R_1 = CH_3, R_2 = H
44 R_1 = CH_3, R_2 = CH_3
45 R_1, R_2 = $CH_2 - (CH_2)_2 - CH_2$
46 R_1, R_2 = $CH_2 - (CH_2)_3 - CH_2$

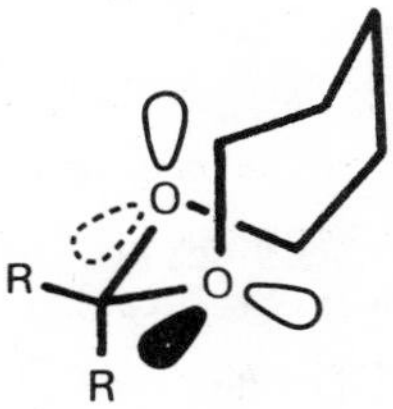

47

Fig. 8

Interestingly, the chair **39** and the twist-boat **40** each have two anomeric effects. The energy difference between the two forms in the simple 2,4-benzodioxepin **42** could be explained by a greater steric interaction in the chair form between the axial hydrogens on C_2, C_4, and C_7. In compound **43**, it appears that the chair form **39** (R_1 = CH_3) with an equatorial methyl is preferred over the twist-boat **40** (R_1 = CH_3) with an isoclinal methyl group because in the latter case, the methyl group has a significant non-bonded repulsive interaction with the methylene group gauche to it. Finally, steric interactions should largely favor the twist-boat **40** over the chair **39** in the disubstituted derivatives **44-46**. In these compounds, there is a severe steric repulsion between the axial alkyl group (R_2) and the hydrogens at C_4 and C_7 when they exist in the chair form **39**. Part of this steric interaction is not present in the twist-boat **40**.

Anet and collaborators (46) as well as Dale and co-workers (47) have also shown the 8-membered ring, 1,3-dioxacyclooctane and some derivatives, to exist only in the boat conformation **47**. Conformer **47** has two anomeric effects and is virtually the same as that found in **40** or in dimethoxymethane.

On the basis of the above results and discussion, the glycosides can now be considered. Efforts have been made previously to evaluate the magnitude of the anomeric effect by undertaking equilibration studies between equatorial and axial isomers at the anomeric center in carbohydrates (48), in monosubstituted 2-alkoxytetrahydropyrans (49, 50) and in more rigid systems (51). The anomeric effect has been evaluated to be of the order of 1.2 to 1.8 kcal/mol from these studies. In these evaluations, the conformation of the OR group in the axial and in the equatorial isomer was not considered; the influence of the exo-anomeric effect was therefore neglected (3). Nevertheless, these studies demonstrated the importance of the anomeric effect.

α and β-Glycosides can adopt conformations $\underline{A}_1$, $\underline{A}_2$, $\underline{A}_3$ and $\underline{E}_1$, $\underline{E}_2$, $\underline{E}_3$ respectively (Fig. 9). The relative proportions of these various conformers should be influenced by the usual steric interactions and by stereoelectronic effects. Conformer $\underline{A}_3$ can be immediately eliminated for steric reasons. Conformer $\underline{A}_1$ has two anomeric effects, conformers $\underline{A}_2$, $\underline{E}_1$, and $\underline{E}_2$ have one each whereas conformer $\underline{E}_3$ has none. Taking into account the usual steric effects and the anomeric effect, this analysis predicts that the relative stability is $\underline{A}_1$ (0 kcal/mol), $\underline{A}_2$ (1.9), $\underline{E}_1$ (0.6), $\underline{E}_2$ (1.5), and $\underline{E}_3$ (2.5). On that basis, the α-isomer exists essentially as conformer $\underline{A}_1$ whereas the β-isomer

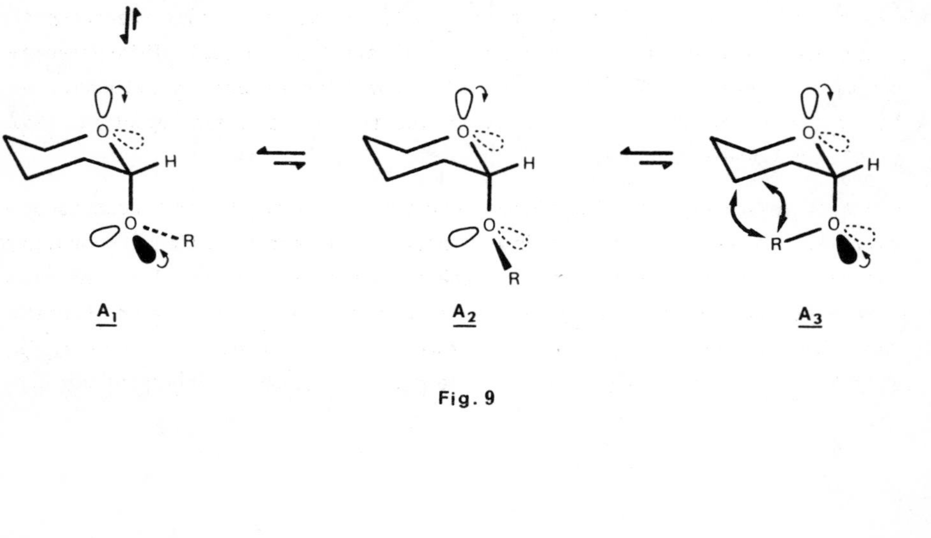

Fig. 9

would be a mixture of E_1 and E_2 where E_1 largely predominates. Also, in a conformationally rigid system, the α-isomer should be formed preferentially under equilibrating conditions, in agreement with experimental results; Eliel and Giza (50) found a difference in stability valued at 0.7-0.8 kcal/ mol in favor of the axial over the equatorial isomer of 1-methoxy-5-methyltetrahydropyran.

Acid equilibration of the equatorial and axial bicyclic acetals **48** and **49** (Fig. 10) has been carried out (52) and the result was compared with those of the monomethyl acetals **50** and **51**, and the **gem**-dimethyl acetals **52** and **53**. Each pair of isomers gave essentially an identical result, i.e. 33% of the equatorial and 77% of the axial isomer (±2%) after equilibration in benzene at 70°C. This corresponds to a ΔG of ≈0.8 kcal/mol, a value close to that predicted for the relative stability of rotamer A_1 over E_1 (0.6 kcal/mol).

Based on steric effects only, the equatorial bicyclic acetal **48** can take the three conformations **54**, **55**, and **56** whereas the axial isomer **49** can exist in the conformations **57** and **58**. The equatorial monosubstituted acetal **50** can take only the conformations **54** and **55** whereas the axial isomer **51** must exist exclusively in the conformation **57**. Finally, the equatorial and axial **gem**-dimethyl acetals **52** and **53** are essentially locked in conformations **54** and **57** respectively. Note that **54**, **55**, and **56** correspond to conformers E_1, E_2, and E_3 whereas **57** and **58** correspond to conformers A_1 and A_2 respectively. The equilibration of each pair of axial and equatorial isomers gave a similar result. Since the isomers **52** and **53** exist only in conformations E_1 (**54**) and A_1 (**57**) respectively, it can be concluded that the population of conformers E_2 (**55**), E_3 (**56**), and A_2 (**58**) must be negligible in the case of the other acetals (49-52). It is known that the anomeric effect becomes less important in polar solvents (49, 50, 53). This phenomenon was observed when the equilibration was carried out in methanol, as the relative percentage of the axial isomer became less important (66% of **49** and **51** and 62% of **53**).

The relative proportion of conformers can be established by nmr spectroscopy provided that the various conformers have sufficiently different coupling constants. This method has, however, the disadvantage that small percentages of some conformers will be difficult to detect. Using this method, Lemieux and co-workers (36) were unable to detect the presence of conformers E_2, E_3, and A_2 in β and α-glycosides. Furthermore, no other experimental methods

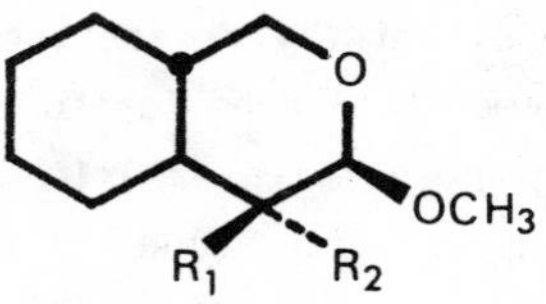

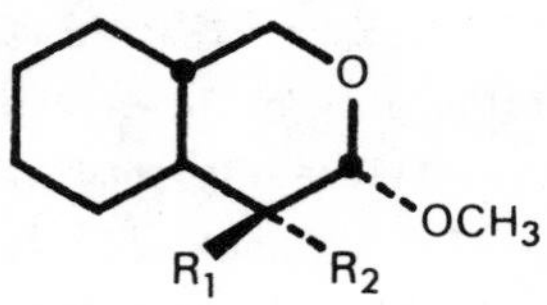

48	$R_1 = R_2 = H$	**49**
50	$R_1 = H, \ R_2 = CH_3$	**51**
52	$R_1 = R_2 = CH_3$	**53**

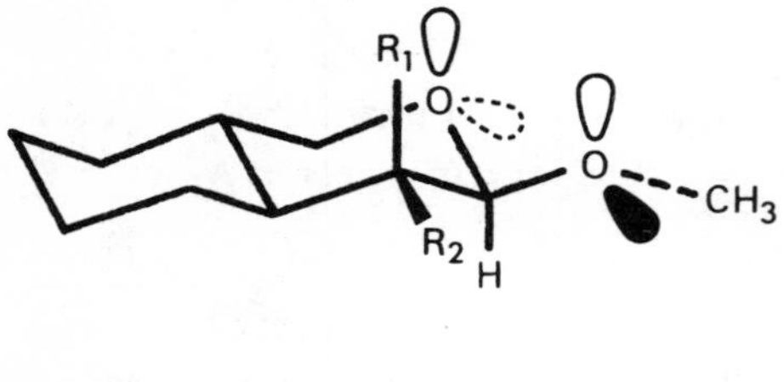

54

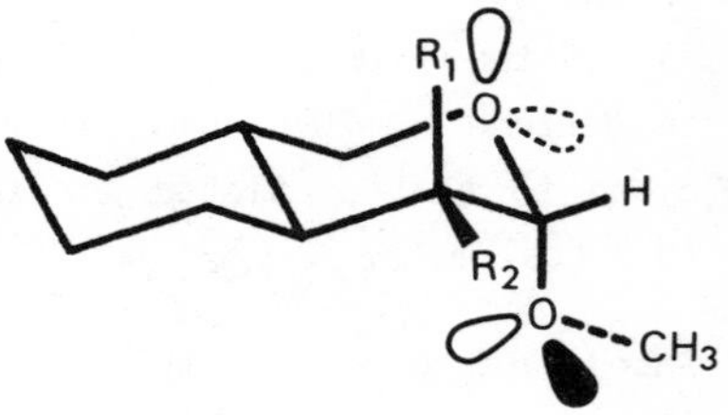

57

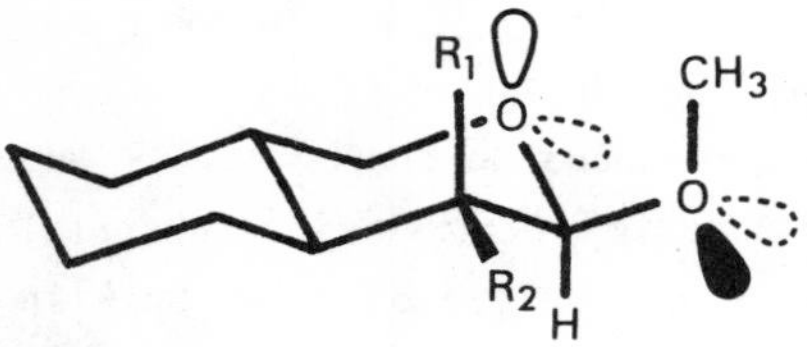

55

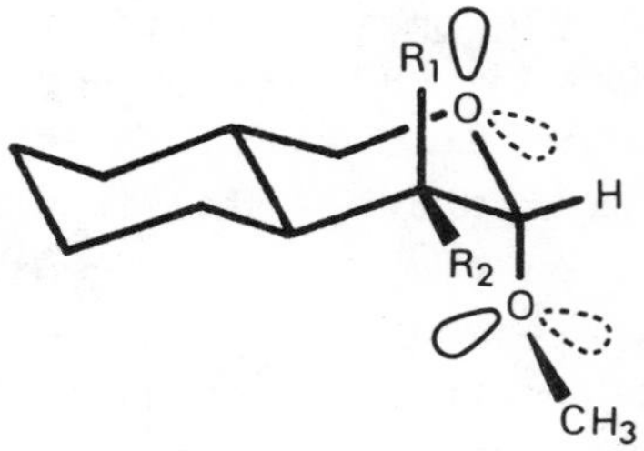

58

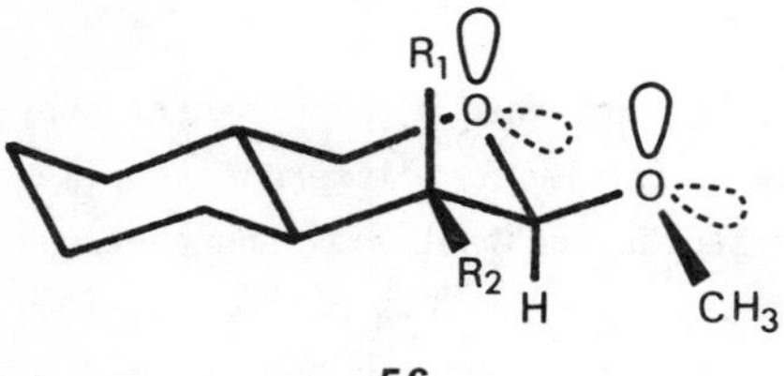

56

Fig. 10

including X-ray (54) and dipole moment studies   (53, 55, 56) indicate that
α and β-glycosides exist in conformations other than $\underline{A}_1$ and $\underline{E}_1$ respectively.

It is rather surprising that conformer $\underline{E}_2$ cannot compete with conformer
$\underline{E}_1$ because they differ by only one gauche form of butane (0.9 kcal/mol).
The experimental results suggest that $\underline{E}_1$ and $\underline{E}_2$ should be separated by over
2 kcal/mol. It was suggested (36) that this unexpected situation could be
mainly the result of a short $C_1 - O_1$ bond which would amplify the steric
interaction in $\underline{E}_2$, because the anomeric effect of an alkoxy oxygen (exo-
anomeric effect) would be larger than that of a tetrahydropyran oxygen
(cf. p. 15). By comparing bond lengths (X-ray) in glycosides, Lemieux has
suggested that the exo-anomeric effect could in fact be stronger. However,
there still remains the possibility that a small percentage of $\underline{E}_2$ exists
in equilibrium with $\underline{E}_1$, but the experimental technique to observe it has
yet to be found (57).

Conformation of mono and dithioacetals

Stereoelectronic effects should also be observed when one or both oxygens
of the acetal function are replaced by another heteroatom such as a sulfur
or a nitrogen atom. Compounds having sulfur atoms will first be considered.

Eliel and Giza (50) have studied the acid equilibration of the isomeric
2-alkylthio 6-methyltetrahydropyrans $\underline{59}$ and $\underline{60}$ (R= CH_3 and $(CH_3)_3C$). They
found about 65% of axial isomer $\underline{59}$ indicating that the monothioacetal function possesses an anomeric effect although it is weaker than that of the
acetal function. Zefirov and Skekhtman (58) have arrived at a similar conclusion by studying 2-phenylthio and 2-ethylthiotetrahydropyran.

$\underline{59}$ $\underline{60}$

2-Alkoxythiatetrahydropyrans were also studied (58, 59). It was found that
the major isomer (90%) had the axial conformation $\underline{61}$ rather than the equatorial conformation $\underline{62}$ when R is a methyl or a propyl group. Also, Perlin
and Nam Shin (60) have carried out the methanolysis of 5-thio-D-galactose

which yielded the α-glycoside **63** almost exclusively. These results suggest that the anomeric effect is appreciable in these compounds. Alkylthiothia-pyrans (**61** and **62**, OR=SR) were also investigated and were shown to exhibit an anomeric effect (61).

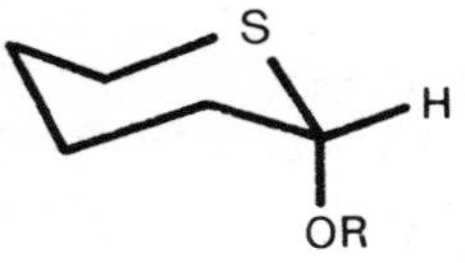

61 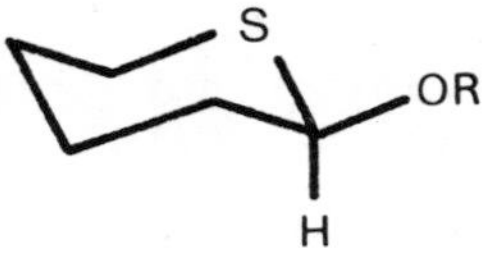   **62** 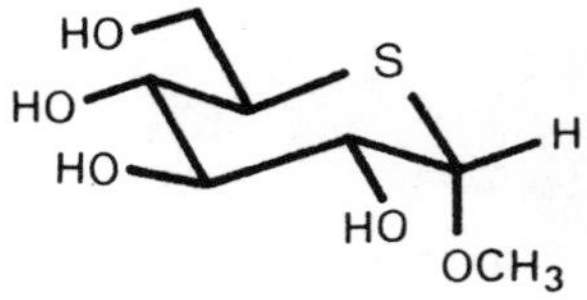   **63**

1,7-Dithiaspiro[5.5]undecane can exist in the three conformations **64A**, **64B**, and **64C** (Fig. 11). Conformer **64A** which has the two sulfurs axially oriented can have two anomeric effects, conformer **64C** which has the two sulfurs equa-torially oriented has no anomeric effects and conformer **64B** with one sulfur axially oriented has only one anomeric effect. Taking into account the ster-ic effects (0.9 kcal/mol for a gauche form of butane and 0.4 kcal/mol for a gauche form of $S-CH_2-X-CH_2$) and an anomeric effect of 1.4 kcal/mol, the relative stabilities **64A**, **64B**, and **64C** are 0, 2.4, and 4.8 kcal/mol. On that basis, this compound must be essentially conformationally rigid existing in the conformation **64A** only. This prediction was verified experi-mentally (25, 62), and further confirmation was found by the study of the acid catalyzed cyclization of ketone dithiol **65**. This compound can give either isomer **66** or **67** which can take two different conformations each.

64A   **64**B   **64**C

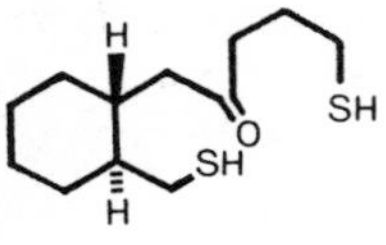

65    **66**    **67**

Fig. 11

The detailed analysis of these conformers showed that **66** exists in a conformation equivalent to **64A** only. Similarly, **67** would also exist in a conformation equivalent to **64B** only. This analysis further predicted that **66** is more stable than **67** by 2.4 kcal/mol and since these isomers can be interconverted by equilibration under acidic conditions, isomer **66** should be the major isomer formed under thermodynamically controlled conditions. Cyclization of **65** under equilibrating conditions gave **66** and **67** in a 98:2 ratio and **66** was shown to exist in a conformation equivalent to **64A**. The fact that a small percentage of **67** was detected at room temperature shows that, at sufficiently low temperature, less than 0.1% of **67** would be present. This result confirms that, at low temperature, 1,7-dithiaspiro[5.5]-undecane exists essentially in the conformation **64A** only. The anomeric effect for a sulfur atom in the dithioacetal function must therefore be of the order of 1.4 kcal/mol.

Molecular models show that 1-oxa-7-thiaspiro[5.5]undecane can take four different conformations **68A-D** (Fig. 12) where **68A** has two anomeric effects, **68B** and **68C** one and **68D** none. Accepting again a value of 1.4 kcal/mol of stabilization for each anomeric effect, and after taking the steric effects into consideration, the relative stabilities were predicted to be 0, 2.4, 2.4, and 4.8 kcal/mol for **68A-D** respectively. Thus, on that basis, 1-oxa-7-thiaspiro[5.5]undecane must be essentially conformationally rigid existing as **68A** only. Again, the experimental results were in complete agreement (25, 62).

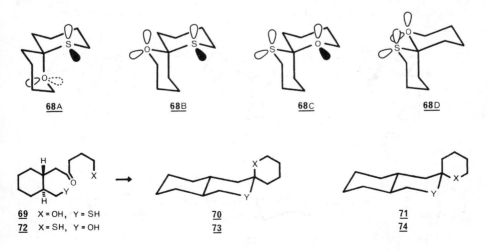

68A	**68**B	**68**C	**68**D

69 X = OH, Y = SH   **70**   **71**
72 X = SH, Y = OH   **73**   **74**

Fig. 12

This analysis was confirmed by the cyclization studies of hydroxyketone thiols **69** and **72**. Compound **69** gave on cyclization the isomer **70** rather than the isomer **71** which has an equatorial oxygen. Similarly, cyclization of **72** gave **73** only; the isomer **74** was not observed. It was further shown that compounds **70** and **73** exist in a conformation equivalent to **68A**, which has two anomeric effects.

The hemiacetal thiol **75** (Fig. 13) gave on acid cyclization an equilibrium mixture of cis and trans monothioacetals **76** and **77** in a 1:1 ratio (63). If a value of 1.7 kcal/mol is accepted for the steric effects in **76** (two gauche forms of SCH_2XCH_2 = 0.8 kcal/mol and one gauche form of butane = 0.9 kcal/mol), the anomeric effect for the ether oxygen in **76** must also be equal to 1.7.

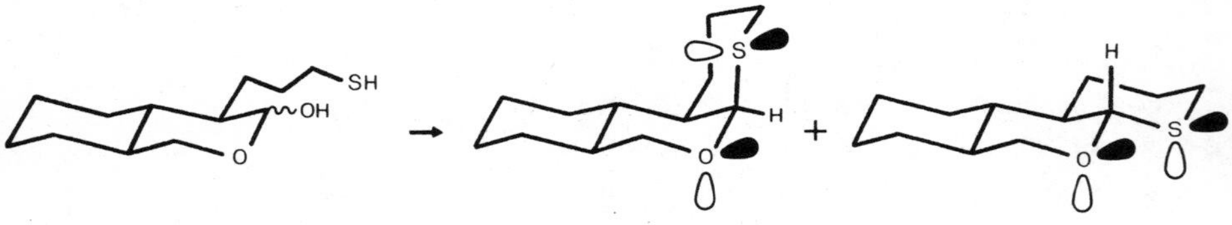

75　　　　　　　　　　　**76**　　　　　　　　　**77**

Fig. 13

Sauriol-Lord and St-Jacques (64) have studied 2,4-benzodithiepin **78** and the substituted derivatives **79-82** (Fig. 14). They found that these compounds exist in the chair form **39** (O=S) (Fig. 8), except for the derivative **82** where 13% of the twist-boat **40** (O=S) was detected. The fact that essentially only the chair form was observed in these compounds would simply be due to the long C — S bonds which attenuate the effects of steric repulsions in conformer **39** (O =S). This is in accord with the fact that in contrast to 2-methoxy-1,3-dioxane which exists as a 7:3 mixture of equatorial conformer **83** and axial conformer **84**, 2-methoxy-1,3-dithiane has been found to exist solely in the axial form **85**.

The same authors have also studied compounds **86** and **87** having a methoxy group (Fig. 15). These compounds could exist either in the chair form with an axial (**88**) or an equatorial methoxy group (**89**), or in the twist-boat **90**. Conformer **88** is disfavored sterically but is favored electronically, whereas the twist boat **90** with a syn clinal methoxy group is favored electronically and does not suffer from severe steric hindrance. It was found

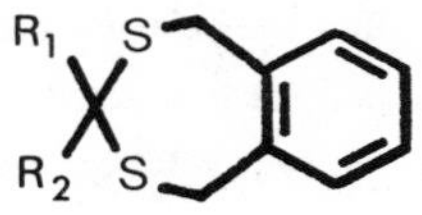

$\underline{78}$ R₁= H, R₂ = H
$\underline{79}$ R₁= CH₃, R₂ = H
$\underline{80}$ R₁= CH₃, R₂ = CH₃
$\underline{81}$ R₁, R₂ = CH₂—(CH₂)₂—CH₂
$\underline{82}$ R₁, R₂ = CH₂—(CH₂)₃—CH₂

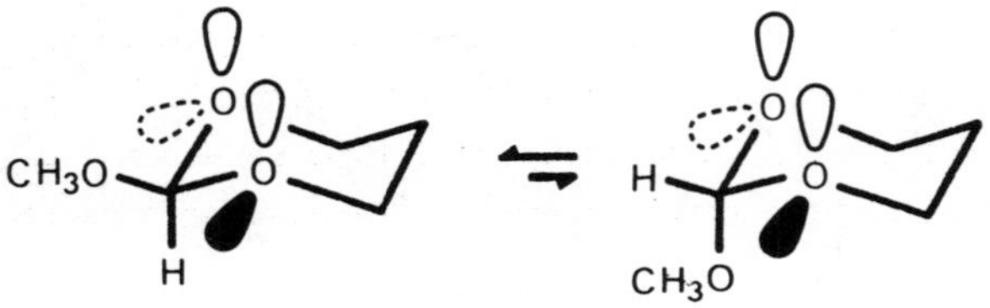

$\underline{83}$ $\underline{84}$ $\underline{85}$

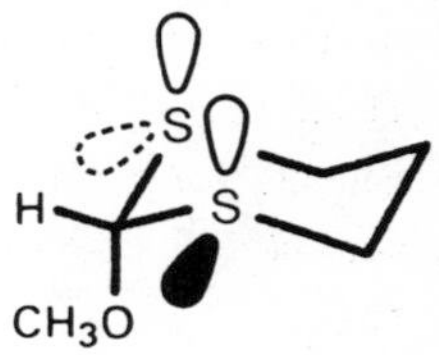

Fig. 14

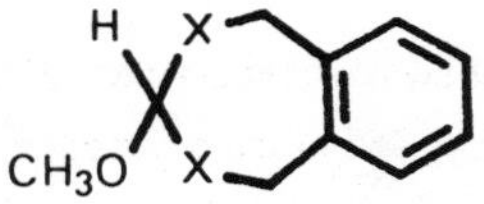

$\underline{86}$ X = O
$\underline{87}$ X = S

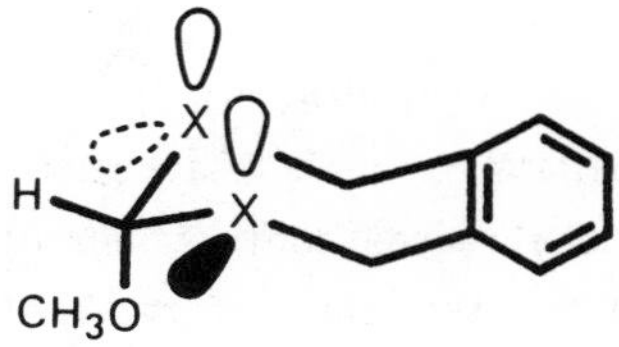

$\underline{88}$

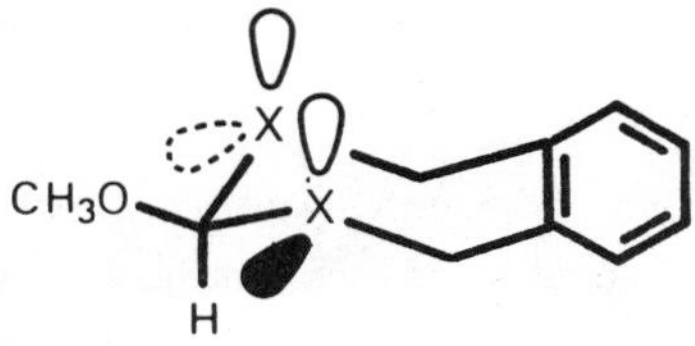

$\underline{89}$ $\underline{90}$

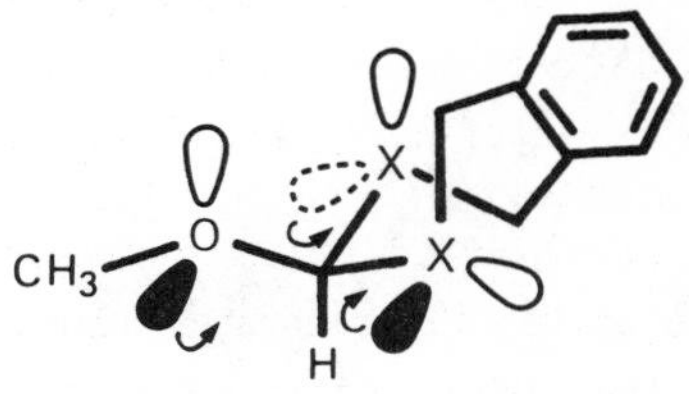

Fig. 15

that contrary to compound **43** which exists in the chair form **88** (X=O and
OCH$_3$= CH$_3$), compound **86** exists in the twist-boat **90** (X=O) which has maximum
electronic delocalization and minimum steric effects. Compound **87** was however found to exist as the chair **88** (X=S) and this is explained by electronic
delocalization due to an axial methoxy group and less important steric repulsion because of the long C—S bond.

Conformation of 1,3-oxazines and 1,3-diazines

The preceding results demonstrate conclusively that the anomeric effect
is important in the case of mono and dithioacetals. The following discussion
will show that this effect is equally important in the case of an acetal
function having one or two nitrogen atoms.

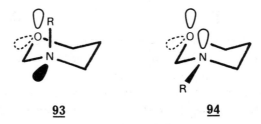

91   **92**

NMR studies at low temperature strongly suggest that, in tetrahydro-1,3-
oxazine, conformer **91** (R=R'=H) makes the major contribution to the equilibrium **91** ⇌ **92** (65). The same conclusion was reached for the case where R=CH$_3$
and R'=H. Two anomeric effects are possible in conformer **91** and only one
in conformer **92**; the greater stability of **91** can therefore be rationalized
by the extra anomeric effect of the nitrogen atom.

93   **94**

N-substituted tetrahydro-1,3-oxazines were also studied. It has been suggested (66) that conformer **93** with the N-alkyl group in the axial orientation,
is more stable than **94** when R is either a methyl or a benzyl group. It was
also found that the contribution made by conformer **95** is appreciable and
it may dominate conformer **96** in tetrahydro-3,4,4,6-tetramethyl-1,3-oxazine.

Furthermore, Riddell and Lehn (67) have been able to show that when R' is a methyl group and R is either a hydrogen atom or a p-nitro benzyl group, conformers **91** and **92** exist in approximately equal proportion.

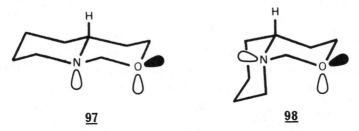

$$\underline{95} \qquad\qquad \underline{96}$$

This last study is quite interesting because it permits an evaluation of the anomeric effect for the nitrogen atom. Conformer **91** with the axial N-methyl group should be less stable than conformer **92** by approximately 1.3 kcal/mol on the basis of steric effects (one gauche form of n-butane, $\approx$0.9 kcal/mol and one gauche form of $CH_3 - N - CH_2 - O$, $\approx$0.4 kcal/mol). The second anomeric effect caused by the equatorial orientation of the nitrogen electron pair in **91** must compensate for the steric effect. An approximate value of 1.3 kcal/mol must therefore be taken for that electronic effect, a value close to that estimated for the oxygen atom of the acetal function.

$$\underline{97} \qquad\qquad\qquad \underline{98}$$

Recently, Crabb, Turner, and Newton (68) have observed that perhydropyrido-[1.3]oxazine exists as a $\approx$9:1 mixture of the trans and the cis forms **97** and **98**. The cis form **98** has two anomeric effects (-2.8 kcal/mol), two gauche forms of butane (1.8 kcal/mol) and one gauche form of n-propyl ether (0.4 kcal/mol) whereas the trans form **97** has only one anomeric effect (-1.4 kcal/mol). On that basis, the trans form **97** should be more stable than the cis form **98** by about 0.8 kcal/mol, in agreement with the experimental result. Katritzky and co-workers (69) have also shown that 1-oxa-3,5-diaza and 1,3-dioxa-5-aza cyclohexane derivatives exist respectively in the conformations **99** and **100**. With an alkyl group in the axial orientation, both conformations gain two anomeric effects.

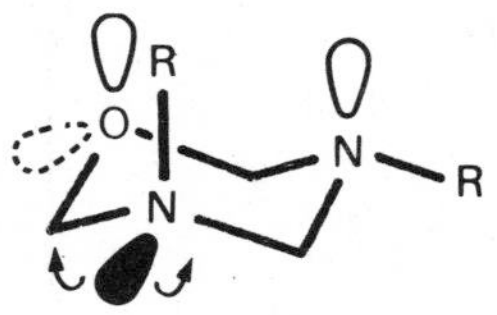

99 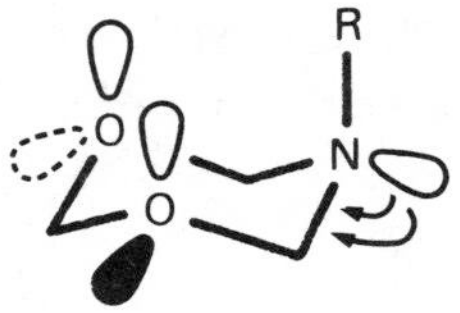   **100**

Interestingly, Allingham et al. (66) have analyzed a series of N-alkyl 5-nitrotetrahydro 1,3-oxazines and concluded that there is a preference for the N-alkyl axial conformer **101** when the alkyl substituent is methyl, ethyl or propyl and a preference for the N-alkyl equatorial conformer **102** when the substituent is isopropyl, cyclohexyl or t-butyl. Thus, **101** is still preferred despite the 1,3-diaxial steric interaction between the nitro group and a primary N-alkyl group. Finally, Katritzky et al. (70) have proposed the conformation **103** as the major one for compounds **104** and **105.**

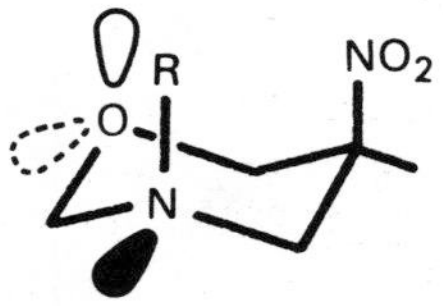

101 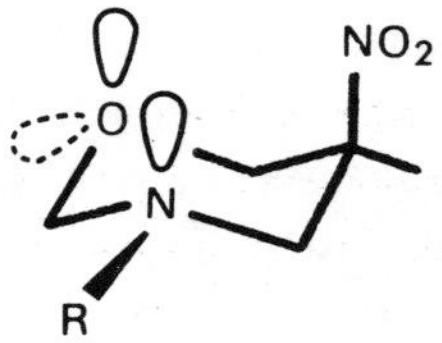   **102**

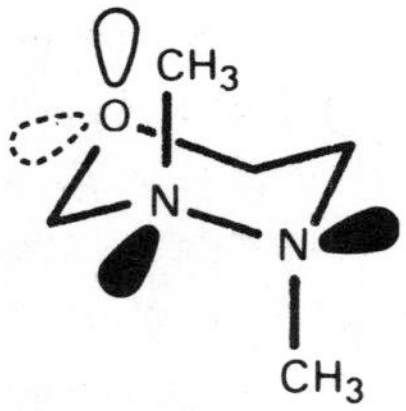

103 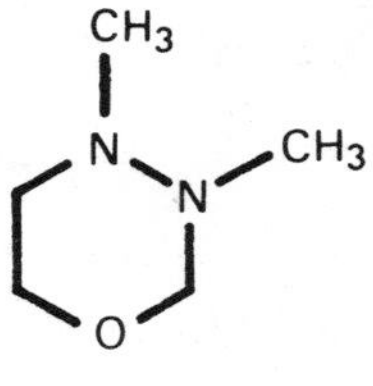   **104** 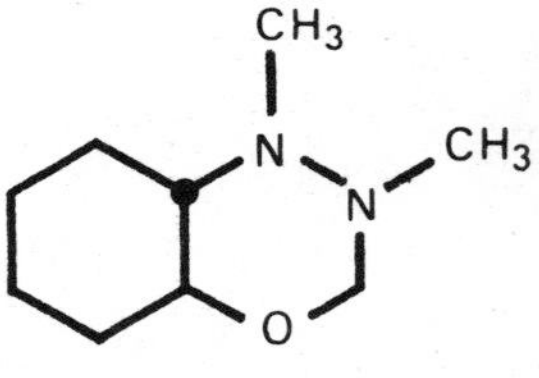   **105**

Several groups (65, 71-75) have studied the 1,3-diazane system **106** (R=R'=H and/or CH$_3$) and found that these compounds exist as a mixture of conformers **107** and **108**. In these studies, the conformation **109** with two anomeric effects was not considered. It is conceivable that when R=R'=H, the most stable conformation is **109** rather than **108**. Also, when R=CH$_3$ and R'=H, conformer **109** with its second anomeric effect to compensate for the steric effect caused by the axial N-methyl group could compete with conformer **108** which has only one anomeric effect. These predictions have not yet been verified experimentally.

106

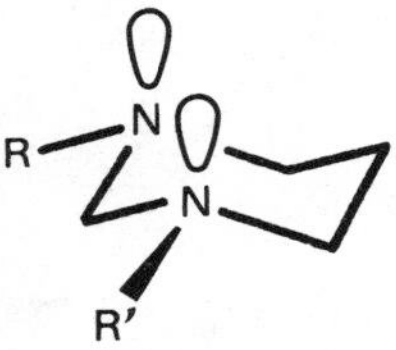

107 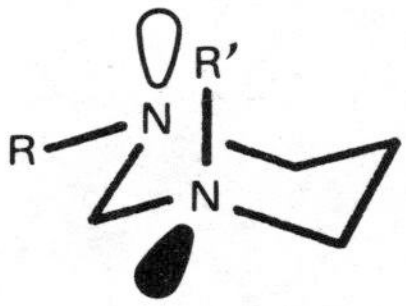   **108**    **109**

Formation and hydrolysis of the acetal function

We have presented experimental results which demonstrate the influence of stereoelectronic effects on the configuration and the conformation of the acetal function. A pertinent question which follows is whether or not these stereoelectronic effects play a similar role in the formation and in the hydrolysis of this functional group.

The formation or the hydrolysis of an acetal function proceeds by the mechanism described in Fig. 16 in which oxonium ions and hemiacetals occur as intermediates. It has also been established (76) that the rate determining step in acetal hydrolysis is generally the cleavage of the $C-O{\raise.5ex\hbox{$\scriptstyle H$}\kern-.1em\raise-.5ex\hbox{$\scriptstyle R$}}$ bond of the protonated acetal **100** to form the oxonium ion **111**. This ion is then rapidly hydrated to yield the protonated hemiacetal **112** which can give the aldehyde product after appropriate proton transfers. It is pertinent therefore to find out if stereoelectronic effects influence the rate determining step (**110** → **111**) of this hydrolysis reaction.

We will first examine what might occur when stereoelectronic effects are taken into account. An analysis of the conformations **D**, **A**, and **C** (Fig. 17) of dimethoxymethane which have two, one, and zero anomeric effects respec-

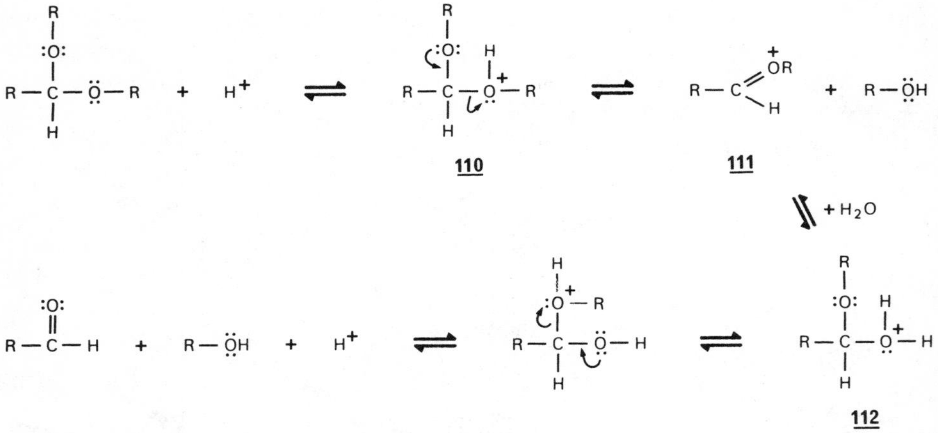

Fig. 16

tively, leads to the following predictions. The $C-OR$ bond in conformer
A should be shorter than a normal $C-O$ ether bond because it has a partial
double-bond character due to the anomeric effect, whereas the $C-O*R$ bond

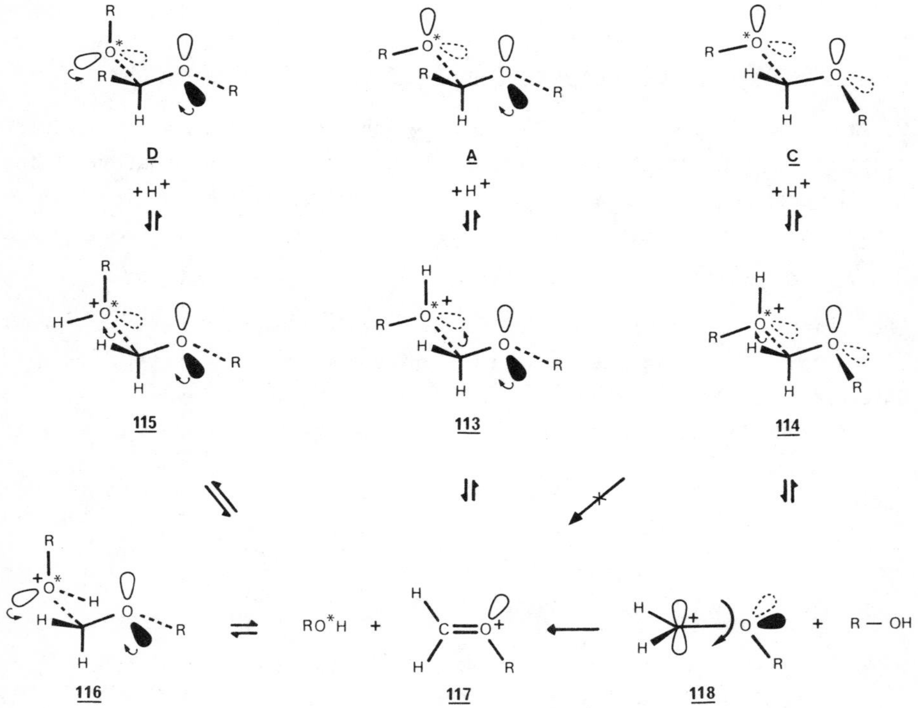

Fig. 17

should be longer than usual because of electron donation from the other oxygen. In conformer $\underline{D}$, both C—O bonds have partial double-bond character; they could be shorter than normal. Since conformer $\underline{C}$ has no anomeric effect, its two oxygens are identical. The C—O bond lengths should be normal with a value intermediate to those predicted for conformer $\underline{A}$.

Bond lengths collected from experimental data in X-ray crystal-structure determination of methyl pyranosides (11) and other products (77) strongly support the above discussion. Oxygen atoms with an electron pair antiperiplanar to a C—O*R bond definitely have shorter C—O bonds whereas the C—O*R bond is longer than normal. Interestingly, theoretical calculations have been carried out (11, 78) and the results are consistent with the above discussion and experimental observations.

As a consequence, stereoelectronic effects should also influence the basicity of the oxygen atoms of the acetal function, thence their relative ease of protonation. For instance, the oxygen atom of the O*—R group should be more basic than the other in conformer $\underline{A}$. In conformer $\underline{C}$ which has no anomeric effect, the two oxygens are completely equivalent, and they should have a basicity intermediary to those of the oxygen atoms of conformer $\underline{A}$. Since conformer $\underline{D}$ has two anomeric effects, both oxygens should be slightly more basic than the OR oxygen of conformer $\underline{A}$.

Consequently, by protonation of the O*R oxygen, the C—O*R bond length should increase to a larger extent in conformer $\underline{A}$ (→$\underline{113}$) than in conformer $\underline{C}$ (→$\underline{114}$) because the non-protonated oxygen of $\underline{113}$ has an electron pair antiperiplanar to the $C-O{\overset{*}{\underset{R}{\lessgtr}}}^{H}$ bond which is not the case with $\underline{114}$. With conformer $\underline{D}$, protonation of the electron pair of the O*R oxygen which is antiperiplanar to the C—OR bond (→$\underline{115}$) should lead to a situation similar to that of $\underline{113}$. On the other hand, protonation of the O*R oxygen can take place on the other electron pair (→$\underline{116}$), and in this case, the $C-O{\overset{*}{\underset{R}{\lessgtr}}}^{H}$ bond length should not increase as much because in $\underline{116}$, the protonated oxygen has still an electron pair antiperiplanar to the C—OR bond. Thus, the stereospecific protonation of one of the two electron pairs of the O*R oxygen in conformer $\underline{D}$ leads to two completely different situations ($\underline{115}$ and $\underline{116}$). This is very interesting because it points out the importance of the stereochemistry of protonation as well as the relative basicity of two electron pairs on the same oxygen atom. For instance in conformer $\underline{D}$, the electron pair of the O*R oxygen which is not engaged in an anomeric effect should be more

basic than the other and on that basis, protonation of **D** should give **116**
in preference to **115**.

An ab initio study on four different conformations of protonated dihydroxy-
methane has been carried out (78, 79). Comparison of these results with
those obtained from the corresponding neutral conformations indicates that
bond length and overlap population show strong stereoelectronic conformational dependence in complete agreement with the above discussion.

We can now examine the cleavage of the $C-O^{*}\overset{H}{\underset{R}{\diagdown}}$ bond in the protonated species **113**, **114**, **115**, and **116**. Intermediates **113** and **115** can undergo cleavage
with the help of an electron pair from the neighboring oxygen atom to give
the oxonium ion **117** and alcohol. In **116**, a similar situation occurs, but
the ejection of RO*H should be more difficult because the oxygen atom of
the leaving group still has an anomeric effect. Conformer **114** cannot break
down with the help of an electron pair to give directly the delocalized
ion **117**; it should give first the high energy undelocalized cation **118**
which can be transformed into the more stable oxonium ion **117** after a rotation of 60° of the C — OR bond. If the energy difference between the proc-
esses **114** → **118** → **117** and that of **115** → **117** or **113** → **117** is such that the former
cannot compete with the others, the acetal hydrolysis will be said to take
place under stereoelectronic control.

As a consequence of the principle of stereoelectronic control, the reverse
process, i.e. the addition of an alcohol to an oxonium ion (e.g. **117**
→ **113**) must take place through a precise pathway. Evidence on this pathway
comes from the X-ray analysis (80, 81) of molecules containing an amino
group and a carbonyl group in close proximity (e.g. **119**) (see also (82)).

119   **120**   **121**

Bürgi, Dunitz, and Shefter (80) made the following observations: the N:···C=O distance was found too long for a bond but much too short for no-bonding, the RRC=O unit (120 → 121) deviates from its usual coplanar geometry and the probable orientation of the electron pair of the approaching nitrogen is assumed (this cannot be observed) to lie close to the local threefold axis of the tertiary amino group. This analysis clearly indicates an interaction between the nitrogen atom and the carbonyl group. The crystal structures leave no doubt that the nucleophile approaches the carbonyl along a vector at an angle close to 109° with the C=O bond. As the nucleophile approaches the carbonyl carbon, the oxygen atom and the alkyl substituents bend out the plane and the carbon oxygen bond length increases (cf. 120 → 121).

SGF-LCGO calculations for the nucleophilic addition of hydride ion to formaldehyde to yield methanolate anion ($H^- + CH_2=O \rightarrow CH_3O^-$) have been carried out (83, 84). Interestingly, the calculated reaction path for this reaction shows striking similarities to that derived from structural correlations for amine addition to a ketone function.

Strong evidence for the reaction path of the addition of a nucleophile to a carbonyl group comes from the X-ray structure of the naphthalene amino ketone 122 (85). In most 1,8-disubstituted naphthalenes, both substituents are splayed outward (cf. 123). In 122, the C—CO bond is splayed outward, but the C—N bond leans inward (cf. 124). This brings the nitrogen into a more favorable position for attack than is possible in the undistorted molecule. Similar observations were made with the amino ester (122, $COCH_3=COOCH_3$) and the amino carboxylic acid (122, $COCH_3=COOH$).

122   **123**   **124**

Capon and Thacker (86) have provided good evidence that the isomerization of methyl α and β-glucopyranosides in acidic methanol proceeds by the ejection of the methoxyl group to give the glucopyranosyl cation. The hydrolysis of these compounds should also proceed in a similar manner to give the same cation which is captured by water to yield the hydrolysis product (87, 88). Since the ejection of a methoxy group can take place with the help of an electron pair in the case of α-glycosides only, the α-anomers should hydrolyze at a faster rate than β-anomers.

Early studies on the relative rate of acetal hydrolysis described in the literature bring no evidence in favor of stereoelectronic control. Feather and Harris (89) have studied 10 anomeric pairs of alkyl glucopyranosides, and found that the β-anomers containing the equatorial methoxy group hydrolyzed more rapidly (1.3-3.2 times faster) than the α-anomers with the axial methoxy group. BeMiller and Doyle (90) have observed similar results by measuring the rate of hydrolysis of 11 anomeric pairs of alkyl glucopyranosides. In the case of aryl glycosides, the α-anomers hydrolyzed slightly more rapidly (91). Van Eikeren (92) has prepared the conformationally rigid model compounds **125** and **126** and found a relative rate of hydrolysis of 1.5 in favor of the axial isomer **125**.

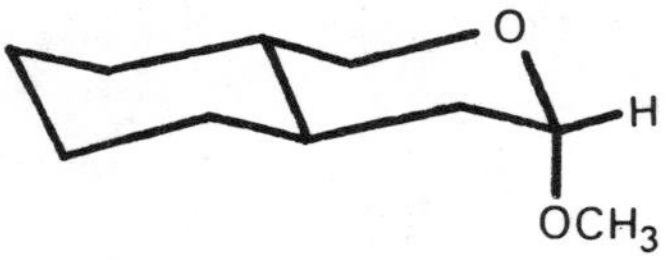

125 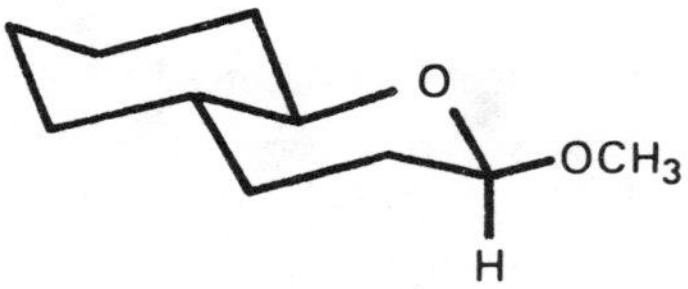   **126**

Chandrasekhar and Kirby (93) have observed that the hydrolysis of the axial p-nitrophenoxy isomer **127** is pH-independent in the pH range 7-10 and found that the rate of this spontaneous hydrolysis is 3.3 times slower than the equatorial isomer **128** under the same conditions. Thus, in a system designed to eliminate all other factors, including the problems of interpretation associated with acid-catalyzed reactions, no evidence was found that acetal cleavage is subject to stereoelectronic control.

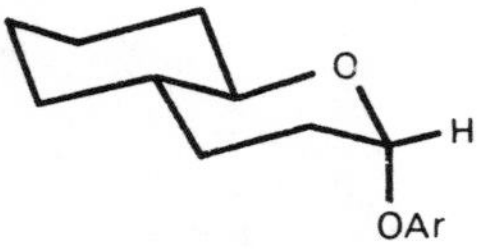

127 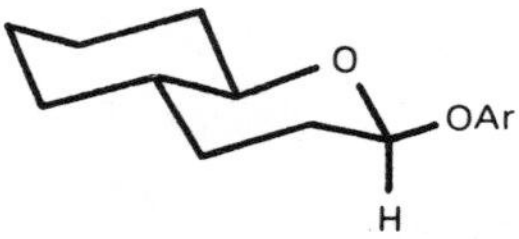   **128**    **129**

A conclusion can therefore be reached that electronic effects are not important in the rate of hydrolysis of acetals, but this is valid only if these
compounds hydrolyze in their ground state conformation. The reality could
well be completely different. For instance, an acetal could prefer to hydrolyze with stereoelectronic control via a higher energy conformation. Thus,
the hydrolysis of 127 and 128 could occur with stereoelectronic control
at a more or less competitive rate. Compound 127 would hydrolyze via its
ground state conformation whereas 128 would hydrolyze via the boat conformation 129. This situation is possible because the enthalpy of activation
for the cleavage of the 4-nitrophenoxy group in 127 and 128 is close to
25 kcal/mol, a much greater value than the barrier for the formation of
the boat conformation 129 which is of the order of 10 kcal/mol (94). Therefore, 129 could lie on the reaction coordinates.

The first experimental evidence that acetal hydrolysis is controlled by
stereoelectronic factors was provided by Kirby and Martin (94). They have
measured the rates of the acid catalyzed hydrolysis (0.1 N HCl) of the
tricyclic acetal 130 which has no electron pair antiperiplanar to the leaving group, and the corresponding cis isomer 131 which does have an electron
pair properly oriented to eject the leaving group. They found that the
cis isomer 131 was rapidly hydrolyzed whereas the trans isomer 130 in which
the conformation of the acetal centre is locked, was still not completely
hydrolyzed after several weeks, indicating a difference in rate of at least
3000.

130   **131**

They also found that at pH 9, 130 was hydrolyzed 3.4 times more slowly
than 131. It was shown, however, that the rate determining steps are not
the same. In 130, it is the C — O bond cleavage whereas in 131, it is the
hydration of the corresponding oxonium ion. Consequently, a direct comparison between the rates of the spontaneous hydrolysis of 130 and 131 cannot be

made. However, **130** can be compared with trans bicyclic acetal **128**. They
both have essentially the same equatorially oriented leaving group in the
ground state and they are hydrolyzed by a rate determining C−OAr cleavage
at pH 9. The only difference is that **128** can adopt the boat form **129** whereas
130 cannot take a similar conformation. Compound **128** at 39°C is hydrolyzed
20 times faster than **130** at 100°C, which is equivalent to a factor of over
10^4. Kirby and Martin (94) concluded that this rate difference is a direct
consequence of the rigid geometry of **130**. Indeed, **130** has no reasonable
accessible conformation with an electron pair antiperiplanar to the leaving
group. The stereoelectronic barrier to the C−O bond cleavage was consequently estimated to be almost 7 kcal/mol.

A further example of stereoelectronic control in acetal cleavage was also
discovered by Kirby and Martin (95) who studied the spontaneous hydrolysis
of the axially oriented p-nitrophenoxy acetals **132** and **133**. Loss of p-nitro-
phenolate from these compounds would generate the oxonium ion **134**, an acetal
with a much better leaving group, i.e. the aldehyde oxygen. If one of the
electron pairs on the oxygen atom of ring A in **134** is in a position to
participate, it should trigger a concerted reaction to form **135** directly.
Such participation is possible only in the case of the cis isomer **133**.
So, the rate of hydrolysis of the trans isomers **132** should be much slower.

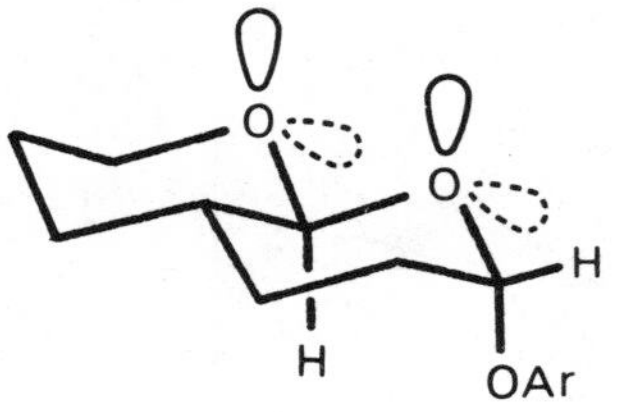

132

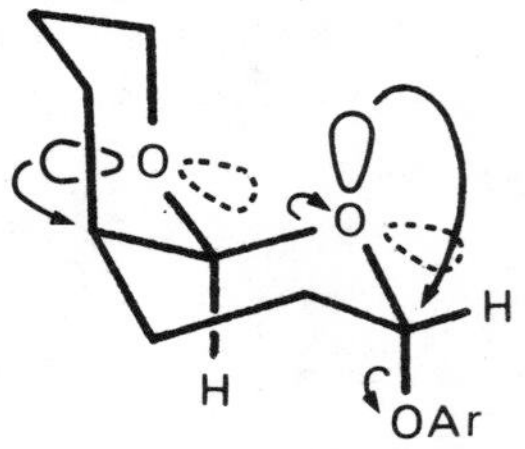

133

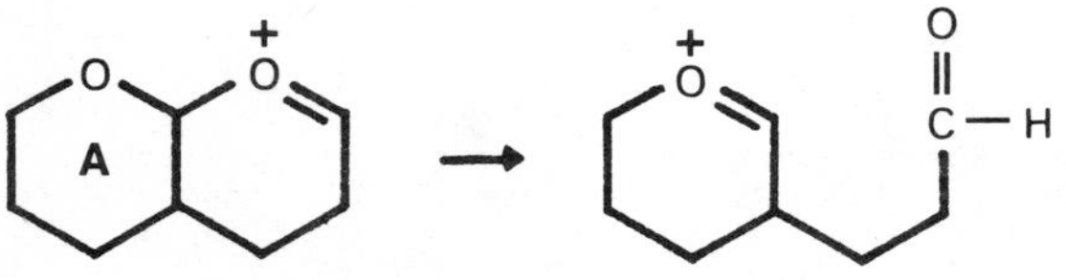

134

135

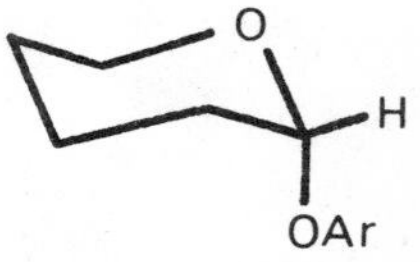

136

The experimental results are consistent with this expectation. The spontaneous hydrolysis of trans isomer **132** is 1380 times slower than that of p-nitrophenoxytetrahydropyran (**136**), while the cis isomer **133** is hydrolyzed ≈7 times more slowly than **136**. The enthalpy barrier associated with stereoelectronic control in the trans isomer **132** was estimated to be 7.2 kcal/mol, a value identical to that obtained for compound **130**.

Following the approach pioneered by Bürgi and Dunitz (80, 96), Jones and Kirby (97) further discovered a linear relationship between bond length and reactivity. They first determined crystal and molecular structures for a series of 2-substituted aryloxytetrahydropyran derivatives (98 - 101). I have previously pointed out that these compounds are hydrolyzed spontaneously by way of an oxonium ion and the crystal structures show that differences in bond length are already present in the ground state. They found that in axially oriented aryloxytetrahydropyrans, the endocyclic $C - O$ bond is significantly shortened and the $C - OAr$ bond of the leaving group lengthened, by an amount which depends on the electronegativity of the O-Aryl oxygen atom. They further noticed that this variation in bond length is very simply related to the rates of hydrolysis of these acetals. Indeed, the rates of hydrolysis show a linear dependence with the pKa of the leaving group when the $C - O$ cleavage is rate determining (102, 103). A plot of bond length against the pKa of the leaving group is equivalent to a plot against the free energy of activation for $C - O$ cleavage in water. This plot shows that the length of the two $C - O$ bonds of the acetal group depend linearly on the pKa of the leaving group. In this manner, they could predict that the $C - OCH_3$ bond length for a typical axial methyl glycopyranoside should be 1.401 Å, a value identical, within experimental error, to the observed mean value of 1.405 Å obtained by X-ray analysis (54).

The first experimental evidence that there is stereoelectronic control in the formation of an acetal function was recently obtained (23) by studying the mild acid cyclization of bicyclic hydroxypropyl acetal **137** (Fig. 18). At room temperature, compound **137** gave only the cis tricyclic acetal **138**; the appearance of a small quantity of the trans tricyclic acetal **139** occurred only after 5 days. On the other hand, when the cis acetal **138** was refluxed under the same conditions, isomerization took place to yield an equilibrium mixture of cis (45%) and trans (55%) acetals **138** and **139**.

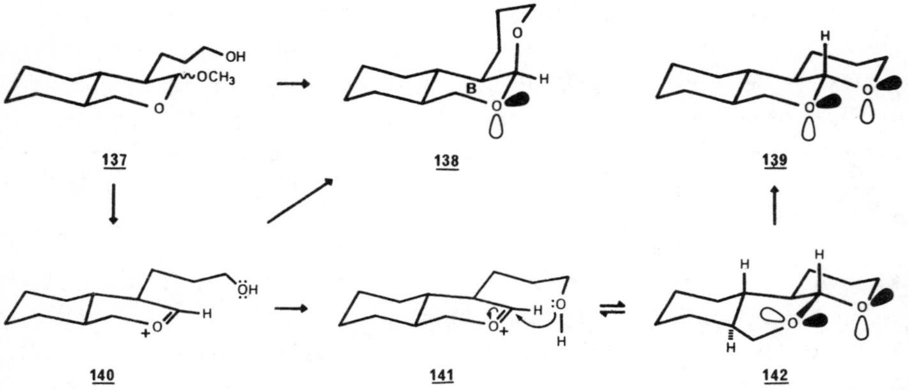

Fig. 18

The specific conversion of 137 into the cis tricyclic acetal 138 is clearly
the result of a kinetically controlled reaction. As a consequence, the
formation of trans acetal 139 must involve a higher energy barrier than
that of cis acetal 138. The first step in the cyclization reaction must
be the loss of methanol from the starting material 137 to give the cyclic
oxonium ion 140. The hydroxyl group has then the choice of attacking the β
or the α-face of the oxonium ion 140 to give the cis or the trans tricyclic
acetals 138 and 139 respectively. It can be readily seen that the β-attack
on 140 can give the cis acetal 138 with stereoelectronic control, because
the oxygen of ring B in 138 has an electron pair oriented antiperiplanar
to the newly formed C − O bond. On the other hand, the α-attack on 140 with
stereoelectronic control cannot yield the trans acetal directly in its
more stable conformation 139. The α-attack (i.e. 141) must first give conformer 142 of the trans acetal having its ring B in a boat form in order
to have the oxygen atom of ring B with an electron pair antiperiplanar
to the newly formed C − O bond. 142 would then undergo a conformational
change to the more stable conformation 139 of the trans acetal. Clearly,
the transition 140 → 141 → 142 → 139 requires more energy than the transition
140 → 138, a conclusion which is in accord with the experimental result.

In a more recent study (104), similar results were obtained with the bicyclic thiol hemiacetal 75 (Fig. 13). Cyclization of 75 under kinetically
controlled conditions gave the cis monothioacetal 76; the trans isomer
77 was not observed. Thus the attack of an SH group on a cyclic oxonium
ion must also take place preferentially with stereoelectronic control.

These recent results obtained with compounds in which the acetal function is locked in a specific conformation clearly demonstrate that stereoelectronic factors are important in connection with the reactivity of acetals. Stereoelectronic effects must therefore play a role in compounds where the conformation of the acetal function is not fixed. On that basis, it seems quite clear that α-glycosides must hydrolyze via their ground state conformation whereas β-glycosides must first assume a boat conformation in order to fulfill the stereoelectronic requirement. Interestingly, van Eikeren (92) in his study on the relative rates of hydrolysis of axial and equatorial bicyclic isomers **125** and **126** has obtained results which indicate that the anomers hydrolyze via different transition states. Moreover, the transition state of the axial anomer would involve more extensive C — O bond breakage than that of the equatorial anomer. This is in complete accord with the above conclusion concerning α and β-glycosides, since it is conceivable that, in the cleavage of a C—OR bond at the anomeric center, the C — OR bond will be longer in the transition state when the molecule exists in a chair rather than a boat form because the chair form is at a lower energy level.

It is well known that in lysozyme-oligosaccharide complexes (105-108), the pyrane ring derived from β-glycosides is distorted from the normal chair conformation toward a half-chair conformation. This distortion raises the energy of the ground state of the substrate and thus lowers the energy of activation for bond cleavage. Thus, in the hydrolysis of β-glycosides by lysozyme, hydrolysis would take place via a distorted substrate and that would be mainly for stereoelectronic reasons.

Chwang, Kresge, and Wiseman (109) have made the interesting observation that the bridged vinyl ether **143** is hydrated (at the bridgehead carbon) more slowly than **144** by a factor of 10^2, even though hydration of vinyl ethers is normally faster than that of analogous olefins by a factor of 10^{5-8}. The sulfur analog (**143**, O=S) was found to be even less reactive than **144**. These remarkable results can be readily explained by the fact that the electron pairs on the oxygen (or sulfur) atom are not properly aligned to give a delocalized oxonium ion (cf. **117**). The oxygen electron pairs are orthogonal to the π system of the double-bond; protonation of **143** must therefore yield a high energy undelocalized cation which is further destabilized by the inductive effect of the oxygen atom (cf. **118**). The overall rate of the hydration reaction is thus considerably reduced.

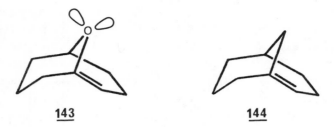

143   **144**

Hydride transfer to cyclic oxonium ion

We should also expect stereoelectronic control when the hydroxyl group
is replaced by another nucleophile in the reaction with cyclic oxonium
ions. A recent report (110) shows that hydride transfer to cyclic oxonium
ion is subject to stereoelectronic control. Tricyclic spiroketal **145** (Fig.
19) undergoes an acid-catalyzed oxidation-reduction reaction to give the
equatorial bicyclic aldehyde **147** stereospecifically. Similarly, spiroketals
148 and **149** gave the corresponding equatorial bicyclic ketone **150**.

145 R = R' = H
148 R = CH₃, R' = H
149 R = H, R' = CH₃

146

147 R = H
150 R = CH₃

151

152

153

154

155

Fig. 19

These results are interpreted by invoking an internal hydride transfer from an alcohol function to the cyclic oxonium ion (i.e. **146**) (111). It can be readily seen that the β-attack on **146**, i.e. **151** can occur with stereoelectronic control yielding **147** and **150** in their most stable conformation **152**. On the other hand, the α-attack on **146**, i.e. **153** must first give conformer **154** having its ring B in a boat form in order to fulfill the electron pair alignment requirement. Conformer **154** would then give the more stable conformation **155** of the corresponding axial isomer. The transformation **153** → **154** → **155** requires more energy than **151** → **152**; thus, the equatorial isomers **147** and **150** are the sole products of these transformations.

This rationalization indicates that internal delivery of a hydride is not a requisite for the observed stereospecificity. Reduction of the oxonium ion with an external hydride reagent should also give equatorially oriented bicyclic ether only. Accordingly (112), reduction of tricyclic spiroketal **145** with sodium cyanoborohydride at pH ≈3-4 yields only the equatorial bicyclic ether alcohol (**147**, CHO = CH$_2$OH). Eliel and co-workers (113) have previously suggested that the orientation of the electron pairs of oxygen atoms influence the course of the reduction of 2-alkoxytetrahydropyran with lithium aluminium hydride-aluminium trichloride.

Oxidation of the C—H bond in acetals

In 1971, it was discovered that ozone reacts in a completely specific fashion with the acetal function derived from an aldehyde to give the corresponding ester and alcohol (114). This reaction proceeds via the insertion of ozone into the C—H bond of the acetal forming a hydrotrioxide intermediate (**156**) which breaks down to yield the reaction products, the ester, the alcohol and singlet oxygen (115, 116). The hydrotrioxide intermediate **156** can be detected at low temperature (115).

$$R - \overset{\overset{\displaystyle OR}{|}}{\underset{\underset{\displaystyle H}{|}}{C}} - OR + O_3 \longrightarrow R - \overset{\overset{\displaystyle OR}{|}}{\underset{\underset{\displaystyle O-O-O-H}{|}}{C}} - OR \longrightarrow R - \overset{\overset{\displaystyle O}{\|}}{C} - OR + ROH + O_2$$

156

In the course of a study on the generality of this reaction (117), it was found that the rates at which different acetals are oxidized vary considerably and this led to the proposal that the reaction is controlled by stereoelectronic factors. Basically, it was proposed that in order for oxidation to proceed, it is required that one non-bonded electron pair on each oxygen atom lie antiperiplanar to the C—H bond of the acetal function.

A study (118, 119) on the reactivity of a series of appropriate chemical models did confirm this postulate. Conformers **A**, **C**, and **F** of an acetal function (Fig. 1, p. 4) possess on each oxygen atom an electron pair oriented antiperiplanar to the C — H bond, conformers **B** and **D** have only one oxygen with an electron pair properly oriented whereas conformer **E** has none. Thus, conformers **B**, **D**, and **E** should be inert and conformers **A**, **C**, and **F** should be reactive towards ozone.

Conformationally rigid β-glycosides were found to be reactive towards ozone. At least one of the rotamers E_1, E_2, and E_3 of a β-glycoside (Fig. 9) which corresponds to conformers **A**, **B**, and **C** respectively of an acetal function should therefore be reactive. Compounds **157** and **158** are rigid model compounds for rotamers E_1 and E_3. These two compounds react with ozone; thus, conformers **A** and **C** are reactive conformers. On the other hand, the unreactive compound **159** is a rigid model for rotamer E_2; thus, conformer **B** is not reactive, as predicted.

157 R = CH$_3$ **158**   **159**

Conformationally rigid α-glycosides were found to be inert toward ozone. Rotamers A_1, A_2, and A_3 of an α-glycoside which correspond to conformers **D**, **B'**, and **E** respectively of an acetal function are consequently unreactive as predicted. Compound **160** can be considered a rigid model for the most stable rotamer A_1 (conformer **D**) of an α-glycoside and it was found unreac-

160 R = CH$_3$ **161**

tive. Finally, 1,3-dioxanes **161** which are rigid models for conformer **F** were smoothly oxidized.

This analysis was supported by the fact that conformationally labile α-glycopyranosides were found to react with ozone. Similar results were also observed with α and β-glycofuranosides as these compounds are not maintained in a rigid conformation.

Interestingly, acyclic (dialkoxy) acetals react with ozone at a much slower rate than cyclic acetals such 1,3-dioxanes **161**. It was proposed (118) that the low rate of oxidation of acyclic acetals is due to the fact that they exist in the conformation **D** (Fig. 1) which is not reactive. Thus, acyclic acetals would first undergo a conformational change from conformer **D** to either conformer **A**, **C**, or **F** before reacting with ozone. The fact that a large percentage of acyclic acetals exist in the non-reactive conformation **D** lowers the effective concentration and the rate of oxidation is affected accordingly. Also, conformer **F** of cyclic acetals might be at a higher energy level, thus more reactive, than the conformations **A** or **C** which can be taken by acyclic acetals.

Taillefer and co-workers (120) have undertaken an interesting kinetic study of cyclic and acyclic acetals. They found that the enthalpy of activation is remarkably low (about 5-7 kcal/mol) for cyclic and acyclic acetals, whereas the entropy of activation was found to be very high and negative, indicating a highly ordered transition state. They also showed that the two different types of acetals behave differently toward ozone. They observed an isokinetic relationship for acyclic acetals and the isokinetic temperature was found to be below the experimental temperature range, a domain where reactivity is dominated by entropy factors. These results contrasted with those obtained for cyclic acetals where the isokinetic temperature falls above the working temperatures, a domain of temperatures where reactivity depends mainly on enthalpy factors. These results and the above rationalization constitute rather strong experimental evidence that a conformational change before oxidation must take place in the case of acyclic acetals.

It is clear from the above results that the oxidation of the acetal function is controlled by stereoelectronic factors. The next step is to try to understand how these electronic effects operate to lower the energy barrier of the oxidation reaction.

The experimental evidence obtained (116, 117) indicates that the central carbon of the acetal function becomes positively charged during the oxidation step. Furthermore, a rather high primary isotope effect ($k^H/k^D = 6.5$) has been measured (121). These results indicate that the reaction mechanism proceeds either via a direct hydride transfer yielding a dialkoxycarbonium ion and a hydrotrioxide ion which would collapse to the hydrotrioxide intermediate ($\underline{162} \rightarrow \underline{163} \rightarrow \underline{165}$) (Fig. 20), or via an insertion of ozone in a 1,3-

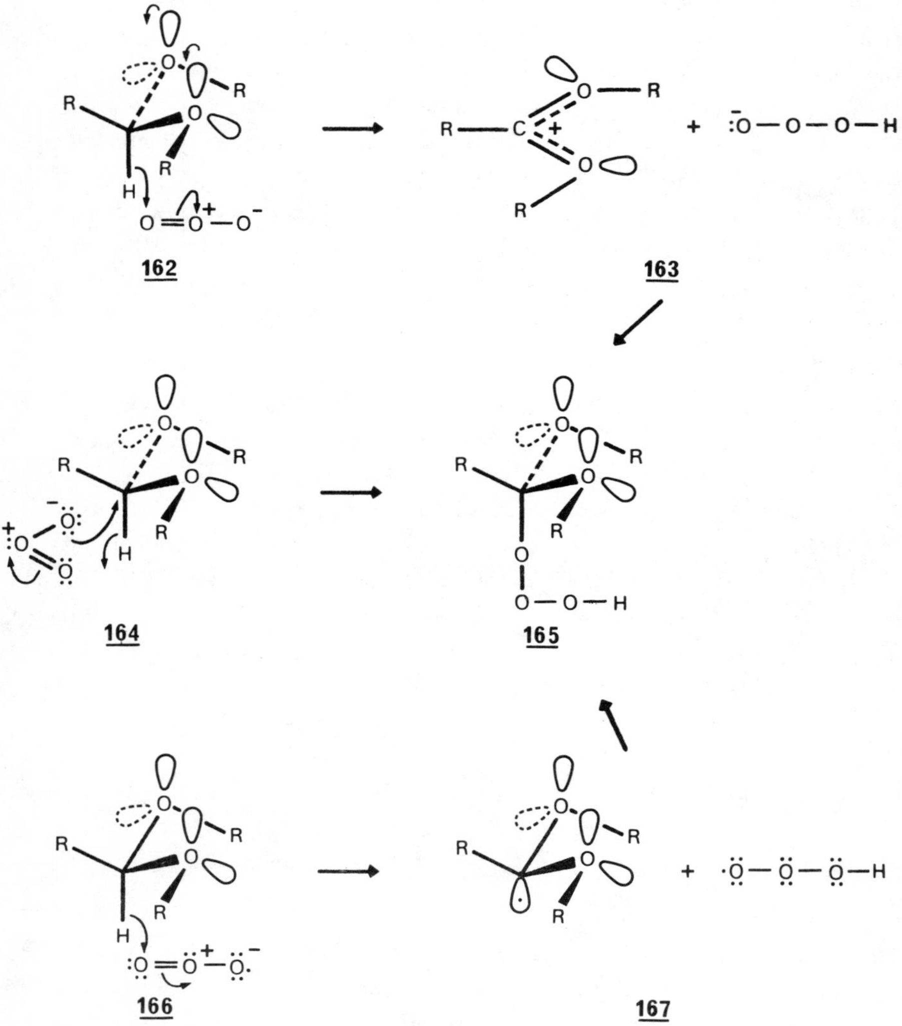

Fig. 20

fashion into the C — H bond of the acetal function (**164** → **165**). If this is the case, the need for an electron pair antiperiplanar to the C — H bond for each oxygen atom of the acetal function becomes clear. Antiperiplanar electron pairs should increase the electron density of the C — H bond which would then be more easily attacked by an electrophile like ozone. An <u>ab initio</u> study carried out by Lehn, Wipff, and Bürgi (78) is in accord with this conclusion. This effect should also stabilize the incipient dialkoxy-carbonium ion **163**, if it is formed. This rationalization is in agreement with the fact that negatively charged hemiacetal oxy anion **168** reacts more rapidly than dialkoxyacetals with ozone (122). The product of the reaction is a mixture of the corresponding carboxylic acid and ester which presumably come from the two possible modes of cleavage of intermediate **169**.

$$
\begin{array}{c}
\underset{\displaystyle \overset{\displaystyle |}{\underset{\displaystyle H}{}}}{R-C-O^-} + O_3
\end{array}
\;\longrightarrow\;
\begin{array}{c}
\overset{\displaystyle OR}{\underset{\displaystyle O-O-O-H}{R-C-O^-}}
\end{array}
\;\longrightarrow\;
\begin{array}{c}
\overset{\displaystyle O}{R-C-OR}
\end{array}
\;+\;
\begin{array}{c}
\overset{\displaystyle O}{R-C-OH}
\end{array}
$$

168   **169**

It is also possible that the formation of the hydrotrioxide intermediate occurs <u>via</u> a radical mechanism (**166** → **167** → **165**). This implies that a radical would be more stable when there are two electron pairs on adjacent oxygens oriented antiperiplanar to it. There is evidence in the literature that this is indeed the case. Bernasconi and Descotes (123) have observed a preference for axial hydrogen abstraction of <u>cis</u> and <u>trans</u> 2,6-dimethoxy-tetrahydropyran by photolysis of benzophenone; the <u>cis</u> isomer **170** was photo-degraded more rapidly than the <u>trans</u> isomer **171**. Hayday and McKelvey (124) found that, at ambient temperature, triplet benzophenone abstracted the axial hydrogen from <u>cis</u>-2-methoxy-4-methyltetrahydropyran (**172**) about 8 times faster than it abstracted the equatorial hydrogen from the <u>trans</u> isomer **173**. Since both compounds gave the same product distribution it was concluded that a common free radical was formed. This was subsequently confirmed (125) by using EPR spectroscopy to identify 2-alkoxytetrahydro-pyran-2-yl radicals generated from conformationally rigid <u>cis</u> and <u>trans</u> precursors. The <u>tert</u>-butoxyl radical was also shown to exhibit a strong (12 fold) preference at room temperature for the removal of the axial over the equatorial hydrogen of the anomeric carbon, with conformationally rigid pairs of 2-methoxy-1,3-dioxanes and 2-methyl-1,3-dioxanes (125).

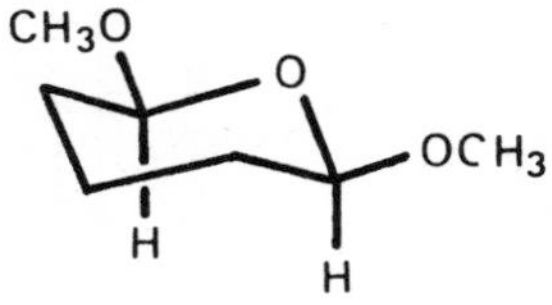

170

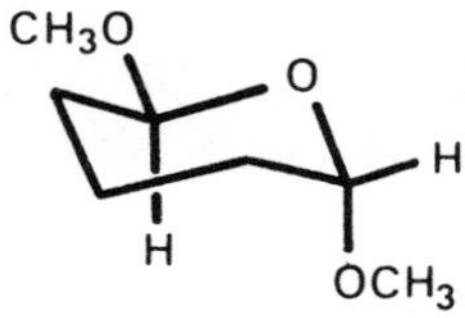

171

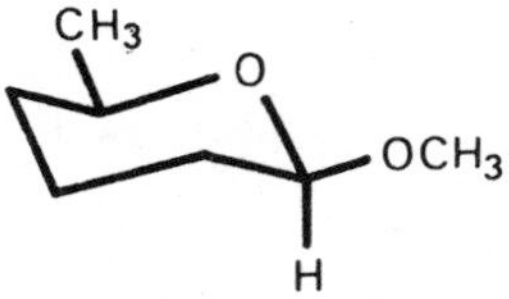

172

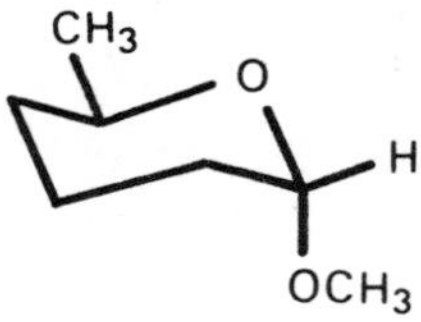

173

In a more recent study on the relative rates of hydrogen atom abstraction by photogenerated tert-butoxyl radical on a variety of cyclic and acyclic ethers, acetals, and orthoformates, Malatesta and Ingold (126) observed a pronounced stereoelectronic effect which produces high rates of abstraction from those C — H bonds which had adjacent oxygen with appropriate electron pair orientation. When the C — H bond had adjacent oxygen with improperly aligned electron pair, abstraction was very much slower. Similar results were observed by Beckwith and Easton (127) with substituted 1,3-dioxanes. The conversion of acetals to lactones by radical abstraction where there would be a stereoelectronic control in bond cleavage has also been reported (128). Remy, Cottier, and Descotes (129) have reported the photolysis of the α and β-isomers **174** and **175**. With the β-anomer **174**, the reaction is complete in 40 h and yields the spiro compound **176**. With the α-anomer **175**, the reaction is slow and does not lead to the formation of a specific product.

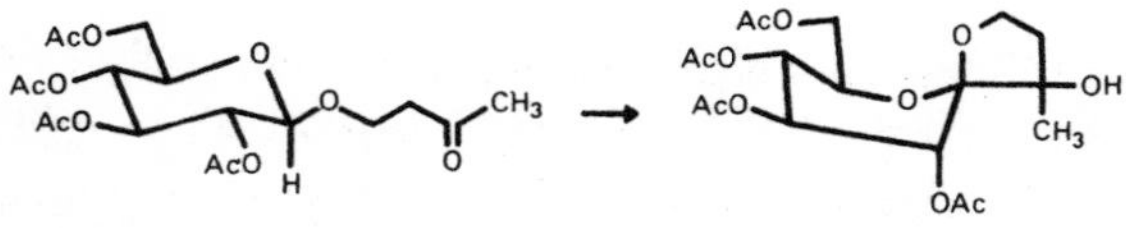

174   **176** 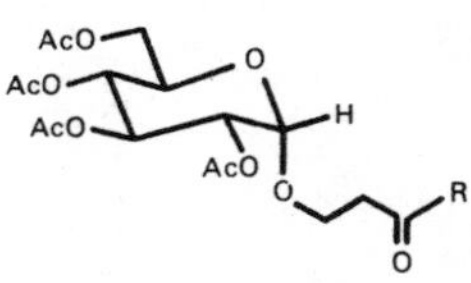   **175**

The oxidation of aldehyde acetals to esters has also been reported with various agents such as N-bromosuccinimide (130-135), triethyloxonium (136, 137) or triphenylmethyl tetrafluoroborate (138). It has not yet been verified if these reactions are controlled by stereoelectronic effects. Angyal and co-workers made an interesting observation in the course of their study (139-143) on the oxidation of acetals derived from carbohydrates with chromium trioxide in acetic acid. They found (139) that α and β-glycosides react differently with this reagent. With β-glycosides, the anomeric hydrogen is first oxidized and the product obtained is the ketoester (177 → 178) whereas, with α-glycosides, the methyl group is slowly oxidized to give the formate (179 → 180). Stereoelectronic effects must again play a role in these two reactions.

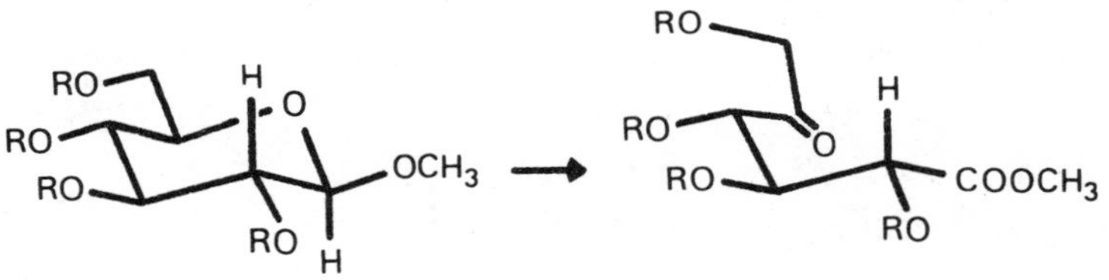

177   178

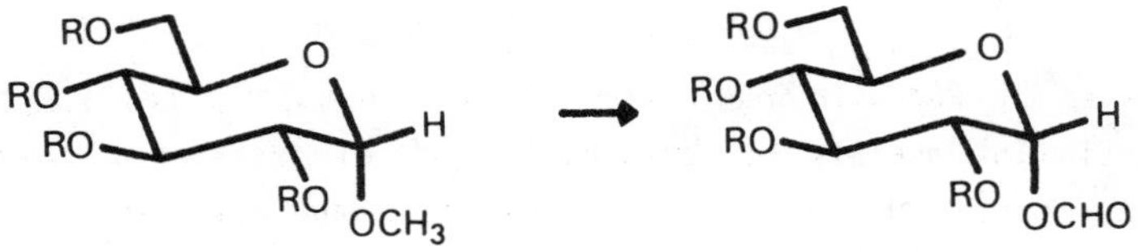

179   180

REFERENCES

(1) Lemieux, R.U.; Koto, S. Tetrahedron 1974, 30, 1933, and references quoted therein.
(2) "Anomeric Effect, Origin Consequences"; Szarek, W.A. and Horton, D., Ed.; ACS Symposium Series no 87; Washington, D.C., 1979.
(3) Zefirov, N.S.; Shekhtman, N.M. Russ. Chem. Rev. 1971, 40(4), 315.
(4) Lemieux, R.U.; Chü, N.J. Abstracts of Papers; Am. Chem. Soc. 1958, 133, 31N.

(5) Lemieux, R.U.; Pavia, A.A.; Martin, J.C.; Watanabe, K.A. Can. J. Chem.
     1969, 47, 4427.

(6) Edward, J.T. Chem. Ind. 1955, 1102.

(7) Kabayama, M.A.; Patterson, D. Can. J. Chem. 1958, 36, 563.

(8) Eliel, E.L. Kem. Tidskr. 1969, 81, NR/6-7, 22.

(9) Wolfe, S.; Whangbo, M.-H.; Mitchell, D. Carbohydr. Res. 1979, 69,
     1 and references quoted therein.

(10) Radom, L.; Hehre, W.J.; Pople, J.A. J. Am. Chem. Soc. 1972, 94, 2371.

(11) Jeffrey, G.A.; Pople, J.A.; Radom, L. Carbohydr. Res. 1972, 25, 117.

(12) David, S.; Eisenstein, O.; Hehre, W.J.; Salem, L.; Hoffmann, R. J.
     Am. Chem. Soc. 1973, 95, 3806.

(13) Baddeley, G. Tetrahedron Lett. 1973, 1645.

(14) Brockway, L.O. J. Phys. Chem. 1937, 41, 185.

(15) Pauling, L. "The Nature of the Chemical Bond"; 3rd Ed.; Cornell Univer-
     sity Press: Ithaca, N.Y., 1960; pp. 314-315.

(16) Altona, C. Ph.D. Thesis, University of Leiden, 1964.

(17) Romers, C.; Altona, C.; Buys, H.R.; Havinga, E. "Topics in Stereochem-
     istry"; Vol. 4; Eliel, E.L. and Allinger, N.L., Eds; Wiley-Interscience:
     New York, 1969; p. 39.

(18) Sweigart, D.A.; Turner, D.W. J. Am. Chem. Soc. 1972, 94, 5599.

(19) Sweigart, D.A. J. Chem. Educ. 1973, 50, 322.

(20) David, S. In reference 2.

(21) Gillespie, R.J. "Molecular Geometry"; Van Nostrand Reinhold Company
     Ltd: London, 1972; p. 6.

(22) Descotes, G.; Lissac, M.; Delmau, J.; Duplan, J. C. R. Hebd. Séances
     Acad. Sci., Ser. C 1968, 267, 1240.

(23) Beaulieu, N.; Dickinson, R.A.; Deslongchamps, P. Can. J. Chem. 1980,
     58, 2531.

(24) Deslongchamps, P.; Rowan, D.D.; Pothier, N.; Sauvé, G.; Saunders,
     J.K. Can. J. Chem. 1981, 59, 1105.

(25) Pothier, N.; Rowan, D.D.; Deslongchamps, P.; Saunders, J.K. Can. J.
     Chem. 1981, 59, 1132.

(26) Evans, D.A.; Sacks, C.E.; Whitney, R.A.; Mandel, N.G. Tetrahedron
     Lett. 1978, 727.

(27) Cresp, T.M.; Probery, C.L.; Sondheimer, F. Tetrahedron Lett. 1978,
     3955.

(28) Hughes, D.H. Tetrahedron Lett. 1978, 3959.

(29) Evans, D.A.; Sacks, C.E.; Kleschick, W.A.; Taber, T.R. J. Am. Chem.
     Soc. 1979, 101, 6789.

(30) Eliel, E.; Knoeber, M.C. J. Am. Chem. Soc. 1968, 90, 3444.

(31) Hintzer, K.; Weher, R.; Schurig, V. Tetrahedron Lett. 1981, 22, 55.

(32) Francke, W.; Reith, W.; Sinnwell, V. Chem. Ber. 1980, 113, 2686.

(33) Fukuyama, T.; Akasaka, K.; Karanewsky, D.S.; Wang, C.-L.; Schmid, G.; Kishi, Y. J. Am. Chem. Soc. 1979, 101, 262.

(34) Dobler, M. "Ionophores and Their Structures"; John Wiley & Sons: New York, 1981.

(35) Cottier, L.; Descotes, G.; Grenier, M.F.; Metras, F. Tetrahedron, 1981, 37, 2515.

(36) Lemieux, R.U.; Koto, S.; Voisin, D. In reference 2.

(37) Exner, O.; Jehlička, V.; Uchytil, B. Collect. Czech. Chem. Commun. 1968, 33, 2862.

(38) Astrup, E.E. Acta Chem. Scand. 1971, 25, 1494.

(39) Uchida, T.; Tadokoro, H. J. Polym. Sci. A-2 1967, 5, 63.

(40) Dale, J. Tetrahedron 1974, 30, 1683.

(41) Borgen, G.; Dale, J. J. Chem. Soc., Chem. Commun. 1974, 484.

(42) Anet, F.A.L.; Cheng, A.K.; Wagner, J.J. J. Am. Chem. Soc. 1972, 94, 9250.

(43) Anet, F.A.L.; Krane, J.; Dale, J.; Daasvatn, K.; Kristiansen, P.O. Acta Chem. Scand. 1973, 27, 3395.

(44) Anet, F.A.L.; Yavari, I. J. Am. Chem. Soc. 1977, 99, 6752.

(45) Blanchette, A.; Sauriol-Lord, F.; St-Jacques, M. J. Am. Chem. Soc. 1978, 100, 4055.

(46) Anet, F.A.L.; Degen, P.J.; Krane, J. J. Am. Chem. Soc. 1976, 98, 2059.

(47) Dale, J.; Ekeland, T.; Krane, J. J. Am. Chem. Soc. 1972, 95, 1389.

(48) Bishop, C.T.; Cooper, F.P. Can. J. Chem. 1963, 41, 2743.

(49) Anderson, C.B.; Sepp, D.T. Tetrahedron 1968, 24, 1707.

(50) Eliel, E.L.; Giza, C.A. J. Org. Chem. 1968, 33, 3754.

(51) Edward, J.T.; Morand, P.F.; Puskas, I. Can. J. Chem. 1961, 39, 2069.

(52) Deslongchamps, P.; Pothier, N.; Ruest, L. Unpublished results.

(53) De Hoog, A.J.; Buys, H.R.; Altona, C.; Havinga, E. Tetrahedron 1969, 25, 3365.

(54) Berman, H.M.; Chu, S.S.; Jeffrey, G.A. Science 1967, 157, 1576.

(55) De Hoog, A.J.; Buys, H.R. Tetrahedron Lett. 1969, 4175.

(56) Gelin, M.; Bahurel, Y.; Descotes, G. Bull. Soc. Chim. Fr. 1970, 3723.

(57) Pothier, N. Ph.D. Thesis, Université de Sherbrooke, 1981.

(58) Zefirov, N.S.; Shekhtman, N.M. Dokl. Akad. Nauk SSSR 1968, 180, 1363.

(59) Zefirov, N.S.; Shekhtman, N.M. Dokl. Akad. Nauk SSSR 1967, 177, 842.

(60) Perlin, A.S. Pure Appl. Chem. 1978, 50, 1401.

(61) Zefirov, N.S.; Blagoveshchenskii, V.S.; Kazamirchik, I.V.; Yakovleva, O.P. Zhur. Org. Khim. 1970, 6, 877.

(62) Deslongchamps, P.; Rowan, D.D.; Pothier, N.; Saunders, J.K. Can. J. Chem. 1981, 59, 1122.

(63) Deslongchamps, P.; Guay, D. Unpublished results.

(64) Sauriol-Lord, F.; St-Jacques, M. Can. J. Chem. 1979, 57, 3221.

(65) Booth, H.; Lemieux, R.U. Can. J. Chem. 1971, 49, 777.

(66) Allingham, Y.; Cookson, R.C.; Crabb, T.A.; Vary, S. Tetrahedron 1968, 24, 4625.

(67) Riddell, F.G.; Lehn, J.M. J. Chem. Soc. B 1968, 1224.

(68) Crabb, T.A.; Turner, C.H.; Newton, R.F. Tetrahedron 1968, 24, 4423.

(69) Baker, V.J.; Ferguson, I.J.; Katritzky, A.R.; Patel, R.; Rahimi-Rastgoo, S. J. Chem. Soc., Perkin Trans 2 1978, 377.

(70) Katritzky, A.R.; Patel, R.C.; Brito-Palma, M.S.F.; Riddell, F.G.; Turner, E.S. Isr. J. Chem. 1980, 20, 150.

(71) Hutchins, R.O.; Kopp, L.D.; Eliel, E.L. J. Am. Chem. Soc. 1968, 90, 7174.

(72) Jones, R.A.Y.; Katritzky, A.R.; Richards, A.C. J. Chem. Soc., Chem. Commun. 1969, 708.

(73) Riddell, F.G.; William, D.A.R. Tetrahedron Lett. 1971, 2073.

(74) Eliel, E.L.; Kopp, L.D.; Dennis, J.E.; Evans, Jr., S.A. Tetrahedron Lett. 1971, 3049.

(75) Jones, R.A.; Katritzky, A.R.; Snarey, M. J. Chem. Soc. B 1970, 131.

(76) Cordes, E.H.; Bull, H.G. Chem. Rev. 1974, 74, 581.

(77) Bürgi, H.-B.; Dunitz, J.D.; Shefter, E. Acta Crystallogr., Sect. B 1974, 30, 1517.

(78) Lehn, J.-M.; Wipff, G.; Bürgi, H.-B. Helv. Chim. Acta 1974, 57, 493.

(79) Wipff, G. Tetrahedron Lett. 1978, 3269.

(80) Bürgi, H.B.; Dunitz, J.D.; Shefter, E. J. Am. Chem. Soc. 1973, 95, 5065.

(81) X-ray Analysis and the Structure of Organic Molecules. J.D. Dunitz. Cornell University Press: Ithaca and London, 1978, pp. 366-384.

(82) Storm, D.R.; Koshland, Jr., D.E. J. Am. Chem. Soc. 1972, 94, 5815.

(83) Bürgi, H.B.; Lehn, J.M.; Wipff, G. J. Am. Chem. Soc. 1974, 96, 1956.

(84) Bürgi, H.B.; Dunitz, J.D.; Lehn, J.M.; Wipff, G. Tetrahedron 1974, 30, 1563.

(85) Schweizer, W.B.; Procter, G.; Kaftory, M.; Dunitz, J.D. Helv. Chim. Acta 1978, 61, 2783.

(86) Capon, B.; Thacker, D. J. Chem. Soc. B 1967, 1010.

(87) Banks, B.E.C.; Meinwald, Y.; Rhind-Tutt, A.J.; Sheft, I.; Vernon, C.A. J. Chem. Soc. 1961, 3240.

(88) Overend, W.G.; Rees, C.W.; Sequeira, J.S. J. Chem. Soc. 1962, 3429.

(89) Feather, M.S.; Harris, J.F. J. Org. Chem. 1965, 30, 153.

(90) BeMiller, J.N.; Doyle, E.R. Carbohydr. Res. 1971, 20, 23.

(91) Capon, B. Chem. Rev. 1969, 69, 407.

(92) van Eikeren, P. J. Org. Chem. 1980, 45, 4641.

(93) Chandrasekhar, S.; Kirby, A.J. J. Chem. Soc., Chem. Commun. 1978, 171.

(94) Kirby, A.J.; Martin, R.J. J. Chem. Soc., Chem. Commun. 1978, 803.

(95) Kirby, A.J.; Martin, R.J. J. Chem. Soc., Chem. Commun. 1979, 1079.

(96) Bürgi, H.B. Angew. Chem. Int. Ed. 1975, 14, 460.

(97) Jones, P.G.; Kirby, A.J. J. Chem. Soc., Chem. Commun. 1979, 288.

(98) Jones, P.G.; Kennard, O.; Chandrasekhar, S.; Kirby, A.J. Acta Crystallogr., Sect. B 1978, 34, 2947.

(99) Jones, P.G.; Kennard, O.; Chandrasekhar, S.; Kirby, A.J. Acta Crystallogr. Sect. B 1978, 34, 3835.

(100) Jones, P.G.; Kennard, O.; Chandrasekhar, S.; Kirby, A.J. Acta Crystallogr. Sect. B 1979, 35, 239.

(101) Jones, P.G.; Kennard, O.; Kirby, A.J.; Martin, R.J. Acta Crystallogr. Sect. B 1979, 35, 242; Ibid.; in the press.

(102) Craze, G.A.; Kirby, A.J. J. Chem. Soc., Perkin Trans. 2 1978, 354.

(103) Fife, T.H.; Brod, L.H. J. Am. Chem. Soc. 1970, 92, 1681.

(104) Deslongchamps, P.; Guay, D. Unpublished results.

(105) Chipman, D.M.; Sharon, N. Science 1969, 165, 454.

(106) Ford, L.O.; Johnson, L.N.; Mechin, P.A.; Phillips, D.C.; Tjian, R. J. Mol. Biol. 1974, 88, 349.

(107) Stryer, L. "Biochemistry"; W.H. Freeman and Company: San Francisco, 1975; p. 165.

(108) Gorenstein, D.G.; Findlay, J.B.; Luxon, B.A.; Kar, D. J. Am. Chem. Soc. 1977, 99, 3473.

(109) Chwang, W.K.; Kresge, A.J.; Wiseman, J.R. J. Am. Chem. Soc. 1979, 101, 6972.

(110) Deslongchamps, P.; Rowan, D.D.; Pothier, N. Heterocycles 1981, 15, 1093.

(111) Woodward, R.B.; Sondheimer, F.; Mazur, Y. J. Am. Chem. Soc. 1958, 80, 6693.

(112) Deslongchamps, P.; Rowan, D.D.; Pothier, N. Can. J. Chem. In press.

(113) Eliel, E.L.; Nowak, B.E.; Daigneault, R.A.; Badding, V.G. J. Org. Chem. 1965, 30, 2441.

(114) Deslongchamps, P.; Moreau, C. Can. J. Chem. 1971, 49, 2465.

(115) Kovac, F.; Plesniscar, B. J. Chem. Soc., Chem. Comm. 1978, 122; J. Am. Chem. Soc. 1979, 101, 2677.

(116) Taillefer, R.J.; Thomas, S.E.; Nadeau, Y.; Fliszar, S.; Henry, H. Can. J. Chem. 1980, 58, 1138.

(117) Deslongchamps, P.; Moreau, C.; Fréhel, D.; Atlani, P. Can. J. Chem. 1972, 50, 3402.

(118) Deslongchamps, P.; Atlani, P.; Fréhel, D.; Malaval, A.; Moreau, C. Can. J. Chem. 1974, 52, 3651.

(119) Deslongchamps, P.; Moreau, C.; Fréhel, D.; Chênevert, R. Can. J. Chem. 1975, 53, 1204.

(120) Taillefer, R.J.; Thomas, S.E.; Nadeau, Y.; Beierbeck, H. Can. J. Chem. 1979, 57, 3041.

(121) Taillefer, R.J.; Thomas, S.E. Unpublished results.

(122) Sundararaman, P.; Walker, E.C.; Djerassi, C. Tetrahedron Lett. 1978, 1627.

(123) Bernasconi, C.; Descotes, G. C. R. Hebd. Séances Acad. Sci., Ser. C 1975, 280, 469.

(124) Hayday, K.; McKelvey, R.D. J. Org. Chem. 1976, 41, 2222.

(125) Malatesta, V.; McKelvey, R.D.; Babcock, B.W.; Ingold, K.U. J. Org. Chem. 1979, 44, 1872.

(126) Malatesta, V.; Ingold, K.U. J. Am. Chem. Soc. 1981, 103, 609.

(127) Beckwith, A.L.J.; Easton, C.J. J. Am. Chem. Soc. 1981, 103, 615.

(128) Rynard, C.M.; Thankaghan, C.; Tidwell, T.T. J. Am. Chem. Soc. 1979, 101, 1196.

(129) Remy, G.; Cottier, L.; Descotes, G. Can. J. Chem. 1980, 58, 2660.

(130) Marvell, E.N.; Joncich, M.J. J. Am. Chem. Soc. 1951, 73, 973.

(131) Rieche, A.; Schmitze, E.; Schade, W.; Beyer, E. Chem. Ber. 1961, 94, 2926.

(132) Prugh, J.D.; McCarthy, W.C. Tetrahedron Lett. 1966, 1351.

(133) Hanessian, S. Carbohydr. Res. 1966, 2, 86.

(134) Hanessian, S.; Plessas, N.R. J. Org. Chem. 1969, 34, 1035, 1045, 1053.

(135) Failla, D.L.; Hullar, T.L.; Siskin, S.B. Chem. Commun. 1966, 716.

(136) Meerwein, H.; Hederick, V.; Morshel, H.; Wunderlich, K. Annalen 1960, 635, 1.

(137) Slomkowski, St.; Penczek, St. Chem. Commun. 1970, 1347.

(138) Barton, D.H.R.; Magnus, P.D.; Smith, G.; Streckert, G.; Zurr, D. J. Chem. Soc., Perkin Trans. 1 1972, 542.

(139) Angyal, S.J.; James, K. Carbohydr. Res. 1970, 12, 147.

(140) Angyal, S.J.; James, K. Aust. J. Chem. 1970, 23, 1209.

(141) Angyal, S.J.; James, K. Aust. J. Chem. 1971, 24, 1219.

(142) Angyal, S.J.; Evans, M.E. Aust. J. Chem. 1972, 25, 1495.

(143) Angyal, S.J.; Evans, M.E. Aust. J. Chem. 1972, 25, 1513.

CHAPTER 3

ESTERS AND RELATED FUNCTIONS

Stereoelectronic effects and the ester function

This Chapter deals with the stereoelectronic effects which control the cleavage of tetrahedral intermediates during the formation or the hydrolysis of an ester. Since these effects are also operative in the ester function itself, a discussion of the functional group will first be presented.

In the ester function, it was proposed that two types of electronic effects referred to as the primary and the secondary electronic effects (1-3), are present. The role and the relative importance of these two types of electronic effects can be understood by analyzing them in relation to the geometry and the relative stability of the two forms, Z and E, of the ester function.

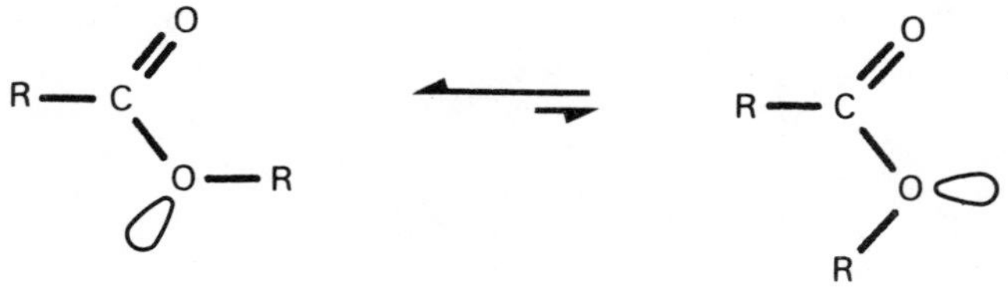

It is well known that esters are planar and that the Z form is much more stable (≈ 3 kcal/mol) than the E form (4). The E form is of course observed in small ring lactones but as soon as the size of the lactone ring allows the Z form to exist, the E form is no longer observed (5). Even in the case of tert-butyl formate, where there must be a strong steric repulsion between the carbonyl oxygen atom and the tert-butyl group in the Z form, this form still predominates ($\approx$90%) (6, 7).

The primary electronic effect is due to the delocalization of electron pairs between the ether oxygen and the carbonyl group of the ester function as expressed by the resonance structure 1, 2, and 3. Resonance structures 1 and 2 show the delocalization of an electron pair between the carbon and oxygen of the carbonyl group (1 ↔ 2) and resonance structures 2 and 3 show that the ether oxygen has one electron pair delocalized towards the same central carbon (2 ↔ 3). The primary electronic effect can therefore be viewed as the result of two n-π* interactions. The three atoms involved can be considered to be sp^2 hybridized and on that basis, the three-dimensional representations 4 and 5 correspond respectively to the Z and E forms of the ester function.

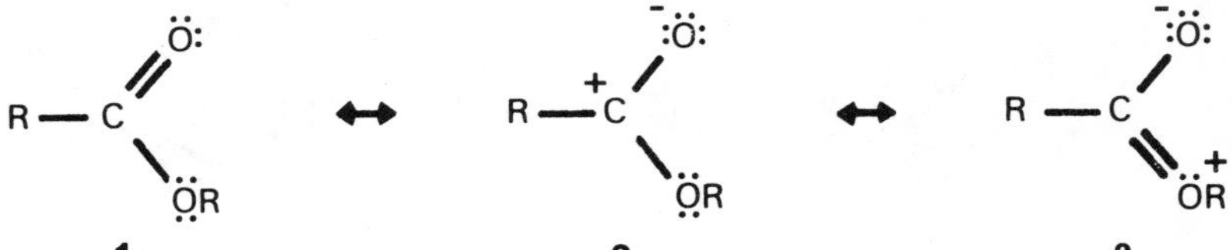

1   **2**   **3**

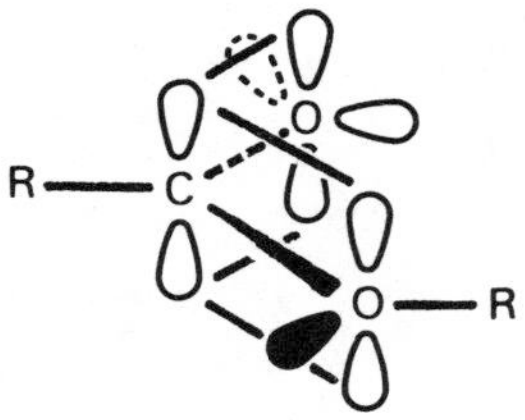

4 (Z) 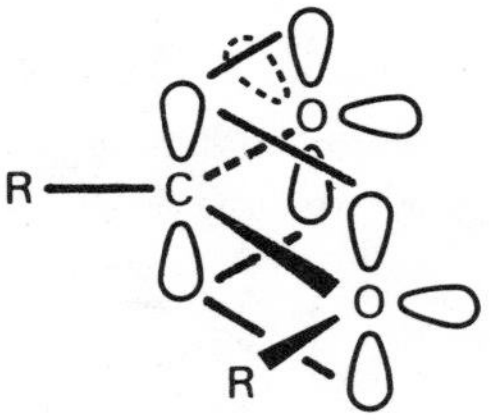   **5 (E)**

The secondary electronic effects in the ester function are essentially similar to the anomeric effect discussed previously for the acetal function, involving an n-σ* interaction. The only difference is that the central carbon is trigonal (sp^2 hybridized) in esters and tetrahedral (sp^3 hybridized) in acetals.

The carbonyl oxygen in both the Z (4) and the E (5) esters has an electron pair oriented antiperiplanar to the C — OR bond, and an n-σ* interaction should therefore exist because this electron pair orbital can overlap with

the antibonding orbital (σ*) of the C—OR bond. In other words, this secondary electronic effect (or anomeric effect) should give a triple bond charac-
ter to the carbonyl group ($\ddot{\text{O}}=$C$-\ddot{\text{O}}$R $\leftrightarrow$ $^{+}\ddot{\text{O}}\equiv$C $^{-}:\ddot{\text{O}}$R). In the Z form (4), there
is the possibility for another secondary electronic effect because, in
this form, the ether oxygen has an electron pair which is oriented antiperiplanar to the C — O σ bond of the carbonyl group. This electron pair
orbital can therefore overlap with the antibonding orbital (σ*) of that
bond. Thus, in Z esters, in addition to the primary electronic effect,
there are two secondary electronic effects which are illustrated in three
dimensions by 6 and in two dimensions by 7. In E esters, besides the primary
electronic effect, there is only one secondary electronic effect as shown
in three dimensions by 8 and in two dimensions by 9.

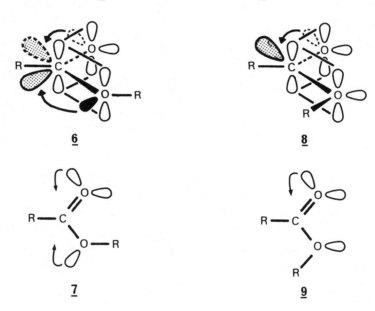

6   8

7   9

The additional stabilizing secondary electronic effect in the Z form might
be larger than that (≈1.4 kcal/mol) observed in acetals, because the carbonyl bond in esters is a more polarized bond than the C — OR bond in acetals
and therefore, the antibonding orbital of the σ C — O bond should be of
lower energy allowing increased overlap. It has been postulated on that
basis (1-3) that the greater stability of the Z form (≈3 kcal/mol) by comparison with the E form in esters would be due mainly to this secondary
electronic effect.

Thus, primary electronic effects (n-π* interaction) form the conjugated system of the ester function whereas secondary electronic effects (n-σ* interaction) are the result of the orientation of non-bonded electron pairs antiperiplanar to the σ C − O bonds of the ester function. Clearly, the primary are energetically more important than the secondary electronic effects and this terminology is justified by the fact that these two effects have their origin in the same chemical principle, orientation in space of electron pairs with resultant electronic delocalization.

There is no direct experimental evidence to show the importance of the secondary electronic effects in the ester function except for the relative stability of the Z over the E form. However, the relative stability of the different forms of dialkoxycarbonium ions can be explained by considering these electronic effects. Since dialkoxycarbonium ions are alkylated derivatives of esters, the result can be used as evidence to support the importance of the secondary electronic effects in the ester function. It is known from X-ray evidence and supported by calculations (8, 9) that dialkoxycarbonium ions like esters are planar and that they can exist theoretically in three different forms, the ZZ (10), the EZ (11), and the EE (12) forms. The two oxygens of 10 each have one non-bonded electron pair antiperiplanar to a polar C − O bond, 11 has one, whereas 12 has none. Thus, 10, 11, and 12 have respectively two, one and zero secondary electronic effects, and on that basis, their relative stability should follow in this order. In the ZZ form (10), there is a severe steric repulsion between the two R groups; thus, with the exception of cases where the two R groups are part of a ring, this form must be eliminated. The EZ form (11) must therefore represent the most stable form of dialkoxycarbonium ions.

| 10 | 11 | 12 |

Ramsey and Taft (10) have provided the first evidence by nmr spectroscopy that the dimethoxycarbonium salt of methyl acetate exists in the EZ form (11) only. It was later found that at -30 to -80°C, there is predominance

(11, 12) of the EZ form, while at 0-38°C, both the EZ and the EE forms exist. The relative percentage of the EE form 12 depends on the nature of the substitutents. When R_1=H and R_2=alkyl, the EZ form predominates to the extent of 70-90%, and when R_1=R_2=alkyl, the EZ form becomes exclusive. Protonated esters (13) and acids (14-17) exist also in the EZ form, preferentially.

Further support for the importance of secondary electronic effects has recently been obtained (2, 18) from a study of the SN_2 displacement by iodide on lactonium salts. It was found that the iodide ion reacts instantly with salts such as 13 to give a mixture of iodoester 14, lactone 15, and alkyl iodide. For example, in the case of 13 (R=CH_3), a mixture of iodoester 14 (R=CH_3) (70%), and δ-valerolactone (15) (30%) was obtained.

14   13   15

These experimental results show clearly that the formation of iodoester (process A) can compete effectively with the formation of lactone and alkyl iodide (process B). This is a priori a surprising result. In process B, two molecules are formed and the ring is not broken while in process A, only one molecule is formed and the ring is cleaved. On that basis, process A should require more energy than process B. Furthermore, in lactonium salts having a methoxy group, iodide displacement on the primary methyl group should be more facile than that on the secondary methylene group of the ring. This factor again favors the formation of lactone, yet important quantities of iodoester are observed. It is therefore clear that there must be a new parameter which operates in process A only, lowering the energy of its transition state, so that it can compete with process B.

Assuming that these lactonium salts exist in the EZ form 11, the two transition states yielding 14 and 15 can be illustrated by 16 and 17 respectively. In 17, the bond to be broken is antiperiplanar to the non polar C_1-C_2 bond whereas in 16, the bond to be broken is antiperiplanar to the C_1-O_1 polar bond. In other words, in 16 the electron pair orbital of the C_5-O_5

16   **17**

bond can be delocalized by an interaction with the antibonding orbital
($\sigma *$) of the polar $C_1 - O_1$ bond whereas in **17**, the electron pair orbital
of the $R - O_1$ bond cannot be delocalized into an antibonding orbital of
a polar $C-O$ bond. Thus, the former process should be favored electronically
over the latter and on that basis, rationalization of the experimental
results becomes clear. These results can therefore be considered as further
support for the importance of secondary electronic effects in the ester
function.

Now that the electronic effects in the ester function have been explained
in detail, we can examine the formation and the cleavage of tetrahedral
intermediates derived from esters by applying the theory of stereoelectronic
control for hydrolytic reactions, which was put forward in 1972 (19). Ac-.
cording to this theory, in order to be favored, nucleophilic attack on
esters must be perpendicular ($\approx 109°$ according to ref. 20) to the plane
at the conjugated system and give rise to a tetrahedral intermediate where
the two oxygen atoms have each a lone pair oriented antiperiplanar to the
newly formed bond. Following this rule, a _Z_ ester (**4**) with a nucleophile
Y must give intermediate **18** (Fig. 1) whereas an _E_ ester (**5**) must give intermediate **19**. Examination of the process **4** → **18** indicates that the conformation
of the _Z_ ester is transposed into the tetrahedral intermediate since the
$C - R$ and the $O - R$ bonds which are antiperiplanar in **4** remain antiperiplanar
in **18**. Similarly, in the process **5** → **19**, the two R groups which are _syn_
in the _E_ ester assume a gauche orientation in the intermediate **19**. At the
same time, the conversions **4** → **18** and **5** → **19** follow the principle of least
motion (21-23).

According to the principle of microscopic reversibility, the reverse process
must follow the same path. In fact, the stereoelectronic theory was first
elaborated by examining this process, _i.e._ the cleavage of tetrahedral

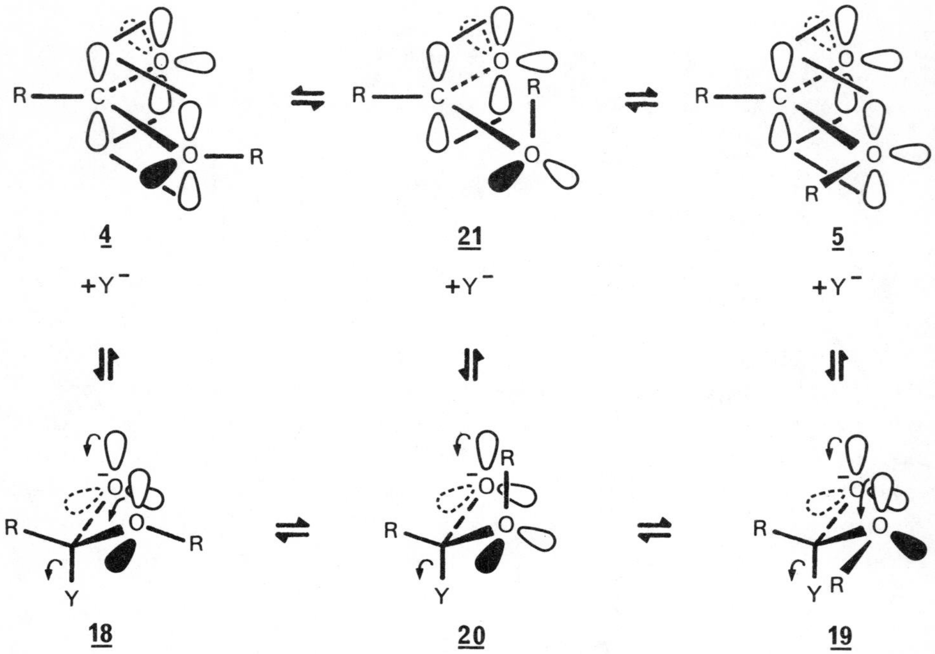

Fig. 1

intermediates. It was defined that the precise conformation of a tetrahedral intermediate must be transposed into the product of the reaction and the cleavage of a C — Y bond is allowed to take place with stereoelectronic control only if the two oxygen atoms of the tetrahedral intermediate have each an electron pair oriented antiperiplanar to the leaving group.

Intermediates **18** and **19** represent two of the three possible gauche conformers for the tetrahedral intermediate, and it is therefore pertinent to analyze the third gauche conformer which corresponds to **20**. Intermediate **20** cannot cleave with stereoelectronic control because the OR oxygen does not have an electron pair properly oriented to eject the leaving group. Ejection of the leaving group in **20** could take place with the help of only one electron pair (from O⁻), but it leads to **21**, a non conjugated "ester" (lacking primary electronic delocalization). Intermediate **21** corresponds to the high energy intermediate which occurs during the thermal isomerization of **Z** and **E** esters. As the energy barrier for such interconversion is at least 15 kcal/mol (4), the energy difference between **21** and **4** (or **5**) must be of that order. It is clear on that basis that the cleavage **20** → **21** is a much higher energy process than the cleavages $\underline{18} \rightarrow \underline{4}$ or $\underline{19} \rightarrow \underline{5}$.*

Intermediate $\underline{18}$ gives the most stable form ($\underline{Z}$) of an ester whereas $\underline{19}$ yields the least stable form ($\underline{E}$). An interesting question which follows is whether or not the energy barrier for the formation of a $\underline{Z}$ ester ($\underline{18} \rightarrow \underline{4}$) is also lower than that for an $\underline{E}$ ester ($\underline{19} \rightarrow \underline{5}$). An answer can be obtained from the analysis of the secondary electronic effects.

Interestingly, there is complete transmission of the electronic effects from the tetrahedral intermediate to the ester function in both processes. In intermediate $\underline{18}$, in addition to the two electron pairs which are antiperiplanar to the leaving group, each oxygen atom also has an electron pair antiperiplanar to a $C-O$ bond. Thus, $\underline{18}$ has four secondary electronic (n-σ*) effects. However, in the cleavage of $\underline{18}$ which yields the $\underline{Z}$ ester $\underline{4}$, two of the secondary electronic effects become primary (n-π* interaction) effects (forming the π system) while the other two remain secondary effects (n-σ*). In intermediate $\underline{19}$, besides the two electron pairs which are antiperiplanar to the leaving group, the negatively charged oxygen also has one electron pair antiperiplanar to the $C-OR$ group. Thus, $\underline{19}$ has three secondary electronic effects and its cleavage gives $\underline{5}$ where two of these secondary (n-σ*) electronic effects become primary (n-π*) while the third one remains a secondary electronic effect (n-σ*).

There is no gain or loss of electronic effects in either cleavage except that two secondary electronic effects (n-σ*) have been transformed into two primary electronic effects (n-π*) and on that basis, the energy barriers should be similar. However, since intermediate $\underline{18}$ has one extra secondary electronic effect by comparison with $\underline{19}$, the energy of the ground state of $\underline{18}$ should be lower than that of $\underline{19}$. Thus, in a situation where a tetrahedral intermediate would exist as a mixture of conformers $\underline{18}$, $\underline{19}$, and $\underline{20}$ which are rapidly equilibrating by rotation, the formation of the $\underline{Z}$ ester $\underline{4}$ would be favored. Consequently, under those conditions not only conformer $\underline{20}$, but also conformer $\underline{19}$ would not be able to compete with conformer $\underline{18}$ yielding the ester having the $\underline{Z}$ form $\underline{4}$. If it is indeed true,

---

*The energy barrier for the ejection of Y in $\underline{20}$ must also be higher than that for the corresponding hemiacetal ($\underline{20}$, $\overline{OR}$=R and Y=OR) because the inductive effect of the oxygen of the OR group in $\underline{20}$ creates a partial positive charge on the central carbon.

the cleavage of that intermediate will be said to take place with primary
and secondary stereoelectronic control.

There are also situations where it is not necessary to have a conformational
change in order to have two possible cleavages in competition. Such a situation is possible in a single tetrahedral intermediate when the leaving
group Y is another OR group. For example, we can consider the three tetrahedral intermediates __22__, __23__, and __24__.

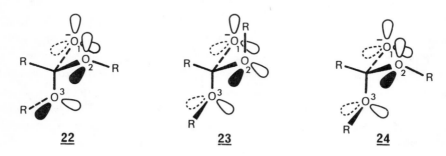

In intermediate __22__, the R group of O_2 is oriented in such a way, that the
cleavage of the $C - O_3$ bond will result in the formation of a __Z__ ester. On
the other hand, as the R group of O_3 is oriented antiperiplanar to the
$C - O_2$ bond, cleavage of the $C - O_2$ bond cannot occur. It is therefore predicted that __22__ will break down to give a __Z__ ester by the cleavage of the
$C - O_3R$ bond. In intermediate __23__, the R group of O_3 is oriented in such
a way to give an __E__ ester by cleavage of the $C - O_2$ bond, and the cleavage
of the $C - O_3$ bond cannot take place because the $R - O_2$ bond is antiperiplanar
to the $C - O_3$ bond.   Thus, __23__ should give an __E__ ester by ejection of the
O_2R group.

The last example, intermediate __24__, is more interesting because it can give
either a __Z__ or an __E__ ester. Indeed, cleavage of the $C - O_3$ bond can yield
a __Z__ ester whereas cleavage of the $C - O_2$ bond would yield an __E__ ester. Now
that the two cleavages are possible in the same conformer, the argument
used previously, __i.e.__ the relative stability of different conformers, cannot
be used to make a prediction in this case. However, it can still be predicted that the cleavage of the $C - O_3$ bond would be favored over that of the
$C - O_2$ bond in __24__ because the former leads to a product (__Z__ ester) having
two secondary electronic effects whereas the latter leads to a product
(__E__ ester) having only one. Consequently, the additional secondary electronic
(n-σ*) effect which lowers the ground state energy of the __Z__ ester product

must also play a role in lowering the energy of the corresponding transition
state which leads to its formation.

We have previously discussed, in the Chapter on acetals, the anomeric (or
secondary electronic) effects as n-σ^* interactions, and their influence
on the leaving group ability of an OR group (cf. p. 32). Applying this
principle to intermediates **22**, **23**, and **24** gives the following results. In
22, O_3 gives one secondary electronic effect (with $C-O_1$) and O_2 gives two
(one with $C-O_1$ and one with $C-O_3$). As a consequence, the $C-O_3$ has one
partial double-bond character whereas the $C-O_2$ bond has two; the O_3R group
is thus a better leaving group than the O_2R group. This is therefore another
argument in favor of the preferred ejection of the O_3R group in intermediate
22. In intermediate **23**, O_2 and O_3 have each one secondary electronic effect;
the O_3R and the O_2R groups must have similar leaving group ability. In intermediate **24**, O_3 has one secondary electronic effect and O_2 has two. The
O_3R group is thus a better leaving group, and its ejection should be easier
than that of the O_2R group. This factor again favors the formation of a
Z ester in preference to an **E** ester from intermediate **24**.

On a more general basis, a hemi-orthoester intermediate such as **25** which
has two different alkoxy groups can yield the two different esters **26** and
27, each having either a **Z** or an **E** conformation. Theoretically, the intermediate **25** can take the nine different gauche conformations described in
Fig. 2, and the predicted stereoelectronically controlled cleavage for each
conformation is indicated in Table 1. Conformer **A** (or **F**) has a conformation
identical to that of intermediate **22**. Similarly, conformer **E** (or **G**) corresponds to **23** and conformer **B** (or **I**) to **24**. Conformer **D** cannot be cleaved;
it is therefore predicted that the energy barrier for its cleavage must
be higher than in the other cases. Conformers **A**, **B**, and **C** represent the
three possible conformers resulting from the attack of methoxide ion on
a **Z** ester. Similarly, conformers **G**, **H**, and **I** are the three possible conformers which can be formed from the reaction of methoxide ion with an **E** ester.

$$R-\overset{\overset{\displaystyle O}{\|}}{C}-OR \ + \ CH_3O^- \ \rightleftharpoons \ R-\overset{\overset{\displaystyle O^-}{|}}{\underset{\underset{\displaystyle OCH_3}{|}}{C}}-OR \ \rightleftharpoons \ R-\overset{\overset{\displaystyle O}{\|}}{C}-OCH_3 \ + \ RO^-$$

    26   **25**   **27**

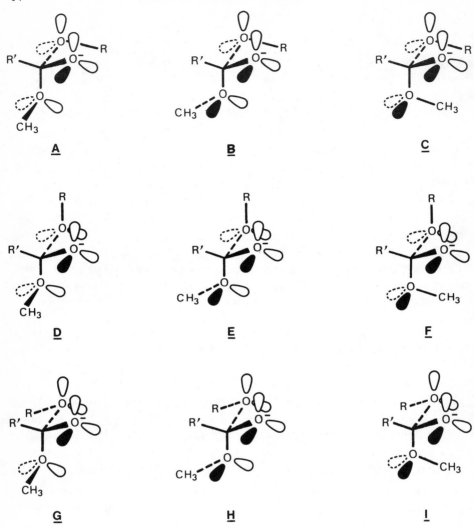

Fig. 2

The correlation between primary (n-π* interactions) and secondary (n-σ*) electronic effects in the cleavage of tetrahedral intermediates has now been explained in detail. Before attempting the rationalization of experiments which could potentially support the importance of the stereoelectronic principle in hydrolytic reactions, the energy barrier for the breakdown of a tetrahedral intermediate relative to that for conformational change must be considered. This is essential in cases where the tetrahedral inter-

TABLE 1 Predicted Cleavages of the Nine Conformers of
Tetrahedral Intermediate **25** Shown in Fig. 2

ester **26**		conformer		ester **27**
Z	⇌	A	⇌̸	
Z	⇌	B	⇌	E
Z	⇌	C	⇌	Z
	⇌̸	D	⇌̸	
	⇌̸	E	⇌	E
	⇌̸	F	⇌	Z
E	⇌	G	⇌̸	
E	⇌	H	⇌	E
E	⇌	I	⇌	Z

mediate might undergo a conformational change prior to break down. As the energy barriers for the breakdown vary with ionic state, which is a function of pH, the various ionic states of tetrahedral intermediates must be taken into consideration.

Hemi-orthoester intermediates have a weakly acidic hydrogen (from the O—H group), and they have electron pairs on the oxygen atoms which can be protonated in acidic medium. They can exist in three different ionic forms T^+, T^o, and T^- depending on the acidity or the basicity of the medium. Since the pKa of a hemi-orthoester is about 10, it will exist in the T^+ and the T^o forms in acidic medium and essentially in the T^o form in neutral. In slightly basic medium (pH ≈8-10), it will exist in the T^o and the T^- forms whereas in stronger basic medium (pH > 11), only T^- should be present.

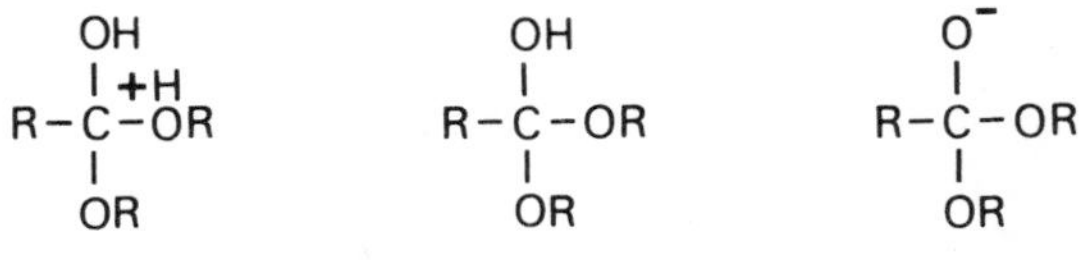

$$
\begin{array}{ccc}
\overset{\displaystyle OH}{\underset{\displaystyle OR}{R-\overset{+H}{C}-OR}} & \quad
\overset{\displaystyle OH}{\underset{\displaystyle OR}{R-C-OR}} & \quad
\overset{\displaystyle O^-}{\underset{\displaystyle OR}{R-C-OR}} \\[2em]
T^+ & T^o & T^-
\end{array}
$$

The energy barrier for conformational change of a given intermediate should not be influenced to a large extent by the ionic state, but the energy barrier for the cleavage should vary; it should be low for the T^+ and the T^- forms and high for the neutral form T°. This is in agreement with the evaluation of Guthrie (24) of the free energies for the formation and the cleavage of hemi-orthoesters in the acid-catalyzed, the uncatalyzed and the hydroxide-catalyzed hydrolysis of methyl formate and methyl acetate.

Capon and collaborators (25-27) have recently demonstrated that a neutral tetrahedral intermediate can exist in solution. The energy barrier for the cleavage of the T° form is consequently much higher than that for molecular rotation. They have observed that acetoxydimethoxymethane (28) in a mixture of acetone and water at -30°C gives acetic acid and the intermediate 29 which can easily be detected by nmr spectroscopy. Disappearance of the intermediate 29 was then observed to give the reaction products. Similar experiments were carried out with 2-acetoxy 1,3-dioxolan (30) and the tetramethyl derivative 31. With the latter, it was possible to obtain a solution in which more than 90% of the starting material was transformed into the intermediate 33. The hemi-orthoester 33 was much easier to detect than 32 because 33 breaks down at a slower rate than 32, and 31 hydrolyzes at a faster rate than 30. McClelland, Kresge, and their co-workers (28) have provided good evidence for the detection of some cyclic hemi-orthoesters in the hydrolysis of the corresponding orthoesters.

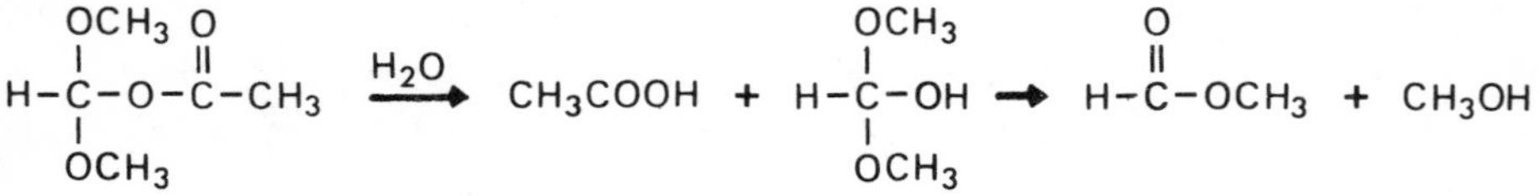

28

29

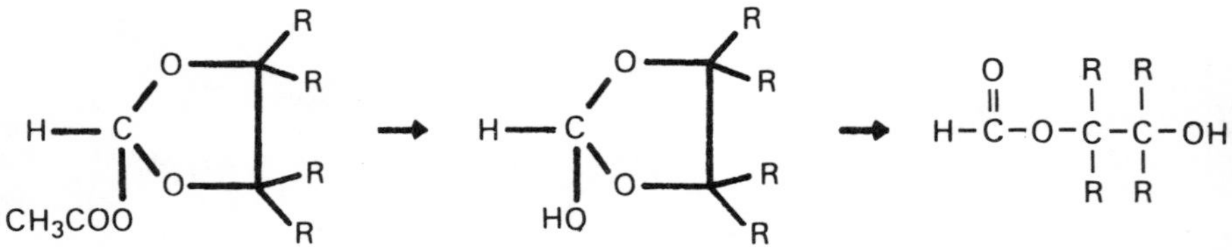

30 R = H
31 R = CH$_3$

32 R = H
33 R = CH$_3$

In acidic medium, the $T°$ form should be present in equilibrium with the T^+ form because hemi-orthoesters are weak bases. Consequently, hemiortho-esters must be allowed to undergo molecular rotation prior to their break-down in this medium. Also, there is no evidence so far that molecular rota-tion can compete with the breakdown of an intermediate in the T^+ ionic form. What we know is that the barrier for cleavage should be definitely lower in the T^+ than in the $T°$ form (24). At pH higher than 11, hemiortho-esters will exist exclusively in the T^- ionic form, and it will be seen that in some cases, the energy barrier for the cleavage of T^- is lower than that of molecular rotation.

Now that the primary and secondary electronic effects have been explained, that the ionic state of tetrahedral intermediates in relation to the energy barrier for breakdown has been examined as well as the possibility that molecular rotation can compete with cleavage, experiments will be presented which show that there is stereoelectronic control in hydrolytic reactions. It should be pointed that it is not easy to find experiments to prove this theory because tetrahedral intermediates have extremely short life times and the knowledge of their stereochemistry cannot be obtained directly. The great advantage of the stereoelectronic theory is that it stresses the importance of the stereochemistry of tetrahedral intermediates, and at the same time, it shows that the orientation in space of the non-bonded electron pairs on heteroatoms is crucial to an understanding of the chemical reactivity of these species. Clearly, rigorous experimental evidence to prove this theory cannot be obtained by a single experiment.

Concurrent carbonyl-oxygen exchange and hydrolysis in esters

Concurrent carbonyl-oxygen exchange and hydrolysis have been demonstrated in the hydrolysis of esters by using oxygen-18 labeling. The oxygen exchange has been taken as experimental evidence for the formation of a tetrahedral intermediate in ester hydrolysis (29, 30). Bender and Heck (31) have also obtained evidence that the intermediate necessary for carbonyl-oxygen ex-change lies on the reaction path of ester hydrolysis; both processes must therefore take place via the same tetrahedral intermediate.

Most substrates which have been studied show concurrent oxygen exchange, but there are exceptions. It has been postulated that reactions in which oxygen exchange is not detected still conform to the general mechanism;

in those cases oxygen exchange does not occur because the ratio of k_3/k_2 is greater than 100 (29, 30, 32-34). Although this explanation is certainly valid in some cases, a more complete rationalization can be obtained on the basis of the stereoelectronic theory.

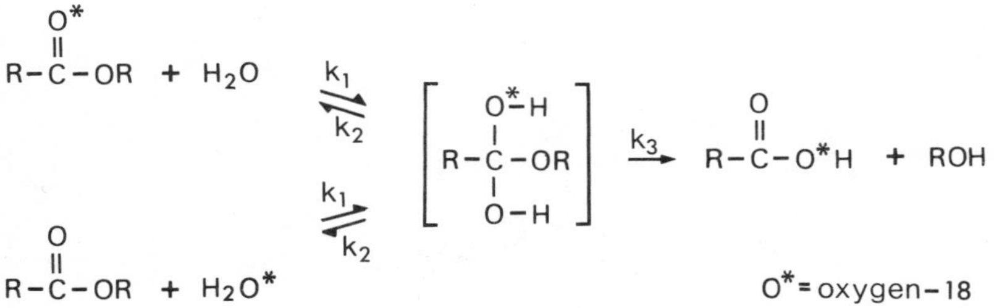

Application of the principle of stereoelectronic control to the hydrolysis of esters under basic conditions leads to the following predictions: <u>Z</u> esters are allowed to undergo carbonyl-oxygen exchange but <u>E</u> esters cannot.

Reaction of an ^{18}O-labeled <u>Z</u> ester (<u>34*</u>) (Fig. 3) with hydroxide ion under stereoelectronic control should give the tetrahedral intermediate <u>35</u>. Proton

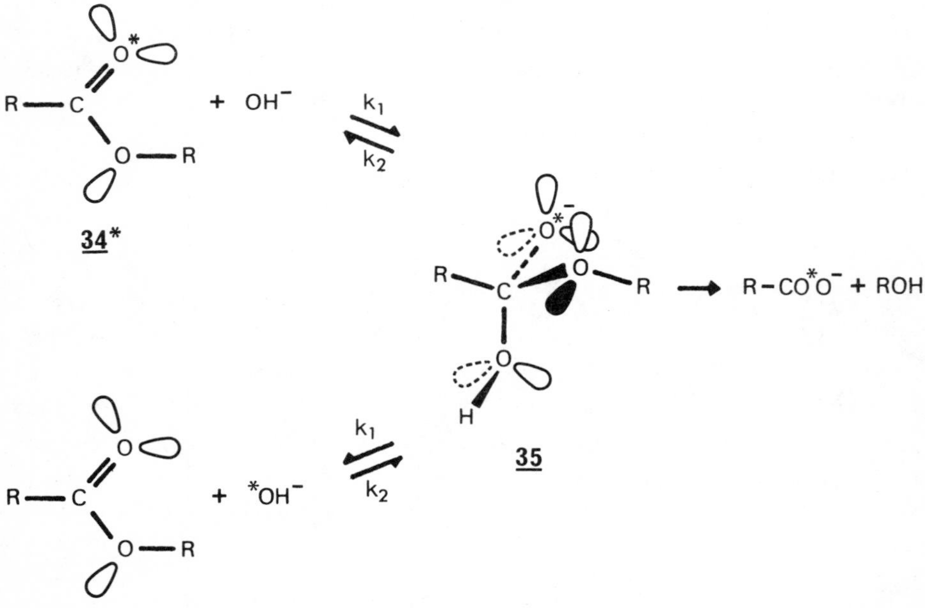

Fig. 3

transfer in the tetrahedral intermediate **35**, assumed to be a very fast process (35), occurs before the cleavage of the tetrahedral intermediate; consequently, an O — H bond is considered equivalent to an electron pair.

It is easy to recognize a plane of symmetry in **35** if the hydrogen of the O — H group is replaced by an electron pair. Examination of the orientation of all the electron pairs in **35** shows that this intermediate can break down in three possible directions to give the starting labeled ester **34***, the products of the hydrolysis reaction, or the unlabeled ester **34** (after appropriate proton transfer). Consequently, the hydrolysis of $\underline{Z}$ esters should always occur with carbonyl-oxygen exchange with the solvent. The extent of exchange will depend on the relative values of k_3 and k_2. This is in agreement with experimental results described in the literature (31-34) for $\underline{Z}$ esters. It should however be pointed out that the fact that $\underline{Z}$ esters do undergo carbonyl-oxygen exchange is in accord, but does not constitute a proof for, the stereoelectronic theory.

The reaction of ^{18}O-labeled $\underline{E}$ ester (**36***) (Fig. 4) with hydroxide ion should give the tetrahedral intermediate **37** which has the required orientation

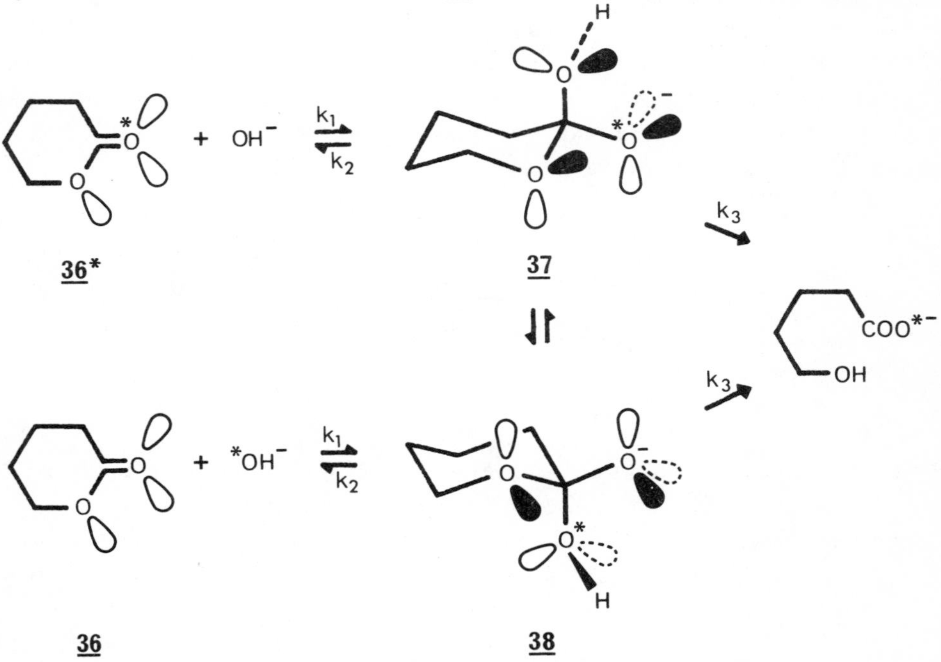

Fig. 4

of electron pairs to break down in two directions yielding either the starting labeled ester **36*** or the hydrolysis product. Indeed, even after appropriate proton transfer, it is not possible to eject the equatorially oriented hydroxyl group ($-*O^- = -*OH$) in **37** because the ring oxygen does not
have an electron pair properly oriented. The only manner by which the formation of unlabeled **E** ester **36** could occur is by a conformational change
of **37** into the new conformer **38**. If the barrier for conformational change
37 → 38 is too high and therefore this process cannot compete with the breakdown of **37**, **E** esters should not undergo carbonyl-oxygen exchange concurrent
with the basic hydrolysis.

It is also possible that the conformational change **37 → 38** does compete
with the breakdown, and carbonyl-oxygen exchange still does not take place
because k_3 would be much larger than k_2. The only scientific argument which
has been advanced to rationalize that k_3 is always greater than k_2 is based
on the secondary electronic effects (1, see also 39). Indeed, k_3 corresponds
to the formation of a **Z** carboxylic acid (equivalent to a **Z** ester) while
k_2 corresponds to the formation of an **E** ester, and as previously discussed,
the secondary electronic effects predict that the formation of an **E** ester
must involve more energy than that of a **Z** ester. Thus, even when conformational change is allowed, **E** esters should not undergo carbonyl-oxygen
exchange. In accord with this prediction, γ-butyrolactone (**39**) (36), D-
gluconolactone (**40**) (37) and isochromane-3-one (**41**) (38) do not undergo
carbonyl-oxygen exchange during their basic hydrolysis.

39   **40**   **41**

As a consequence of stereoelectronic control, it is possible to predict
the direction of attack of a hydroxyl ion on a conformationally rigid lactone
(40). For instance, we can consider the two possible modes of attack (α and
β) of hydroxide ion on the bicyclic lactone **42**. In both cases, the two
oxygens of the lactone function must each develop an electron pair antiperiplanar to the new C—OH bond in the corresponding orthoacid. On that basis,
a β-attack leads to the chair form **43** and an α-attack must give the less

stable boat form **44**. Clearly, the energy difference between the two transition states leading to **43** and **44** must be more than 3 kcal/mol, and since
these two modes of attack are competing, the reaction of hydroxide ion
on the β face of **42** (axial attack) must be the favored process. Thus, stereoelectronic control provides not only an explanation for the non-exchange
of the carbonyl-oxygen during the hydrolysis of lactones, it also predicts
which face of a lactone function will react with hydroxide ion. Consequently, a more precise knowledge of the reaction mechanism can be obtained.

44   **42**   **43**

There is yet no direct experimental evidence which demonstrates that one
particular face of a conformationally rigid lactone is more vulnerable
than the other to a nucleophilic attack. However, rigorous evidence has
been obtained from the study of the reactivity of the methyl and ethyl
lactonium salts **45** and **46** with alkoxide ion (1, 3, 41). Indeed, the reaction
of **45** with tetradeuterated methanol (CD_3OD), ethanol and chloro-2-ethanol
gave specifically only the bicyclic orthoesters **47**, **48**, and **49** respectively.
Similarly, the reaction of the ethyl lactonium ion **46** with methanol gave
the bicyclic orthoester **50** exclusively, indicating again an exclusive
β-attack (axial) by the alkoxide ion on the lactonium salt.

45   **47** R = CD_3
                                48 R = C_2H_5
                                49 R = $ClCH_2CH_2$

46   **50**

Returning to lactones, it is interesting to compare the relative rates of hydrolysis in isomeric lactones such as **51** and **52** which are obtained from the Baeyer-Villiger oxidation of norbornanone (42). Attack by hydroxide ion should give the orthoacids **53** and **54** respectively. The orthoacid **54** has two 1,3-diaxial steric interactions between the axial OH group and the methylene groups of the bridge. Such interactions do not exist in ortho-acid **53**, so formation of **53** should demand less energy than that of **54**. Indeed, the hydrolysis of a mixture of **51** and **52** was carried out (37) with sodium hydroxide and lactone **51** was found to hydrolyze at a much faster rate. This difference in the rate of hydrolysis has been used to separate mixtures of isomeric lactones (43-46).

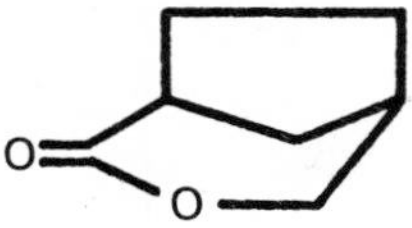

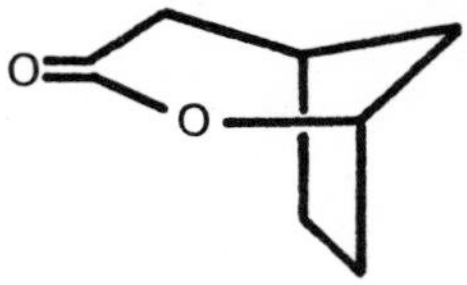

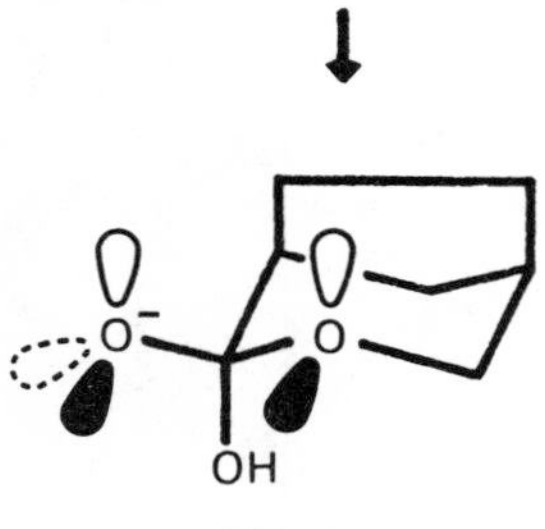

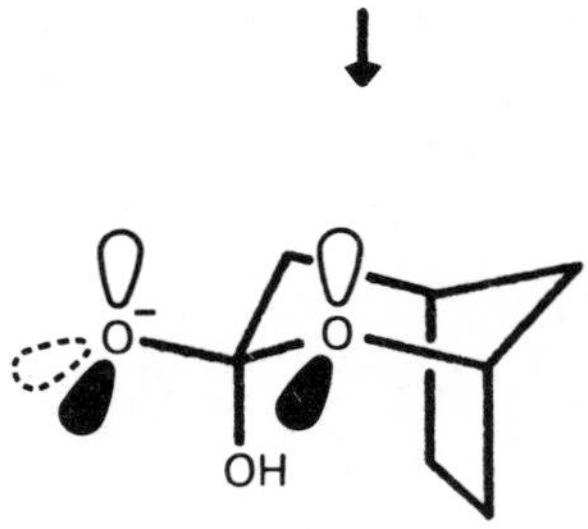

Hydrolysis of cyclic orthoesters

The formation of esters from the mild acid hydrolysis of orthoesters pro-ceeds through the formation of a hemi-orthoester tetrahedral intermediate as described by the following equation (47-53).

$$\underset{\substack{|\\OR}}{\overset{\substack{OR\\|}}{R-C-OR}} + H_2O \ \xrightarrow{H^+}\ \underset{\substack{|\\OR}}{\overset{\substack{OH\\|}}{R-C-OR}} + R-OH \ \longrightarrow\ \overset{\substack{O\\\parallel}}{R-C-OR} + 2\,R-OH$$

hemi-orthoester

In a cyclic orthoester such as **55** (Fig. 5) when the two alkoxy groups are different, there is the possibility of forming three different hemi-orthoesters (**56**, **57**, and **58**) which can lead to three different esters, the two hydroxyesters **59** and **60** and the lactone **61**. Thus, there is a possibility that some specific hemi-orthoesters will be generated which will lead to the preferential formation of one of the ester products. The mild acid hydrolysis of orthoesters is therefore a potential method to test the principle of stereoelectronic control in the formation and cleavage of hemi-orthoester tetrahedral intermediates.

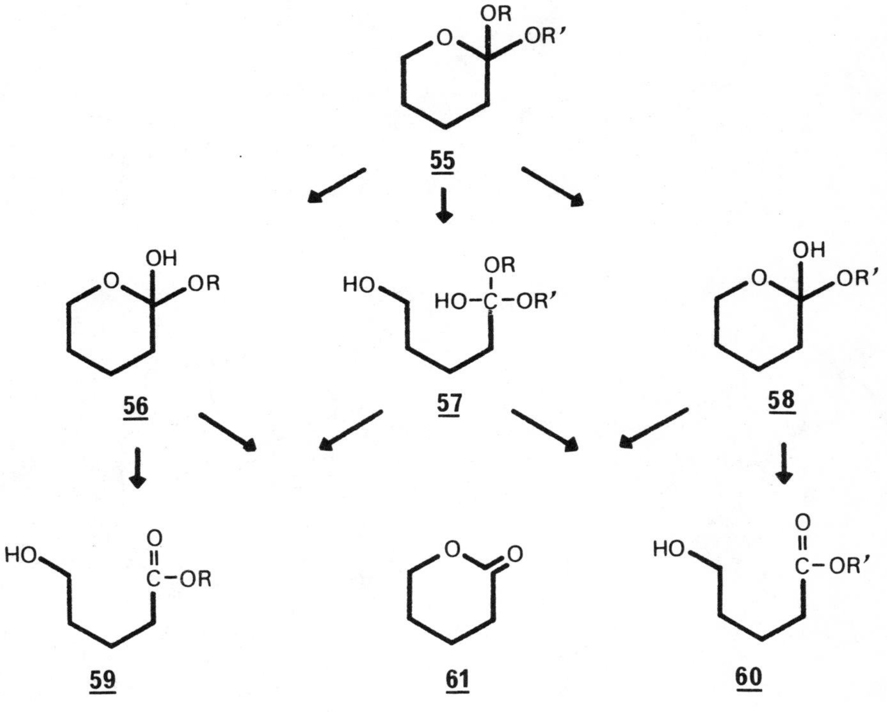

Fig. 5

A cyclic orthoester such as **55** in which the two alkoxy groups are identical will first be examined. There are nine gauche conformers which are theoretically possible for this type of cyclic orthoester and they are described in Fig. 6. The next task is to analyze each of these conformers on the basis of the principle of stereoelectronic control.

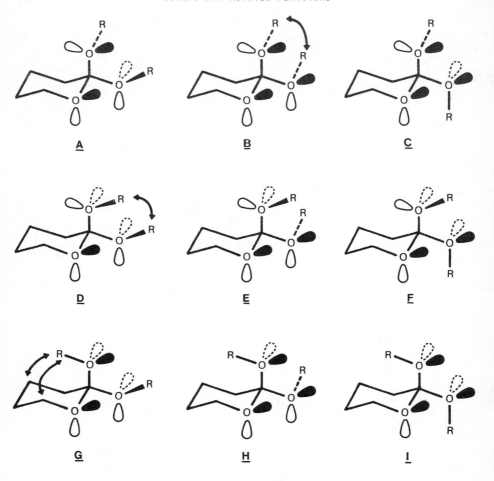

Fig. 6

The hydrolysis should proceed through the most energetically favored conformers which can expel one alkoxy group with the help of primary and secondary
electronic effects after appropriate protonation. There is a severe steric
interaction between the two R groups in conformers **B** and **D**; the population
of these conformers at equilibrium will be very small. Conformers **G**, **H**, and
I can also be ruled out by utilizing a similar argument; the R group of
the axial alkoxy group has a severe steric interaction with the two methylene groups (C_3 and C_5) of the ring. The remaining four conformers **A**, **C**, **E**,
and **F** do not have strong steric interactions and their cleavage can be
analyzed by considering first the primary stereoelectronic effects (i.e.

an electron pair on each oxygen atom oriented antiperiplanar to the leaving group).

On that basis, conformers **A** and **E** can break down by the loss of the axial alkoxy group. Conformer **F** can undergo a cleavage via the fission of the carbon-oxygen bond of the ring but conformer **C** cannot break down (no primary electronic effect). Conformer **C** must therefore be unreactive and this prediction was verified experimentally (38, 50) by studying the reactivity of the tricyclic orthoester **62** (54-56) which is a perfect rigid model for conformer **C**, as shown by X-ray analysis (57). Indeed, compound **62** was found completely stable under the mild acid conditions that are normally used for the hydrolysis of other cyclic orthoesters. Thus, conformer **C** is a remarkably unreactive conformer which must be eliminated.

62

The reactivity of the three remaining conformers **A**, **E**, and **F** can now be analyzed by taking into account the <u>secondary electronic effects</u>. In the cleavage of the axial OR group in conformer **A**, the ring and the equatorial oxygens have each an electron pair oriented antiperiplanar to the leaving group (giving a primary electronic effect: $n-\sigma^* \rightarrow n-\pi^*$). The equatorial oxygen has also an electron pair oriented antiperiplanar to the $C-O$ bond of the ring oxygen; one secondary $(n-\sigma^*)$ electronic effect will therefore help this cleavage which should yield the <u>EZ</u> lactonium ion **63**. By comparison, conformer **E** can eject the axial alkoxy group with the help of the primary $(n-\sigma^* \rightarrow n-\pi^*)$ stereoelectronic effect only, and it will give the <u>EE</u> lactonium ion **64**. Conformer **F** can undergo the cleavage of the $C-O$ bond of the ring with the help of the primary stereoelectronic effect and of one secondary electronic effect yielding the <u>EZ</u> dialkoxycarbonium ion **65**. Thus, on the basis of secondary electronic effects, cleavages of **A** and **F** are favored over that of conformer **E**. Finally, the cleavage of conformer **F** by comparison with that of conformer **A** must be a higher energy process because in the reaction $\underline{A} \rightarrow \underline{63}$, there is formation of two molecules ($\underline{63}$ and alcohol) and the ring is not broken whereas in the reaction $\underline{F} \rightarrow \underline{65}$, the ring is cleaved and only one molecule is formed. Thus, in the last process, the internal return ($\underline{65} \rightarrow \underline{F}$) might be important. The above analysis predicts that a cyclic orthoester should undergo hydrolysis via conformer $\underline{A}$ only.

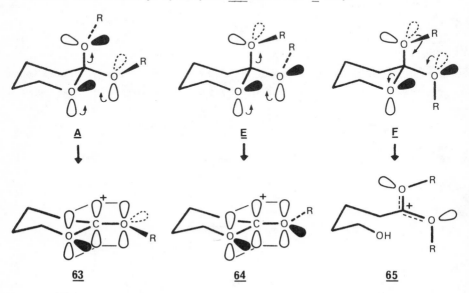

A   E   F

63   64   65

The next operation consists in the analysis of the hydration of lactonium ion $\underline{63}$ and the subsequent breakdown of the resulting hemi-orthoester. A stereoelectronically controlled attack of water on the lactonium ion $\underline{63}$ must take place on the β face yielding the tetrahedral conformer $\underline{66}$ (Fig. 7). As previously discussed for the case of lactones (cf. p. 70), a reaction with water on the α face of $\underline{63}$ would result in a tetrahedral intermediate having a boat conformation, and this process is therefore eliminated.

Under mild acidic conditions, tetrahedral intermediates exist in the T° and the T$^+$ form and conformational change prior to the breakdown can take place (cf. p. 66). Thus, in systems where the tetrahydropyran ring cannot easily undergo a chair inversion, hydration of $\underline{63}$ will first give $\underline{66}$ which will then yield a mixture at equilibrium of rotamers $\underline{66}$, $\underline{67}$, and $\underline{68}$. In cases where the tetrahydropyran ring can easily undergo a chair inversion, there should be an equilibrium mixture of the six possible conformers $\underline{66-71}$. The relative population of $\underline{71}$ will be negligible because this conformer has a strong steric interaction between the R group and the ring. It should also

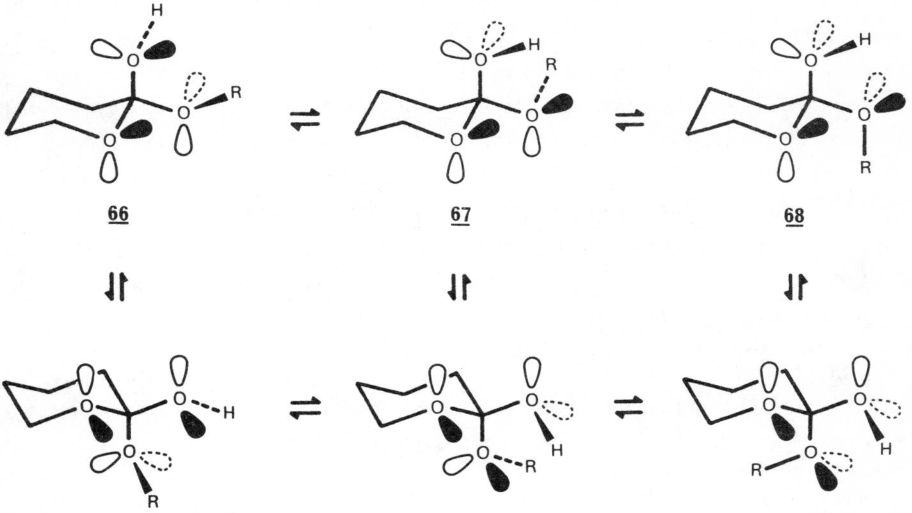

Fig. 7

be pointed out that the exact orientation of the hydrogen atom of the OH group in the six conformers can be neglected because proton-exchange is a fast process.

By taking into consideration the primary and the secondary electronic effects, the predicted cleavages of the six conformers are the following. Conformer **67** cannot break down (no primary stereoelectronic control); it is therefore unreactive and must be eliminated. Conformer **66** can yield a hydroxy-ester having a **Z** conformation whereas conformer **68** can produce a hydroxy-ester which should have an **E** conformation. Interestingly, lactone formation from **66-68** cannot take place with primary stereoelectronic control. Thus, in cases where the tetrahydropyran ring is conformationally rigid, the hydrolysis must take place preferentially <u>via</u> conformer **66**, yielding only the hydroxy-ester product in the **Z** conformation.

Intermediate **69** can either yield a **Z** (hydroxy-ester) or an **E** (lactone) ester. Intermediate **70** can only yield an **E** ester (lactone) whereas intermediate **71** can produce two **E** esters, the hydroxy-ester having an **E** conformation and the lactone. Thus, primary stereoelectronic effects allow the cleavage of intermediates **69-71** to produce either the hydroxy-ester or the lactone product. However, the cleavage of __69__ to yield a __Z__ hydroxy-ester is favored by one secondary electronic effect. On that basis, it is predicted that only hydroxy-ester should be observed as hydrolysis product with conformationally labile cyclic orthoesters. On the other hand, it should also be pointed out that by comparison with the formation of lactone and alcohol, the formation of hydroxy-ester is not favored due to the reversibility of ring opening. If this factor is as important as one secondary electronic effect, lactone formation could compete with that of the hydroxy-ester. On that basis, a mixture of hydroxy-ester and lactone should be obtained with conformationally labile cyclic orthoesters.

The first results reported (41) showed that the mild acid hydrolysis of the five cyclic orthoesters __72-76__ (R=C_2H_5) gave the corresponding hydroxy-ester as the sole product of the reaction. It was found later by Capon and Grieve (58) that the hydrolysis of orthoester __73__ (R=CH_3 or C_2H_5) gave a mixture of hydroxy-ester ($\approx$70%) and lactone (30%). The hydrolysis of orthoesters __72-76__ was subsequently repeated (59). It was confirmed that orthoester __73__ indeed gave a $\approx$7:3 mixture of hydroxy-ester and lactone. A similar result was observed with orthoester __72__ but the other three orthoesters __74-76__ gave exclusively hydroxy-ester as previously reported.

__72__   __73__ R' = H   __75__   __76__
                     __74__ R' = CH$_3$

Compounds __74__ and __75__ cannot easily undergo a chair inversion because the methyl and the isopropyl groups of __74__ and __75__ would become axially oriented and would develop a strong 1,3-diaxial interaction with the axial alkoxy group. A similar situation would also occur in the corresponding hemi-orthoester intermediates; compounds __74__ and __75__ can therefore be considered equivalent to the conformationally rigid bicyclic orthoester __76__. Consequently, the results obtained with compounds __74__, __75__, and __76__ confirm the above conclusion that a conformationally rigid orthoester should give exclusively the hydroxy-ester product.

The results obtained with compounds 72 and 73 show that in orthoesters which are conformationally labile, the hydroxy-ester is still the major product but the formation of lactone can occur to some extent. Thus, it can be concluded that whenever a tetrahedral intermediate can exist as a mixture of conformers 66-71, the formation of an E ester (from 69, 70, or 71) will be able to compete (due to an entropy factor) to some extent with the formation of Z ester (from 66 or 69).

More precise information concerning the course of events in the acid hydrolysis of orthoesters was obtained from the study of the four bicyclic orthoesters 77-80 which have two different alkoxy groups. Each orthoester yielded exclusively the hydroxy-ester resulting from the ejection of the axial alkoxy group. Thus, 77, 78, and 79 afforded the same hydroxy methyl ester 81 whereas orthoester 80 furnished the hydroxyl ethyl ester 82. The reverse process which occurs under basic conditions, i.e. the addition of alkoxide ion to the corresponding bicyclic lactonium salt, has already been described (cf. p. 71) and it was shown to take place with the same specificity.

77 R′ = CD$_3$, R = CH$_3$ 81 R = CH$_3$
78 R′ = C$_2$H$_5$, R = CH$_3$ 82 R = C$_2$H$_5$
79 R′ = ClCH$_2$CH$_2$, R = CH$_3$
80 R′ = CH$_3$, R = C$_2$H$_5$

These results demonstrate that conformer F is definitely not involved in the course of the hydrolysis reaction. For example, if conformer F, i.e. 83 of orthoester 78 is examined on the basis of the stereoelectronic principle, it must yield a mixture of the two hydroxy-esters 81 and 82. Indeed, conformer 83 must produce first the open-chain dialkoxycarbonium ion 84 which after hydration would give the acyclic tetrahedral intermediates 85 and 86. Since internal rotation is allowed in 85 and 86, they would then give a series of different conformers which should fragment to give a mixture of the hydroxy-esters 81 and 82.

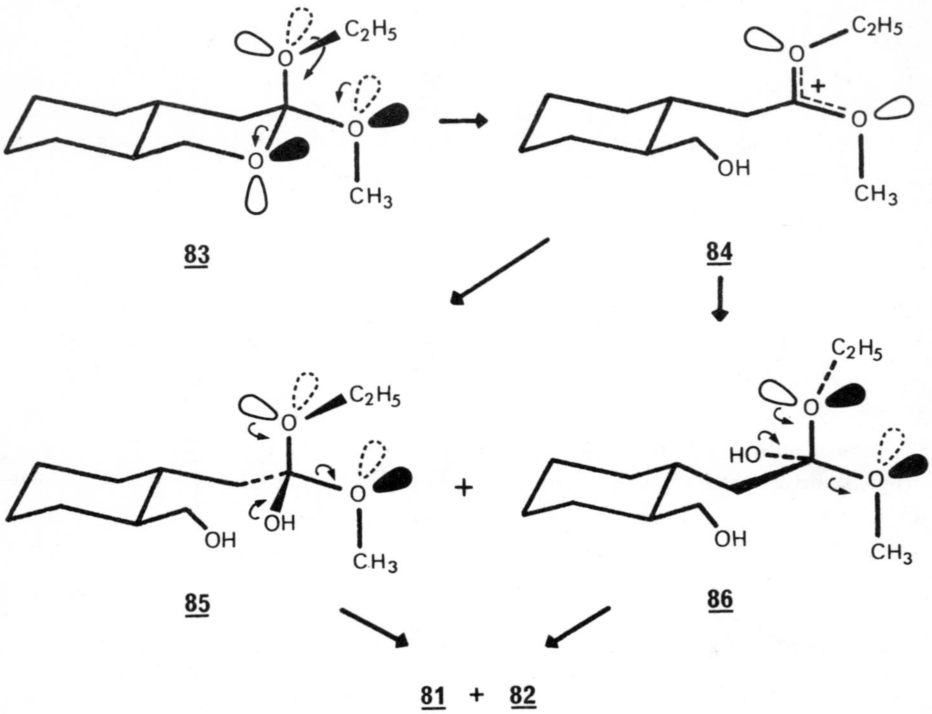

$$\underline{81} + \underline{82}$$

An interesting experimental result was observed in the study of the mild acid hydrolysis of the cis and the trans bicyclic orthoesters **87** and **88** (60). The cis orthoester **87** gave under kinetically controlled conditions the dihydroxy methyl ester **89** whereas the trans orthoester **88** produced directly the hydroxy-lactone **90** under the same experimental conditions. These results can be explained on the basis of the principle of stereoelectronic control.

The cis bicyclic orthoester **87** can exist in the two different conformations **91** and **92**. The conformation **92** corresponds to that of the unreactive tricyclic orthoester **62**, i.e. conformer **C**; it can therefore be eliminated. Conformer **91** can undergo the cleavage of the axial C—O bond with stereoelectronic control to produce the lactonium ion **93** which after hydration will give the hemi-orthoester **94**. Since the chair inversion in **94** is not favored because the hydroxyalkyl side chain would have to take the axial orientation, it is expected that **94** would give the dihydroxy methyl ester **89** preferentially.

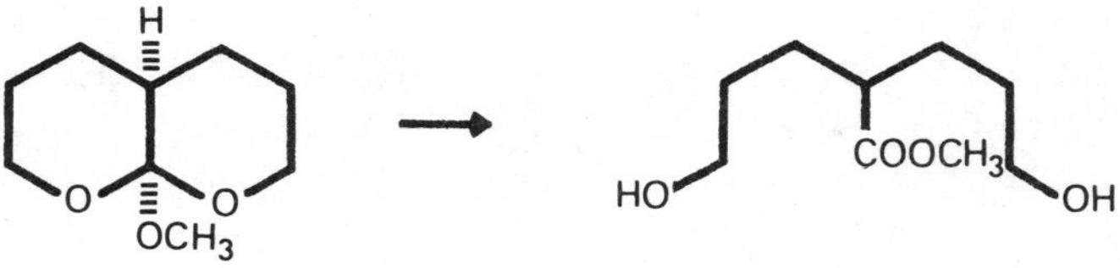

87 89

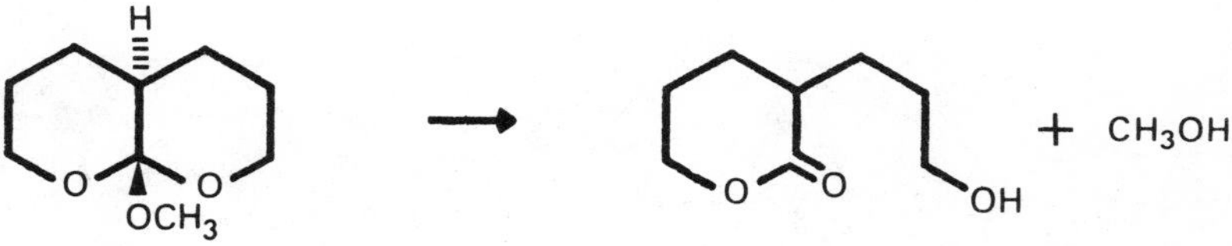

88 90

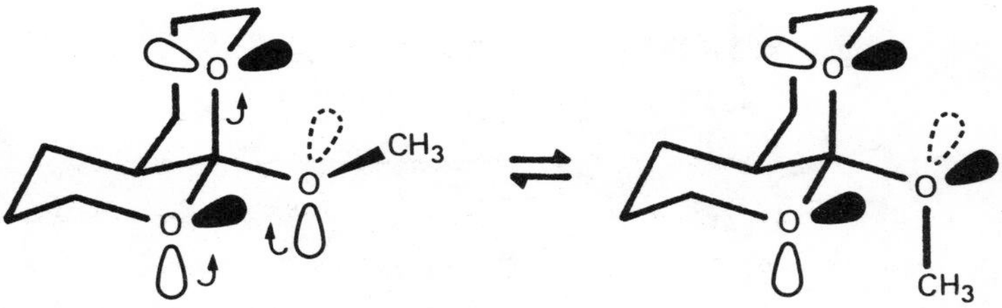

91 92

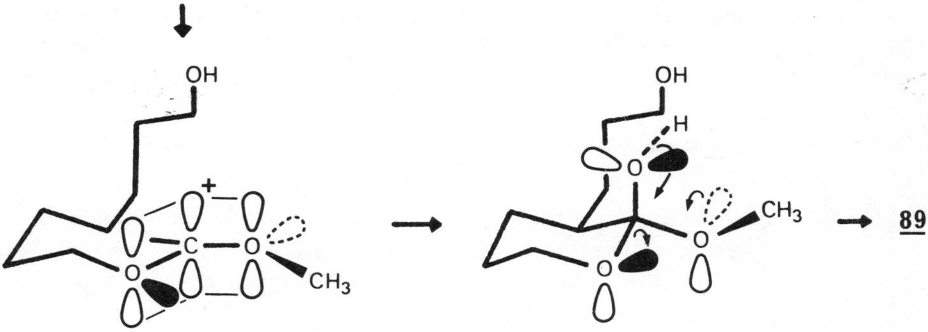

93 94

The trans bicyclic orthoester 88 must exist in the conformation 95 and the primary stereoelectronic effects permit the ejection of the methoxy group only, yielding the bicyclic lactonium ion 96 which after hydration (→97) and cleavage will give the hydroxy-lactone 90 only. The process 97 → 90 cannot occur with stereoelectronic control; its energy barrier must therefore be higher than in other cases. Kaloustian and Khouri (61) have shown that the reaction of sodium methoxide with the bicyclic salt 96 gave the trans orthoester 95 specifically.

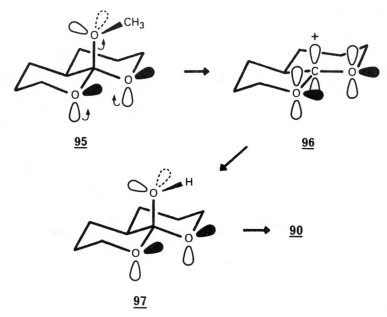

King and Allbutt (60, 62) have described the stereoselective hydrolysis of the dioxolenium ion 98 derived from a trans-decalin. Mild acid hydrolysis of 98 afforded almost entirely the axial ester product 99 with only a trace (<0.5%) of the equatorial ester 100. A similar result was obtained by using mild basic conditions. The authors further established by equilibration studies that the equatorial ester 100 is the most stable isomer, demonstrating that the formation of the axial isomer 99 is subject to kinetic rather than thermodynamic control. Essentially identical results were produced with two other dioxolenium salts derived from steroids.

The same authors have also carried out the hydrolysis of orthoesters 101. When R=CH$_3$, C$_2$H$_5$ or C$_6$H$_5$, they found again an almost exclusive formation of

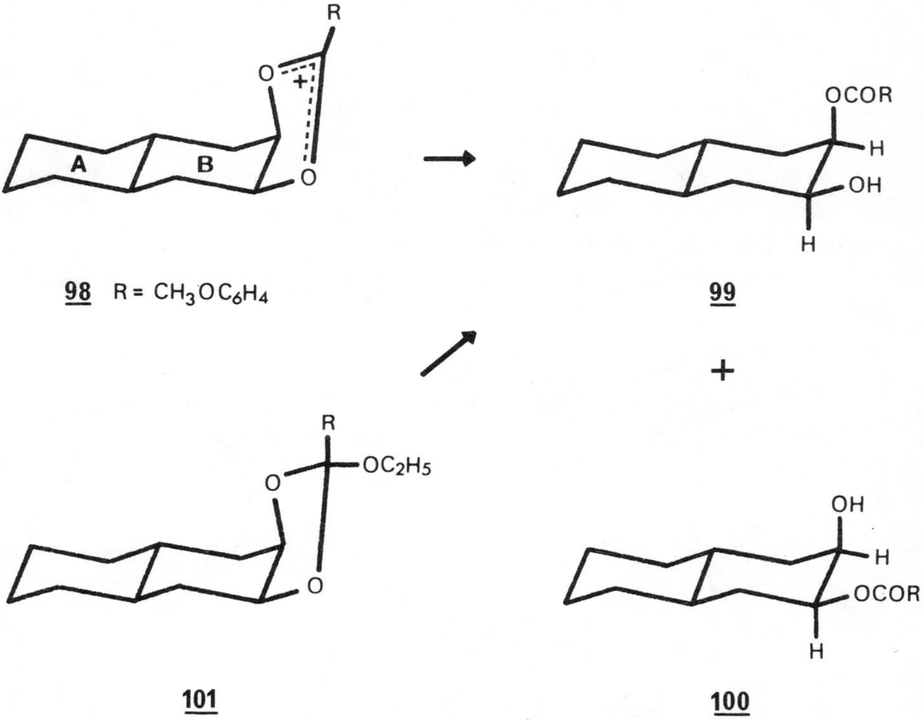

98 R = $CH_3OC_6H_4$ **99**

+

101   **100**

the axial ester **99** in preference to the equatorial ester **100**. When R=H, they obtained a mixture of 60% axial ester and 40% equatorial ester under kinetically controlled conditions. The orthoester **101** (R=CH_3, C_2H_5 or C_6H_5) yielded the same result as the dioxolenium salt **98**, and this is not surprising as **98** is believed to be an intermediate in the hydrolysis of **101**. Other workers have also observed the formation of axial esters in preference to equatorial esters (63-66).

King and Allbutt have explained their results by invoking either a preferential protonation of the equatorial oxygen in the corresponding hemiorthoester (60, 62) or a combination of steric and stereoelectronic factors (60). It is highly probable that both explanations are valid because the greater ease of protonation of the equatorial oxygen is a direct consequence of the stereoelectronic effects (which make that oxygen atom more basic (cf. p. 31)). It should be further pointed out that these workers were the first to report that stereoelectronic effects might play a role in hydrolysis reactions. However, the generality and the importance of the principle of stereoelectronic control was not fully recognized.

According to the present theory, these reactions can be explained in the following manner. Since dioxolenium ions are essentially planar (8, 9), the chair form of ring B of salt **98** must be distorted towards a half-chair. Hydration of **98** with stereoelectronic control must take place from the α face to give the half-chair hemi-orthoester **102** (Fig. 8), because the steric hindrance between the incoming water molecule and ring B must inhibit

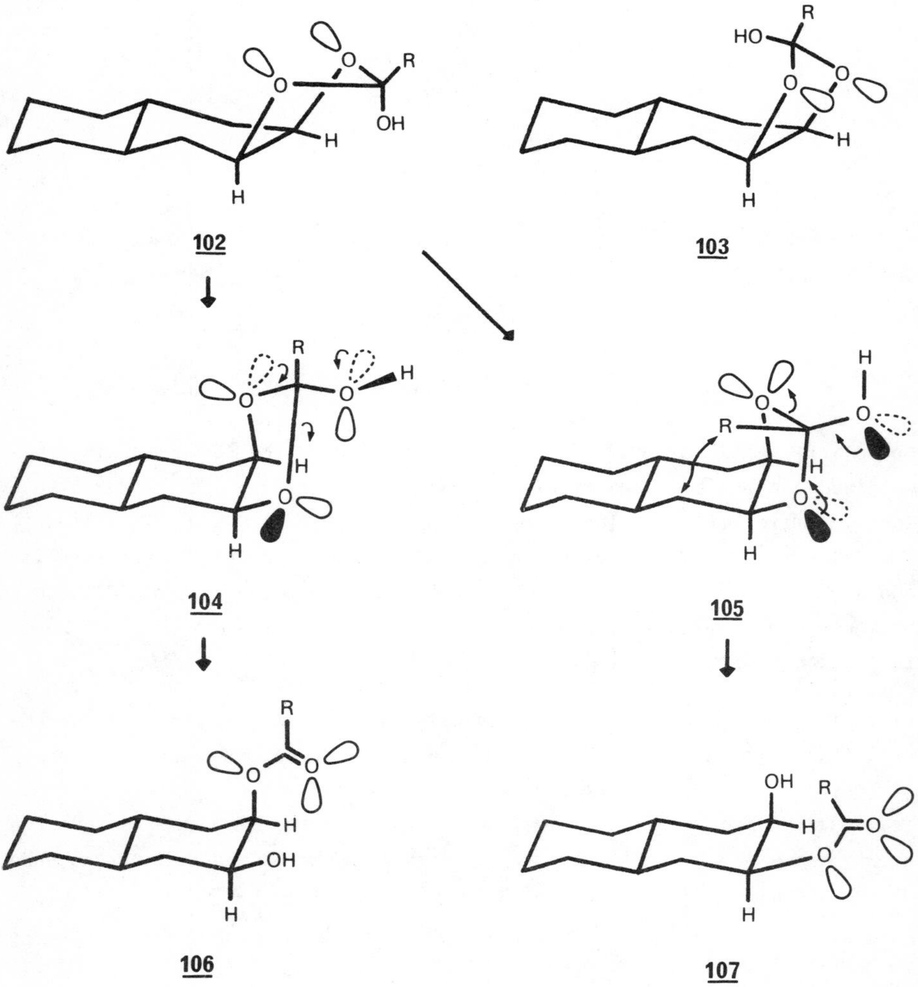

Fig. 8

the formation of hemi-orthoester 103. Once the hemi-orthoester 102 is formed, it can then take the conformations 104 or 105 where the tension caused
by the half-chair of ring B is released. Conformers 104 and 105 can break
down with the help of the primary stereoelectronic effects to give the
axial ester 99 and the equatorial ester 100 respectively.

The R group in 105 is in a 1,3-diaxial arrangement with a methylene group
of ring B (cf. arrow); thus when R is a large group, conformer 104 will be
favored over 105, and the axial ester will be preferentially formed. However, when R is a hydrogen atom, 104 and 105 will exist as an equilibrium
mixture and a mixture of axial and equatorial esters 99 and 100 will be
formed. This is essentially the steric and stereoelectronic argument used
by King and Allbutt (62). There is an additional reason for the preferential
formation of axial ester which is simply that the cleavage of 104 gives
an axial ester where the ester function will be produced in the Z conformation 106 whereas the cleavage of 105 will form an equatorial ester 107
which has the less stable E conformation.

Ozonolysis of tetrahydropyranyl ethers

The relationship between the conformation of the acetal function and its
reactivity towards ozone has been described in detail (cf. p. 41). It was
shown that the insertion of ozone into the C$-$H bond of the acetal function
to form the corresponding hydrotrioxide tetrahedral intermediate is subject
to stereoelectronic control. This section deals with the next step of this
reaction, i.e., the decomposition of the hydrotrioxide tetrahedral intermediate to yield the ester product. Experimental results will be presented to
show that this step is also controlled by stereoelectronic effects. These
results can therefore be used as evidence for the principle of stereoelectronic control in ester formation.

In the course of the studies on the ozonolysis of acetals (67-69), it was
observed that the reaction of ozone with simple tetrahydropyranyl ethers
as well as with conformationally rigid β-glycosides always gave the corresponding hydroxy-esters exclusively, under kinetically controlled conditions, lactone formation was never observed. For example, ozonolysis of
2-alkoxytetrahydropyran 108 under acetylating conditions (O_3 in Ac_2O-AcONa)
gave alkyl 5-0-acetyl pentanoate 109 exclusively. Similarly, oxidation,
of methyl 2,3,4,6-tetra-0-acetyl-β-D-glucopyranoside 110 yielded the corre-

sponding penta-0-acetyl aldonic acid methyl ester <u>111</u>. Similar results
were obtained with tetrahydrofuranyl ethers. Ozonolysis of 2-methoxytetra-
hydrofuran (<u>112</u>) gave only methyl-4 0-acetyl-butanoate (<u>113</u>); the corresponding γ-butyrolactone was not detected. Likewise, methyl 2,3,5-tri-0-
acetyl α and β-D-ribofuranoside (<u>114</u> and <u>115</u>) were both converted into
methyl 2,3,4,5-tetra-0-acetyl-D-ribonate (<u>116</u>).

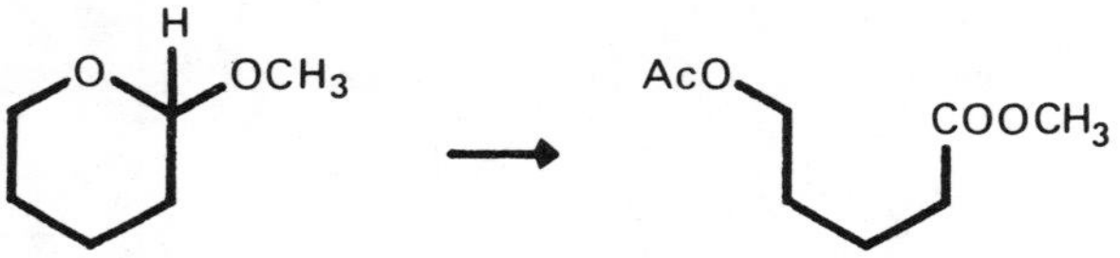

<u>108</u>   <u>109</u>

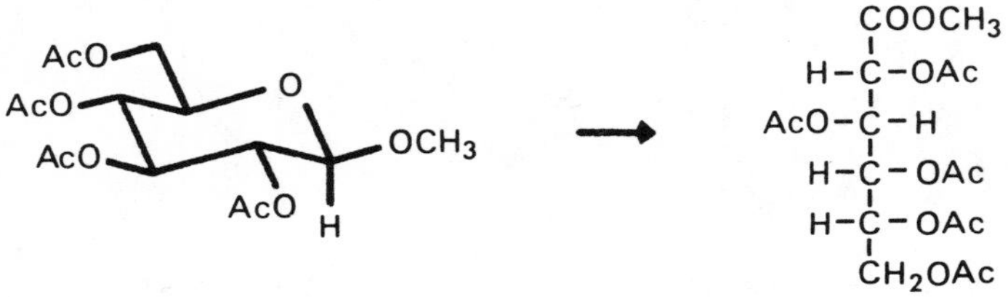

<u>110</u>   <u>111</u>

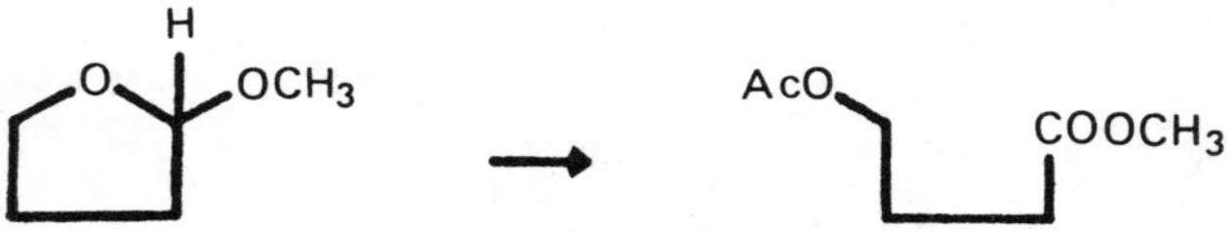

112   113

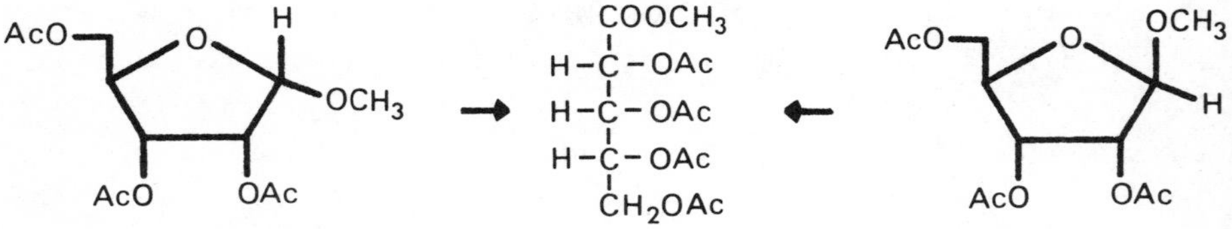

<u>114</u>   <u>116</u>   <u>115</u>

These results demonstrate that the hydrotrioxide intermediate **118** which
is formed from a cyclic acetal such as **117** is cleaved in a completely specific manner, yielding the hydroxy-ester **119** in preference to the lactone
120.

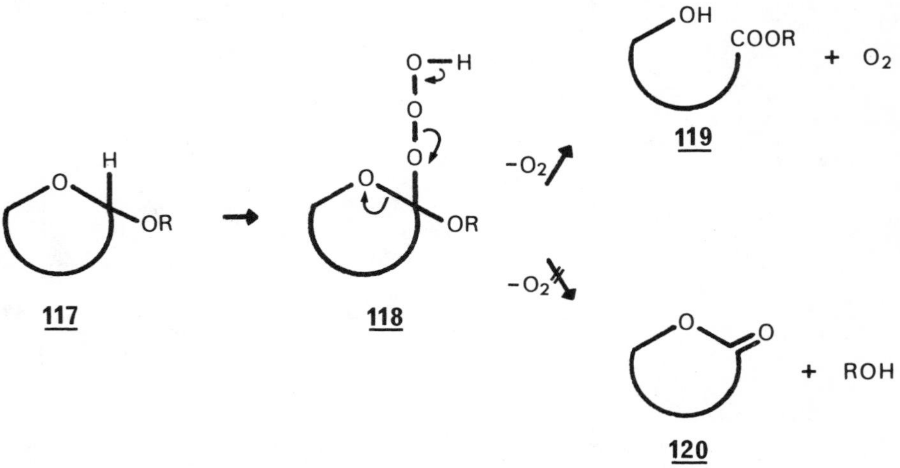

These results are very similar to those obtained from the mild acid hydrolysis of the cyclic orthoesters described in the preceding section. The difference is that in the ozonolysis reaction, the tetrahedral intermediate
which is formed bears a hydrotrioxide group instead of a hydroxyl group.
Also the fragmentation must take place on the T° neutral form of the hydrotrioxide intermediate (70), and it is likely that the hydroxyl group forms
a hydrogen bond with the ring oxygen in order to assist its departure,
forming molecular oxygen and the hydroxy-ester (**119**). Except for these
differences, the fragmentation of the hydrotrioxide intermediate remains
essentially the same as that for the hemi-orthoester intermediate, and
must therefore be governed by the same electronic factors.

Consequently, with conformationally rigid tetrahydropyranyl ethers which
have an alkoxy group in the equatorial configuration (as in β-glycosides),
or with conformationally mobile compounds, the reaction of ozone must take
place with either conformer **121** or conformer **122** (Fig. 9). The resulting
hydrotrioxides are **124**, **125**, and **126** when the tetrahydropyran ring is rigid,
and an equilibrium mixture of the six conformers **124-129** when that ring
can easily undergo a chair inversion. The $HO_2 - O$ bond of the hydrotrioxide
intermediate will be cleaved during the formation of the ester function;

that bond is therefore equivalent to an electron pair and its precise orientation in space does not need to be considered now.

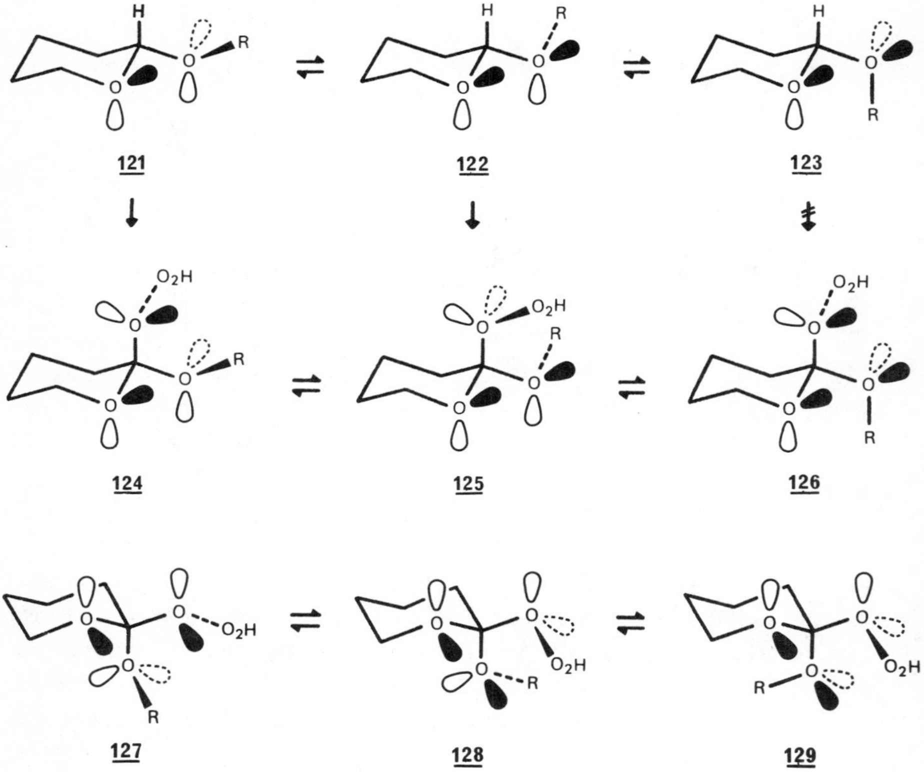

Fig. 9

We have discussed (cf. p. 77) the preferred cleavages of the six hemiortho-ester conformations corresponding to the hydrotrioxide intermediates 124-129. It is not necessary to repeat this discussion here, except to mention that, on this basis, only conformers 124 and 127 can be cleaved with primary and secondary stereoelectronic control to produce the hydroxy-ester with the ester function in a Z conformation. Thus, when the tetrahydropyran ring is locked conformationally, fragmentation must take place from intermediate 124 and if this ring is mobile, the cleavage will occur from either 124 or 127. If it is a requisite that the hydroxyl group must form a hydrogen bond with the leaving group (70), then the true intermediate for the fragmentation of 124 would correspond to 130 whereas that of 127 would be either 131 or 132.

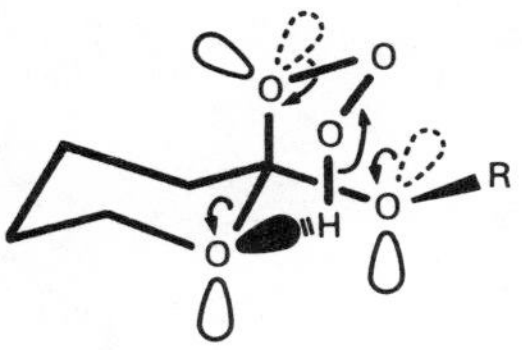

130 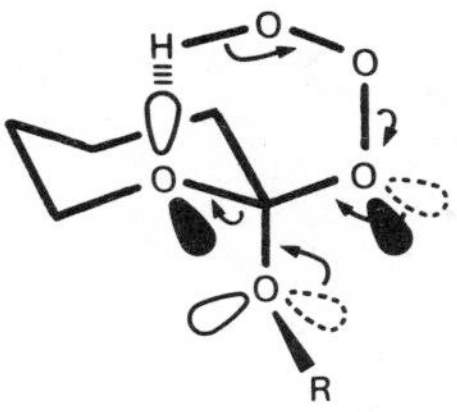   **131** 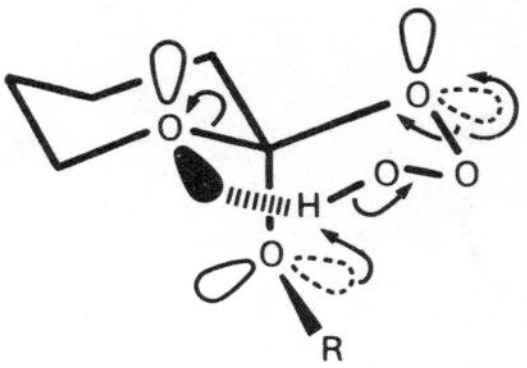   **132**

Ozone oxidation of the trans-decalin diol benzylidene **133** has been carried
out (71). Under kinetically controlled conditions, it produces the axial
benzoate **134** in preference to the more stable equatorial benzoate **135**.
Similar results were obtained with an analogous case derived from cholestane-
2β,3β-diol. These results are essentially identical to those obtained by
King and Allbutt (60, 62) in their study on the hydrolysis of dioxolane
orthoesters and dioxolenium salts (cf. p. 82), and can therefore be explained in the same manner. These results further confirm that the oxidation of
acetals by ozone produces an intermediate which behaves like the hemiortho-
ester tetrahedral intermediate which is formed in the hydrolysis of orthoesters.

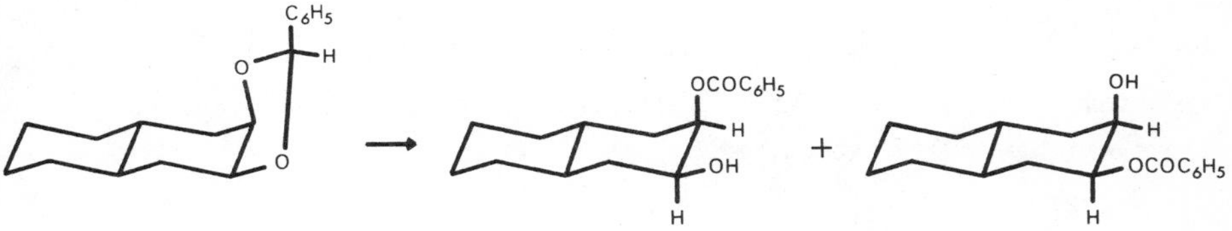

133   **134**   **135**

The ozonolysis of tetrahydropyranyl alcohol was also carried out (72) and
it proceeds smoothly to give the hydroxy-acid in essentially quantitative
yield. This compound **136** must give intermediate **137** and then a mixture
of **137** and **138**, which fragment to give the hydroxy-acid **139** (a Z carboxylic
acid) instead of δ-valerolactone. This result further demonstrates that k_3
must be larger than k_2 in the case of tetrahedral intermediates derived
from lactones, a conclusion which was reached previously in the study of
the carbonyl-oxygen exchange during the hydrolysis of lactones (cf. p. 70).

It is appropriate to point out here that the development of the theory
of stereoelectronic control has had its origin in the study of the ozonolysis of the acetal function (67, 68). It was first demonstrated that there

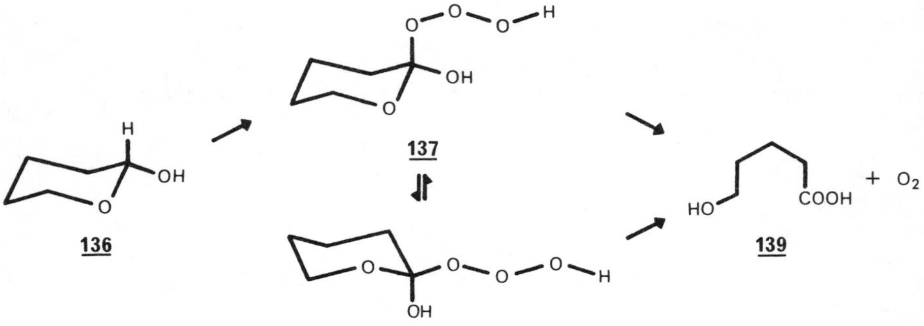

is a direct relationship between the orientation of the electron pairs relative to the C — H bond and the reactivity of the acetal function toward ozone. As a second step, it was postulated that this reaction proceeds via the formation of a hydrotrioxide tetrahedral intermediate. As a third step, it remained to explain the specific decomposition of such an intermediate to a hydroxy-ester. This led to the postulate that the orientation of the electron pairs, which plays a key role in the reaction of ozone with the C — H bond of the acetal function, might also play an equivalent role in the cleavage of these tetrahedral intermediates. Finally, as it was realized that the hydrotrioxide intermediate **118** is equivalent to a hemi-orthoester, the tetrahedral intermediate normally observed in ester hydrolysis, the importance of the principle of stereoelectronic control in hydrolytic reactions was then fully recognized.

Cleavage of vinyl orthoesters

The study of the cleavage of the axial and the equatorial vinyl bicyclic orthoesters **140** and **141** (Fig. 10) with potassium permanganate was reported recently (1, 3). Permanganate reacts with the vinyl orthoester double-bond yielding first **142** and then the tetrahedral intermediate **143**. On that basis,

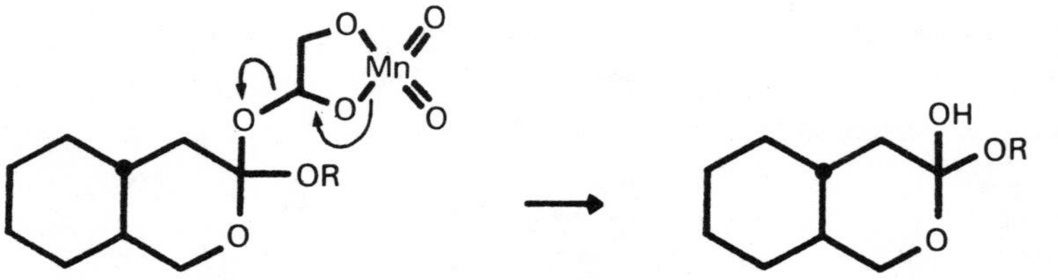

140 must produce the tetrahedral intermediate 144 with an axial OH group whereas 141 must give 145 with an equatorial OH group. Consequently, the configuration of the tetrahedral intermediate is determined only by that of the vinyl orthoester from which it originates.

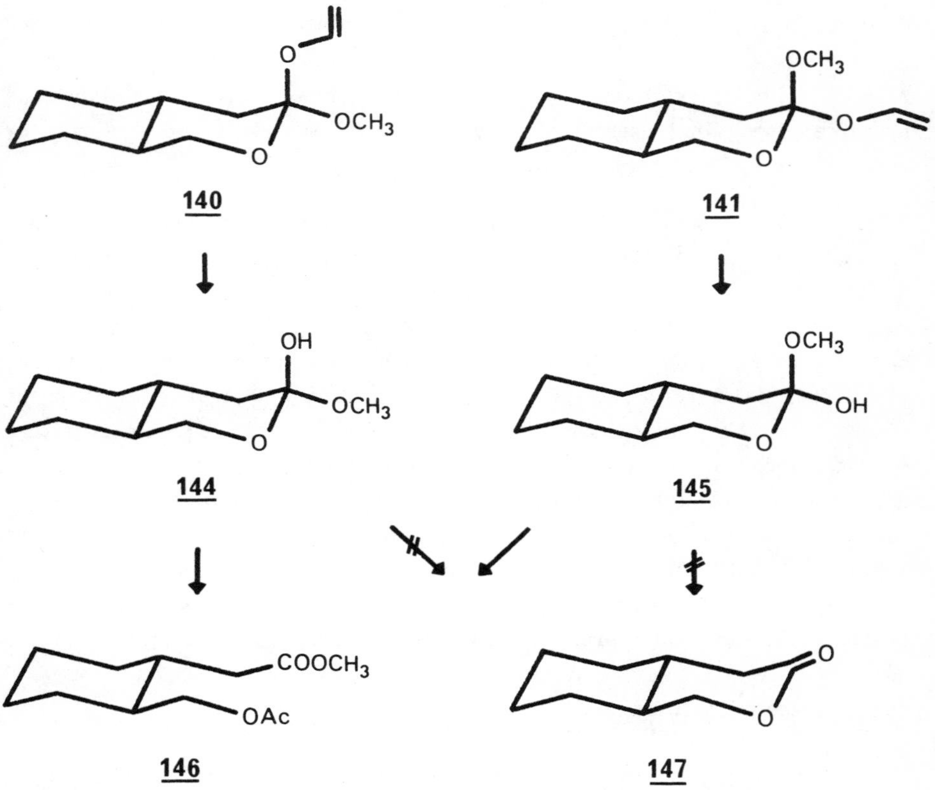

Fig. 10

In previous studies, i.e. concurrent carbonyl-oxygen exchange in the hydrolysis of esters, acid hydrolysis of orthoesters and oxidation of acetals by ozone, the configuration of the tetrahedral intermediate was determined by the application of the principle of stereoelectronic control. There could be some ambiguity in these experiments as the theory of stereoelectronic control is used to predict both the stereochemistry of the tetrahedral intermediate as well as its breakdown. The oxidation cleavage of vinyl orthoesters can therefore be considered a more powerful experimental technique in that respect because the configuration of the hemi-orthoester is determined independently of the stereoelectronic theory. It has also the advantage that the behavior of a tetrahedral intermediate having either an axial or an equatorial hydroxyl group can be observed in separate experiments.

The permanganate oxidation of axial and equatorial vinyl orthoesters **140** and **141** was carried out in a buffered solution (pH = 10) mixed with acetonitrile. The reaction mixture was then esterified with acetic anhydride and pyridine. Both vinyl orthoesters gave an identical result: >95% of acetoxy ester **146** and <5% of bicyclic lactone **147**.

These results can again be explained on the basis of the stereoelectronic theory. Compound **140** must give a mixture of conformers **148**, **149**, and **150**. We have previously discussed (cf. p. 77) that none of these conformers can break down with primary stereoelectronic control to give the lactone, and that only conformer **148** can yield the hydroxy-ester in the Z conformation. Thus, the formation of hydroxy-ester from **140** must come from the cleavage of **148**. The equatorial vinyl orthoester **141** must produce a mixture of conformers **151** and **152**. It has also been previously discussed that stereoelectronic effects predict that conformer **151** can either give a Z (hydroxy-ester) or an E (lactone) ester whereas conformer **152** can only give the lactone product. Since the hydroxy-ester is the product of the reaction, this result demonstrates again the importance of the secondary stereoelectronic effects which predict that a Z ester will be formed preferentially to an E ester. Thus formation of hydroxy-ester from **141** must come from the cleavage of **151**.

148   149   150

151   152

Cleavage of tetrahedral intermediates containing a sulfur atom

Kaloustian and Khouri (73-76) have studied the cleavage of hemi-orthothiol ester tetrahedral intermediates generated by the reaction of dialkoxycarbonium ions with hydrosulfide anion in aprotic solvent. The results obtained show that cleavage of the hemi-orthothiol ester intermediate is subject to stereoelectronic control.

$$
\underset{OR}{\overset{OR}{R-C^+}} + {}^-SH \longrightarrow \underset{OR}{\overset{OR}{R-C-SH}} \longrightarrow \overset{S}{R-C-OR} + ROH
$$

The authors (73) have first observed that cyclic dialkoxycarbonium ions **153** and **154** (R=CH$_3$ or C$_6$H$_5$) as well as acyclic dialkoxycarbonium ion **155** (R=CH$_3$ or C$_2$H$_5$) reacted with sodium hydrosulfide (NaSH) to give monothioesters **159** and **160** and thionobenzoate **161** respectively. These results show that the hemi-orthothiol esters **156**, **157**, **158** must be formed as intermediates in these reactions. The salts **153** and **154** (R=H) behave in a similar fashion (nmr and tlc analyses), but isolation of the products (**159** and **160**, (R=H)) was thwarted by their high reactivity.

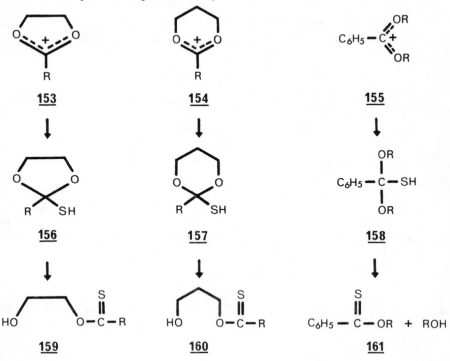

Secondly, it was observed by the same authors (61) that the reaction of
bicyclic 1,3-dioxolenium salt **162** with sodium hydrosulfide gave a mixture
of the two isomeric hydroxy thionoacetates **164** and **165** in a 1.5:1 ratio.

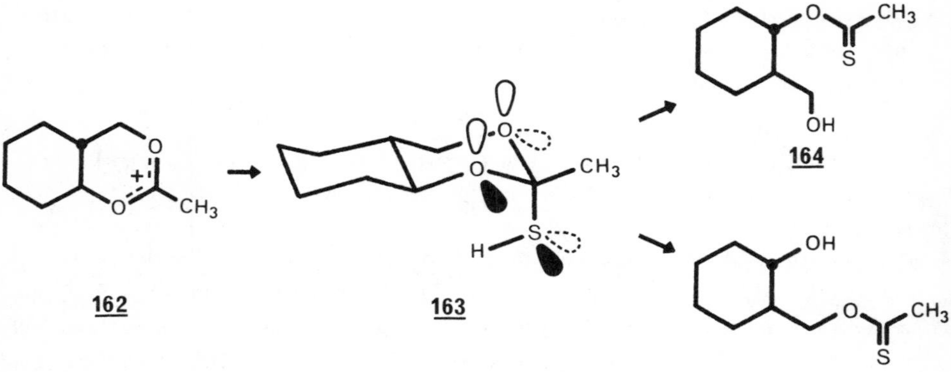

On the other hand, the same reaction with the bicyclic lactonium salt **166**
(R = H or CH$_3$) yielded the hydroxy thionoester **168** exclusively; the other
a priori possible product, i.e. the thionolactone **169**, was not produced.
Following the principle of stereoelectronic control, the salt **162** must react
with sodium hydrosulfide to give the axial hemi-orthothiol ester **163**. The
specific formation of methyl orthothio ester **163** (SH=SCH$_3$) from the reaction
of ion **162** with methyllithium mercaptide (CH$_3$SLi) confirmed the configuration of **163**. The intermediate **163** can break down with stereoelectronic
control to yield the two hydroxy thionoesters **164** and **165**. The reaction of
ion **166** with sodium hydrosulfide must produce the intermediate **167** which
can only break down to yield the hydroxy thionoester **168** because the cleavage of **167** to furnish the thionolactone **169** cannot take place with stereo-

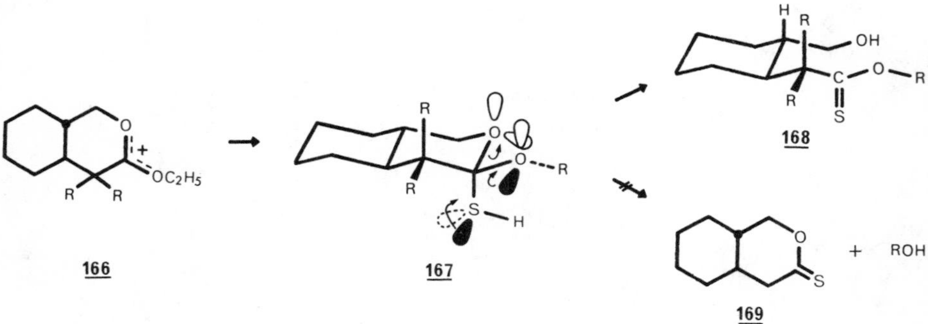

electronic control. This rationalization of Kaloustian and Khouri based on the principle of stereoelectronic control is completely consistent with their experimental results.

The same workers (61) have also studied the reaction of bicyclic salt **170** with sodium hydrosulfide. To their surprise, they isolated a mixture of hemi-orthothiol esters **171** and **172** in preference to the hydroxy thionolactone **173**. They further showed that **171** and **172** are rapidly interconverted. The reluctance of **172** to give hydroxy thionolactone **173**, despite stereoelectronic assistance, could be explained on kinetic and/or thermodynamic grounds. Kaloustian and Khouri favored the kinetic explanation because other hydroxythionoesters do not exist in the tetrahedral form. This writer favors the thermodynamic explanation. Compound **173** has a thionolactone rather than a thionoester functional group. It is possible that, as in the case of lactones by comparison with esters, thionolactones might be more reactive than thionoesters, and on that basis, the equilibrium would be in favor of **171** and **172** in preference to **173**. The interconversion **171** ⇌ **172** cannot occur with stereoelectronic control unless considerable twisting of one of the rings in **171** can take place. The interconversion **171** ⇌ **172** must therefore have a relatively high energy barrier.

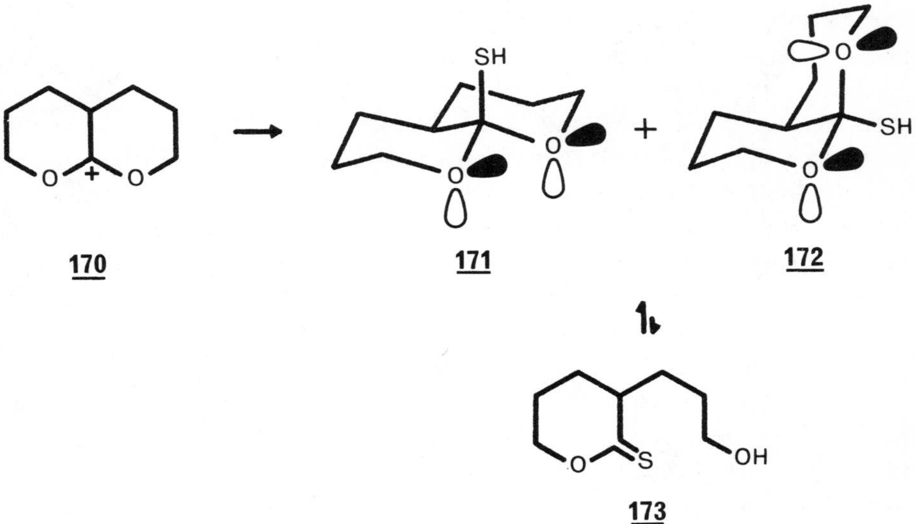

170   **171**   +   **172**

173

Kaloustian and Khouri (74) have also observed that the sodium salts **174** (n=2 and 3) of the hemi-orthothiol esters **156** and **157** are stable tetrahedral intermediates. These insoluble salts were produced by reacting the dialkoxycarbonium ions **153** and **154** with sodium sulfide (Na_2S). These stable insoluble tetrahedral species were fully characterized by the following chemical interconversions. Reaction of **174** with methyl iodide or trimethyl oxonium tetrafluoroborate gave the cyclic methyl orthothioesters **175** (n=2 and 3) which were also produced from the reaction of methyllithium mercaptide (CH_3SLi) with the cyclic dialkoxycarbonium ions **153** and **154**. Treatment of hemi-orthothiolate ester anions **174** (n=2 and 3) with water gave the hydroxy thionoesters **176** (n=2 and 3). Finally, treatment of hydroxy thionoester **176** (n=2 and 3) with sodium hydride in acetonitrile regenerated the tetrahedral salts **174** (n=2 and 3). These results show clearly that sulfur containing tetrahedral intermediate sodium salts **174** are more stable than the corresponding thionoester sodium alcoholates (**176**, $OH=O^-Na^+$).

Finally, the reaction of lactonium ions **177-181** with sodium hydrosulfide in acetonitrile at 0°C was also studied (75, 76). Each ion gave a mixture of the corresponding thionolactones **182-186** and hydroxy thionoesters **187-191**. With the exception of **191**, hydroxy thionoesters **187-190** underwent on mild acid catalysis some degree of lactonization which was then followed by rapid decomposition. These results suggest that **187-191** are primary products resulting directly from the breakdown of the corresponding tetrahedral intermediate. They further indicate that the formation of thionolactone could compete with the formation of hydroxy thionoesters. However, temperature studies of the sulfhydrolysis of salt **180** revealed that the cleavage products **185** and **190** were formed in the ratio of 0:100 at -78°C,

20:80 at -42°C and 47:53 at 0°C. Thus, at low temperature, the cleavage of the hemi-orthothiol ester derived from __180__ occurs by preferential cleavage of the endocyclic C — O bond. The breakdown of this intermediate would therefore take place with stereoelectronic control as in the case of conformationally rigid hemi-orthothiol ester __167__.

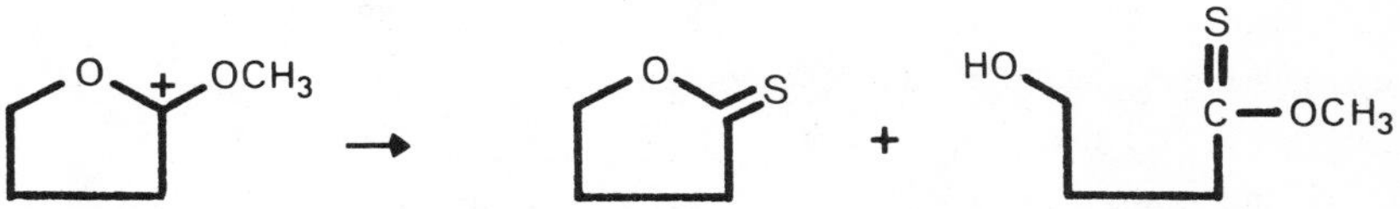

__177__   →   __182__   +   __187__

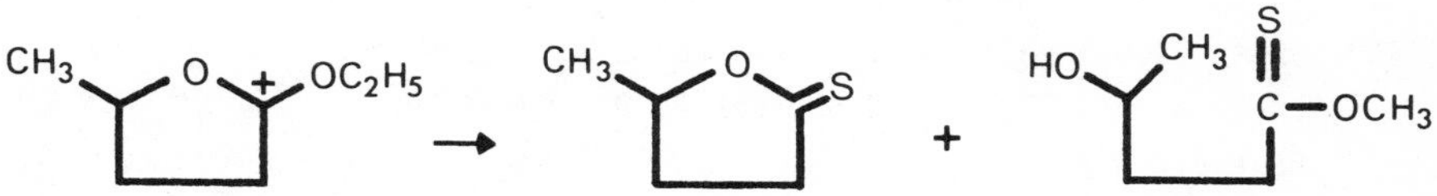

__178__   →   __183__   +   __188__

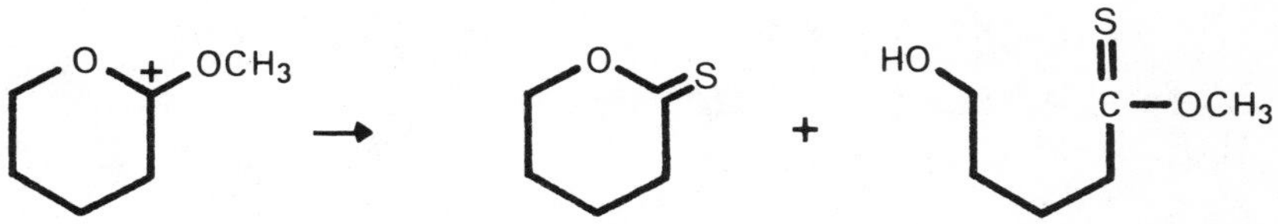

__179__   →   __184__   +   __189__

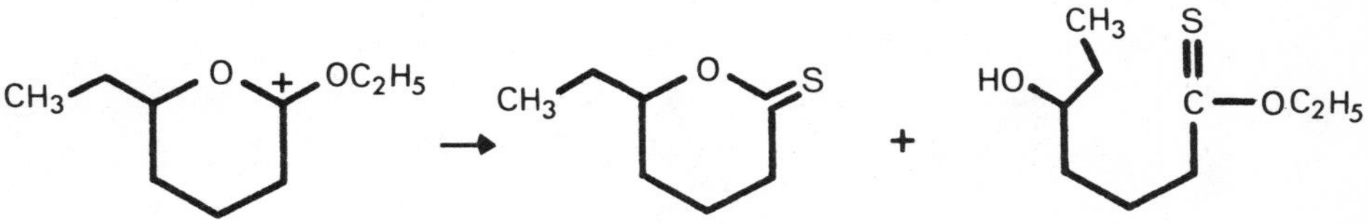

__180__   →   __185__   +   __190__

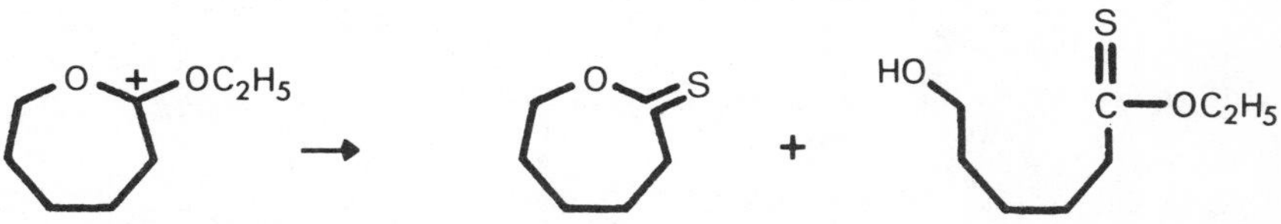

__181__   →   __186__   +   __191__

REFERENCES

(1) Deslongchamps, P. Heterocycles 1977, 7, 1271.

(2) Beaulieu, N.; Deslongchamps, P. Can. J. Chem. 1980, 58, 164.

(3) Deslongchamps, P.; Beaulieu, N.; Chênevert, R.; Dickinson, R.A. Can.
     J. Chem. 1980, 58, 1051.

(4) Jones, G.I.L.; Owen, N.L. J. Mol. Struct. 1973, 18, 1.

(5) Huisgen, R.; Ott, H. Tetrahedron 1959, 6, 253.

(6) Oki, M.; Nakanishi, H. Bull. Chem. Soc. Jpn. 1970, 43, 2558.

(7) Nakanishi, H.; Fujita, H.; Yamamoto, O. Bull. Chem. Soc. Jpn. 1978,
     51, 214.

(8) Paulsen, H. Pure Appl. Chem. 1975, 49, 69.

(9) Paulsen, H.; Dammeyer, R. Chem. Ber. 1976, 109, 1837.

(10) Ramsey, B.G.; Taft, R.W. J. Am. Chem. Soc. 1966, 88, 3058.

(11) Borch, R.F. J. Am. Chem. Soc. 1968, 90, 5303.

(12) Dusseau, Ch.H.V.; Schaafsma, S.E.; Steinberg, H.; De Boer, Th.J. Tetra-
     hedron Lett. 1969, 467.

(13) Olah, G.A.; O'Brien, D.H.; White, A.M. J. Am. Chem. Soc. 1967, 89,
     5694.

(14) Hogeveen, H.; Bickel, A.F.; Hilbers, C.W.; Mackor, E.L.; MacLean,
     C. J. Chem. Soc., Chem. Commun. 1966, 898.

(15) Brookhart, M.; Levy, G.C.; Winstein, S. J. Am. Chem. Soc. 1967, 89,
     1735.

(16) Olah, G.A.; White, A.M. J. Am. Chem. Soc. 1967, 89, 3591.

(17) Hogeveen, H. Recl. Trav. Chim. Pays Bas 1968, 87, 1313.

(18) Baldwin, J.E. Unpublished results. Personal communication.

(19) Deslongchamps, P.; Atlani, P.; Fréhel, D.; Malaval, A. Can. J. Chem.
     1972, 50, 3405.

(20) Bürgi, H.B.; Dunitz, J.D.; Shefter, E. J. Am. Chem. Soc. 1973, 95,
     5065.

(21) Rice, F.O.; Teller, E. J. Chem. Phys. 1938, 6, 489; 1939, 7, 199.

(22) Hine, J. J. Org. Chem. 1966, 31, 1236.

(23) Altmann, J.A.; Tee, O.S.; Yates, K. J. Am. Chem. Soc. 1976, 98, 7132.

(24) Guthrie, J.P. J. Am. Chem. Soc. 1973, 95, 6999.

(25) Capon, B.; Gall, J.H.; Grieve, D.McL.A. J. Chem. Soc., Chem. Commun.
     1976, 1034.

(26) Capon, B.; Grieve, D.McL.A. J. Chem. Soc., Perkin Trans. 2 1980, 300.

(27) Capon, B.; Ghoshi, A.K.; Grieve, D.McL.A. Acc. Chem. Res. 1981, 14,
     306.

(28) Ahmad, M.; Bergstrom, R.G.; Cashen, M.J.; Kresge, A.J.; McClelland, R.A.; Powell, M.F. J. Am. Chem. Soc. 1977, 99, 4827.

(29) Bender, M.L. Chem. Rev. 1960, 60, 53.

(30) Johnson, S.L. "Advances in Physical Organic Chemistry" Vol. 5; Gold, V., Ed; Academic Press: London, 1967; pp. 237-331.

(31) Bender, M.L.; Heck, H.d'A. J. Am. Chem. Soc. 1967, 89, 1211.

(32) Bender, M.L.; Thomas, R.J. J. Am. Chem. Soc. 1961, 83, 4189.

(33) Shain, S.A.; Kirsch, J.F. J. Am. Chem. Soc. 1968, 90, 5848.

(34) Bender, M.L.; Matsui, H.; Thomas, R.J.; Tobey, S.W. J. Am. Chem. Soc. 1961, 83, 4193.

(35) Eigen, M. Angew. Chem. Int. Ed. 1964, 3, 1.

(36) McClelland, R.A.; Somani, R.; Kresge, A.J. Can. J. Chem. 1979, 57, 2260.

(37) Long, F.A.; Friedman, L. J. Am. Chem. Soc. 1950, 72, 3692.

(38) Pocker, Y.; Green, E. J. Am. Chem. Soc. 1973, 95, 113; 1974, 96, 166.

(39) Deslongchamps, P.; Cheriyan, U.O.; Guida, A.; Taillefer, R.J. Nouv. J. Chim. 1977, 1, 235.

(40) Deslongchamps, P. Tetrahedron 1975, 31, 2463.

(41) Deslongchamps, P.; Chênevert, R.; Taillefer, R.J.; Moreau, C.; Saunders, J.K. Can. J. Chem. 1975, 53, 1601.

(42) Meinwald, J.; Frauenglass, E. J. Am. Chem. Soc. 1960, 82, 5235.

(43) Ficini, J.; Krief, A. Tetrahedron Lett. 1970, 1397.

(44) Russo, R.; Lambert, Y.; Deslongchamps, P. Can. J. Chem. 1971, 49, 531.

(45) Reynolds, D.P.; Newton, R.F.; Roberts, S.M. J. Chem. Soc., Chem. Commun. 1979, 1150.

(46) Newton, R.F.; Roberts, S.M. Tetrahedron 1980, 36, 2163.

(47) Cordes, E.H. "Progress in Physical Organic Chemistry" Vol. 4; Streitwieser, Jr., A. and Taft, R.W., Eds; Interscience: New York, 1967, pp. 1-44.

(48) Cordes, E.H.; Bull, H.G. Chem. Rev. 1974, 74, 581.

(49) Fife, T.H. Acc. Chem. Res. 1972, 5, 264.

(50) Bouab, O.; Lamaty, G.; Moreau, C.; Pomares, O.; Deslongchamps, P.; Ruest, L. Can. J. Chem. 1980, 58, 567.

(51) Chiang, Y.; Kresge, A.J.; Salomaa, P.; Young, C.I. J. Am. Chem. Soc. 1974, 96, 4494.

(52) Lam, P.W.K.; McClelland, R.A. J. Chem. Soc., Chem. Commun. 1980, 883.

(53) Bouab, O.; Moreau, C.; Zeh Ako, M. Tetrahedron Lett. 1978, 61.

(54) Beaulieu, N.; Deslongchamps, P. Can. J. Chem. 1980, 58, 875.

(55) McElvain, S.M.; Degginger, E.R.; Behun, J.D. J. Am. Chem. Soc. 1954, 76, 5736.

(56) McElvain, S.M.; McKay, Jr. G.R. J. Am. Chem. Soc. 1955, 77, 5601.

(57) Banyard, S.H.; Dunitz, J.D. Acta Crystallogr., Sect. B 1976, 32, 318.

(58) Capon, B.; Grieve, D.McL.A. Tetrahedron Lett. 1982, 23, 4823.

(59) Deslongchamps, P.; Beaulieu, N. Unpublished results.

(60) King, J.F.; Allbutt, A.D. Tetrahedron Lett. 1967, 49.

(61) Kaloustian, M.K.; Khouri, F. J. Am. Chem. Soc. 1980, 102, 7579.

(62) King, J.F.; Allbutt, A.D. Can. J. Chem. 1970, 48, 1754.

(63) Buchanan, J.G.; Fletcher, R. J. Chem. Soc. 1965, 6316.

(64) Perlin, A.S. Can. J. Chem. 1963, 41, 399.

(65) Note added in Proof p 1759. Julia, S.; Lorne, R. C. R. Hebd. Séances Acad. Sci., Ser. C 1969, 268, 1617.

(66) Lemieux, R.U.; Driguez, H. J. Am. Chem. Soc. 1975, 97, 4069.

(67) Deslongchamps, P.; Moreau, C. Can. J. Chem. 1971, 49, 2465.

(68) Deslongchamps, P.; Moreau, C.; Fréhel, D.; Atlani, P. Can. J. Chem. 1972, 50, 3402.

(69) Deslongchamps, P.; Atlani, P.; Fréhel, D.; Malaval, A.; Moreau, C. Can. J. Chem. 1974, 52, 3651.

(70) Kovac, F.; Plesničar, B. J. Chem. Soc., Chem. Commun. 1978, 122; J. Am. Chem. Soc. 1979, 101, 2677.

(71) Deslongchamps, P.; Moreau, C.; Fréhel, D.; Chênevert, R. Can. J. Chem. 1975, 53, 1204.

(72) See p 1306 in reference 1.

(73) Khouri, F.; Kaloustian, M.K. Tetrahedron Lett. 1978, 5067.

(74) Khouri, F.; Kaloustian, M.K. J. Am. Chem. Soc. 1979, 101, 2249.

(75) Kaloustian, M.K.; Khouri, F. Tetrahedron Lett. In Press.

(76) Khouri, F. Ph.D. Thesis, Fordham University, New York, N.Y., 1981.

CHAPTER 4

AMIDES AND RELATED FUNCTIONS

Stereoelectronic effects and the amide function

This Chapter deals with the stereoelectronic effects which control the cleavage of tetrahedral intermediates during the formation or the hydrolysis of the amide function (1-4). These electronic effects will be examined in the amide function first.

In amides, the nitrogen electron pair is n-π conjugated with the carbonyl group and this electronic delocalization is normally expressed by resonance structures $\underline{1}$, $\underline{2}$, and $\underline{3}$. As a result, the amide function is essentially planar and it is assumed that the three atoms (C, N, and O) of this function are sp^2 hybridized. The amide function can be illustrated in three dimensions by structure $\underline{4}$. The electronic distribution can also be viewed as the result of the delocalization of two n electron pairs, one from the oxygen atom and one from the nitrogen atom (cf. $\underline{1}$ and $\underline{3}$ versus $\underline{2}$) and on that basis, it is referred to here as the primary electronic delocalization of the amide function.

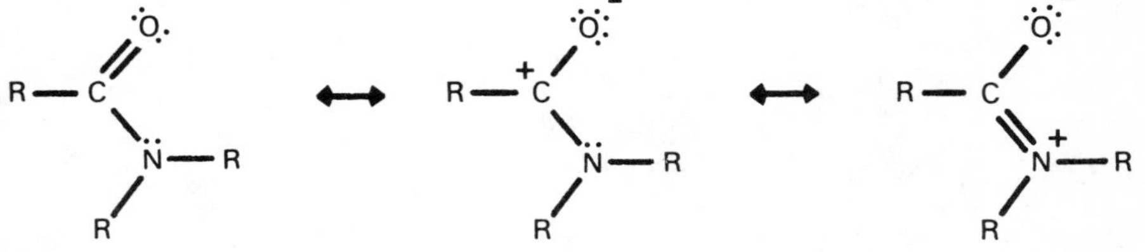

$$\underline{1} \qquad\qquad \underline{2} \qquad\qquad \underline{3}$$

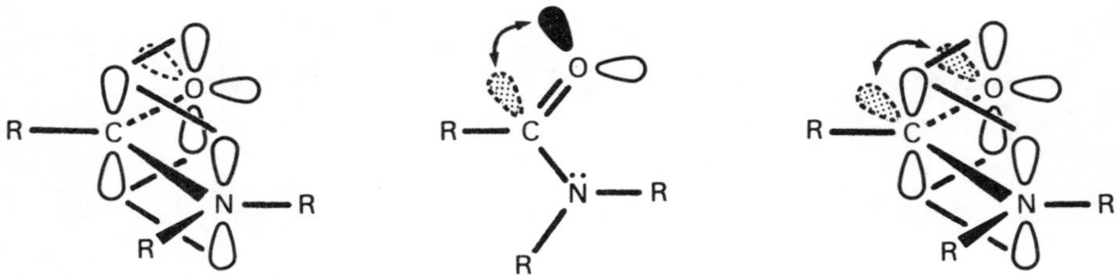

Furthermore, the oxygen atom of the carbonyl group in the amide function has an electron pair oriented antiperiplanar to the polar C—N bond; there is therefore an electronic delocalization caused by the overlap of that oxygen electron pair orbital with the antibonding orbital of the C—N sigma bond (σ^*) as shown in two dimensions by structure 5 and in three dimensions by structure 6. This additional n-σ^* delocalization is referred to here as a secondary electronic delocalization. Thus, amides are similar to E esters because they both have the primary electronic effect and one secondary electronic effect. This is in contrast with Z esters which have two secondary electronic effects besides the primary electronic effect.

As in the case of esters, formation of tetrahedral intermediates from amides must take place with stereoelectronic control (1). Under these conditions a nucleophile Y^- must make an approach almost perpendicular (i.e. with a ≈109° angle (5)) to the plane of the conjugated system of the amide, giving a tetrahedral intermediate which has an electron pair on both the oxygen and the nitrogen atom oriented antiperiplanar to the newly formed C — Y bond (4 → 7). Note that the R'—N bond remains antiperiplanar to the C — R bond in 7 and that the R" group of the nitrogen atom which was syn to the R group in 4 becomes gauche to it in 7. Consequently, the conversion 4 → 7 follows the principle of least motion (6-8).

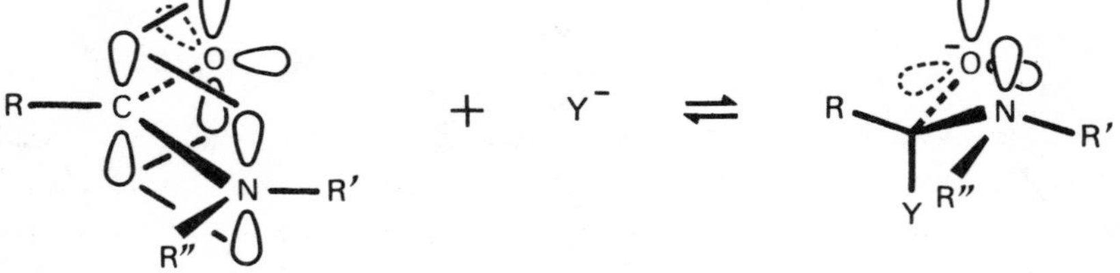

The principle of microscopic reversibility predicts that the reverse process must follow the same path which is indeed stereoelectronically allowed: the oxygen atom in $\underline{7}$ has two secondary electronic effects (n-σ*) (one electron pair of the oxygen atom is antiperiplanar to the C — N bond while the other is antiperiplanar to C — Y bond) and the nitrogen has one (the nitrogen electron pair is antiperiplanar to the C — Y bond). Thus, there are three secondary electronic effects (n-σ*) in $\underline{7}$ and by the ejection of Y⁻ to form $\underline{4}$, two of these (due to the two electron pairs antiperiplanar to the C — Y bond) have been transformed into primary electronic effects (n-π*) in the product $\underline{4}$. The third secondary electronic effect remains a n-σ* interaction in the product. The ejection of Y⁻ can therefore take place with the help of the primary and one secondary electronic effects.

In cases where Y is an alkoxy group, there is the possibility of forming either an ester or an amide function, and the proportion of each will depend on the conformation of the tetrahedral intermediate. The nine different gauche conformers for such a hemi-orthoamide tetrahedral intermediate are shown in Fig. 1, and the stereoelectronically controlled cleavages are described in Table 1.

TABLE 1   Predicted Cleavages of the Nine Conformers of Hemi-orthoamide

amide		conformer		ester
″	⇌	A	↮	
″	⇌	B	⇌	$\underline{E}$
″	⇌	C	⇌	$\underline{Z}$
	↮	D	↮	
	↮	E	⇌	$\underline{E}$
	↮	F	⇌	$\underline{Z}$
	↮	G	↮	
	↮	H	⇌	$\underline{E}$
	↮	I	⇌	$\underline{Z}$

A detailed examination of the cleavage in conformers $\underline{A}$, $\underline{B}$, and $\underline{C}$ follows. Interestingly these three are the only conformers generated directly from reaction of alkoxide ion on a tertiary amide, with stereoelectronic control. Cleavage of conformer $\underline{A}$ can only lead to the tertiary amide, as the ejection of the amino group cannot occur with the help of the primary electronic effect (the O — R bond is antiperiplanar to the C — N bond). Conformer $\underline{B}$

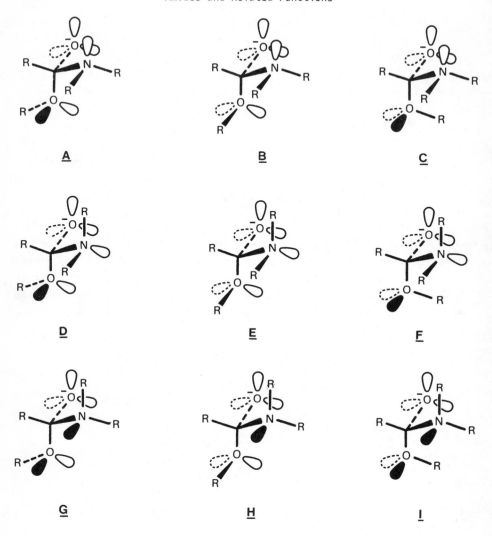

Fig. 1

can give either the tertiary amide or an **E** ester whereas conformer **C** can yield either the tertiary amide or a **Z** ester.

In addition to the primary electronic effect, in conformer **B**, the ejection of the amino group and the OR group occur each with the help of one secondary electronic effect, while in conformer **C**, the cleavage of the amino group occurs with two secondary electronic effects and that of the OR group

with one. In conformer **B**, both cleavages are consequently equally favored
electronically but since the OR group is a better leaving group, the formation of a tertiary amide should be favored. In conformer **C**, the ejection of
the amino group is electronically favored by the additional secondary electronic effect. On that basis, the ejection of the amino group by comparison
with the OR group should be easier.

The secondary electronic effect can also influence the ability of a leaving
group (cf. p. 32). The OR oxygen has one secondary electronic effect in
conformer **B** and two in conformer **C** while the amino group has only one in
both conformers. Thus, the OR oxygen has one double character in conformer
B and two in conformer **C**. The OR group is thus a poorer leaving group in
conformer **C** than in conformer **B**. This factor favors again the ejection
of the amino group in conformer **C** by comparison with conformer **B**. Consequently, in conformer **C**, the secondary electronic effect makes the OR group
a poorer leaving group and, at the same time, this effect favors the ejection of the amino group electronically. On that basis, it is possible to
rationalize why an a priori poorer leaving group (R_2N^-) can be ejected
in preference to a better leaving group.

The ionic state of the hemi-orthoamide tetrahedral intermediate must also
be considered (9-12). In acidic medium, the intermediate will exist in
the protonated form T^+, in slightly basic medium (near the pKa of $T°$, pH ≈
10), it will exist as a mixture of $T°$ and $T^±$, and in basic medium (pH
> 11), it will exist as T^-.

$$
\begin{array}{cccc}
\text{OR} & \text{OR} & \text{OR} & \text{OR} \\
| & | & | & | \\
R-C-NR_2 \quad H+ & R-C-NR_2 \quad H+ & R-C-\ddot{N}R_2 & R-C-\ddot{N}R_2 \\
| & | & | & | \\
\text{OH} & \text{O}_- & \text{OH} & \text{O}_- \\
\\
T^+ & T^± & T° & T^-
\end{array}
$$

The energy barrier for the breakdown of $T^±$ is much lower than that of $T°$;
in $T^±$, the ammonium group is a good leaving group and the negatively charged
oxygen should ease its ejection. Since the rate of proton transfer is a
fast process (13), the interconversion $T° \rightleftharpoons T^±$ will be so fast that the
cleavage of the neutral form $T°$ need not to be considered since it will not
take place.

It is expected that the T^+ and $T^\pm$ ionic forms will always eject the amino group because the protonated nitrogen atom does not have an electron pair available to eject the alkoxy group and, the ammonium group is a much better leaving group than the alkoxy group. Ejection of the alkoxy group in competition with the amino group can occur in the T^- ionic form, where the cleavage of the $C-N$ bond will take place only if the nitrogen electron pair is hydrogen bonded with the solvent (cf. 8) (14-18). Under such conditions, the amino group can leave as a secondary amine and the ejection of the high energy amide ion ($R_2\ddot{N}\mathbf{:}^-$) is avoided.

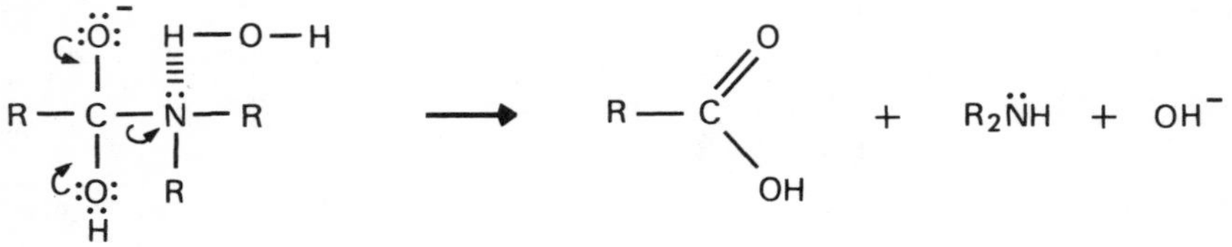

8

The relative energy barrier for stereochemical change from one tetrahedral intermediate to another and that for a stereoelectronically controlled cleavage must also be taken into consideration when analyzing experimental results. Evidence will be presented that in certain cases (specially when the tetrahedral intermediate exists in the T^- ionic form) conformational change of the intermediate does not compete with its cleavage.

Lehn and Wipff (19, 20) have reported an ab initio quantum chemical study of aminodihydroxy methane ($CH(OH)_2\ddot{N}H_2$) which indicates that there are marked stereoelectronic effects. They found that when there are two electron pairs antiperiplanar to the $C-Y$ polar bond, the $C-Y$ bond becomes long and weak. Also, the elongation of the $C-Y$ bond is more pronounced when the atom Y does not have an electron pair antiperiplanar to a polar bond. For exam-

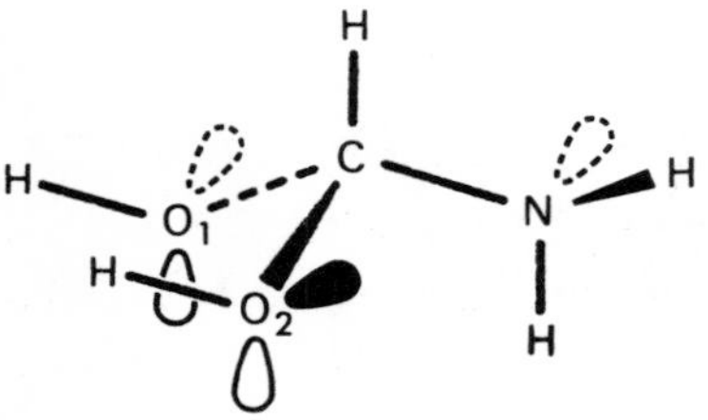

9 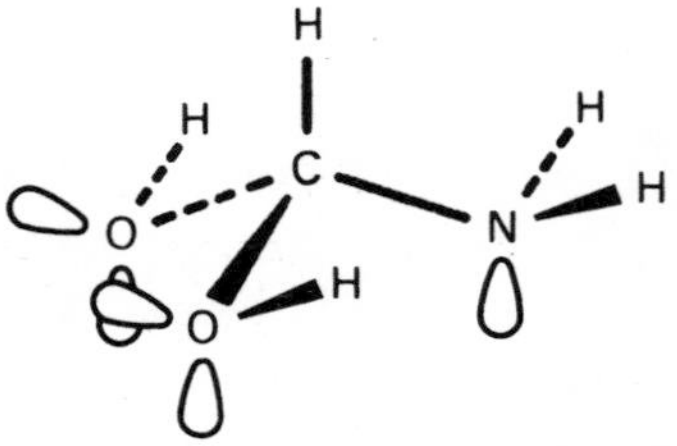   **10**

ple, the C — N bond is short and strong in **9** and long and weak in **10**; also, the $C - O_2$ bond in **9** is long and weak.

In another theoretical study (21) they have compared the neutral form T° of a hemi-orthoamide with the T^+ and the T^- ionic forms. Protonation leads to a marked and selective lengthening and weakening of the C — N bond, whereas the C — O bonds are slightly shortened and strengthened. The changes in C — N bond length show strong stereoelectronic conformation dependence; the forms where the C — N bond is antiperiplanar to two electron pairs have a very long and weak C — N bond. The C — O bond properties are also conformation dependent. They concluded that by comparison with the neutral T° form, the T^+ ionic form should lead to preferential C — N bond cleavage with higher reactivity and higher conformational selectivity. Also, deprotonation of an OH group lengthens and weakens the C — N bond. Both the C — N and the C — OH bonds should be cleaved in the T^- ionic form more easily than in the neutral T° form.

A very interesting observation was made by Dunitz and co-workers (22) in the crystal structure analyses of N,N-dimethyl-8-hydroxynaphthalene 1-carboxamide **11** and the corresponding methoxy derivative **12**. The amide function is perpendicular to the aromatic ring, and is splayed outward while the C — OR bond is inward, _i.e._ toward the carbonyl group (cf. **13**). The carbonyl naphthalene bond is bent in such a way to allow a better alignment of the oxygen nucleophile toward the carbonyl amide (5). There is also a small but significant pyramidalization of the carbonyl group carbon as well as the amide nitrogen but in the opposite direction; the carbonyl carbon atom is closest to the nucleophilic oxygen atom while the nitrogen atom displacement is away from it as illustrated in **14**. This result is in complete agreement with the principle of stereoelectronic control in hydrolytic reactions.

11 R = H
12 R = CH₃

13

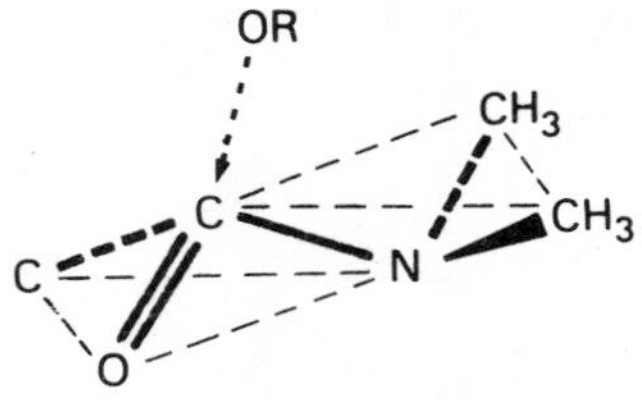

OR

14

Further experimental evidence supporting the principle of stereoelectronic control in the cleavage of hemi-orthoamide tetrahedral intermediates has been obtained from studies on the carbonyl-oxygen exchange during the basic hydrolysis of amides, and from the hydrolysis of imidate salts. These experiments are described next.

Carbonyl-oxygen exchange concurrent with hydrolysis in amides

Carbonyl-oxygen exchange has been observed in the course of the basic hydrolysis of primary amides (23, 24). The exchange, observed by using ^{18}O-labeling ($O* = {}^{18}O$), occurs via a tetrahedral hemi-orthoamide intermediate and the extensive exchange observed was explained by the fact that k_2 is larger than k_3 because an OH group is a much better leaving group than an NH_2 group.

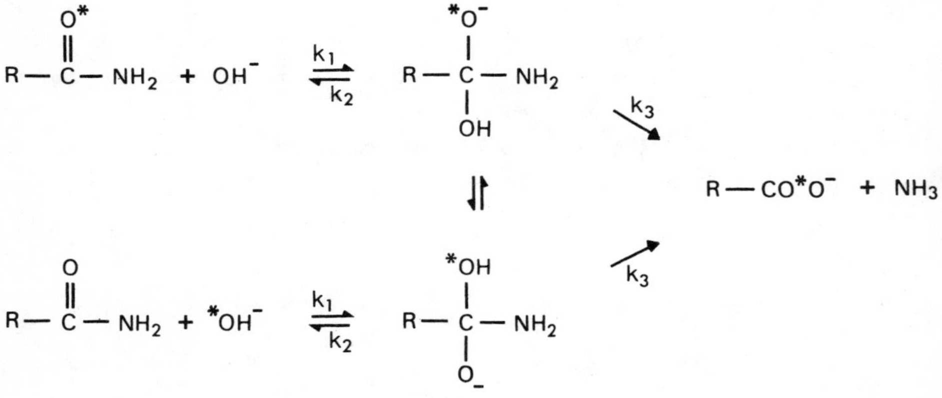

We will see that this technique can be used to demonstrate the importance of the principle of stereoelectronic control in tetrahedral intermediates derived from amides. Primary, secondary, and tertiary amides as well as N — H and N-alkyllactams will be examined.

The stereoelectronically controlled reaction of hydroxide ion with the ^{18}O-labeled primary amide **15*** (R'=H) (Fig. 2) should form tetrahedral conformer **16** (R'=H) specifically. It is assumed that proton transfer on the two oxygens can take place prior to the breakdown of intermediate **16** (R'=H). The same assumption is also made for the proton transfer on the nitrogen. The conversion **16** (R'=H) → **17** (R'=H) is therefore allowed. The proton transfer on the nitrogen can occur with the solvent via the following process. Interme-

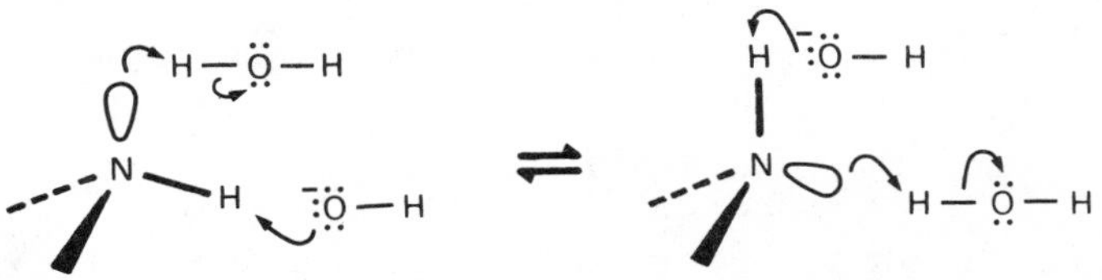

diate **16** (R'=H) has proper electron pair orientation (primary electronic effect) to give the starting amide **15*** (R'=H) and the hydrolysis products or the unlabeled amide **15** (R'=H) via **17** (R'=H). As k_2 is larger than k_3 (OH being a better leaving group than NH_2), primary amides should exhibit significant exchange, a prediction which is supported by experiments (23, 24).

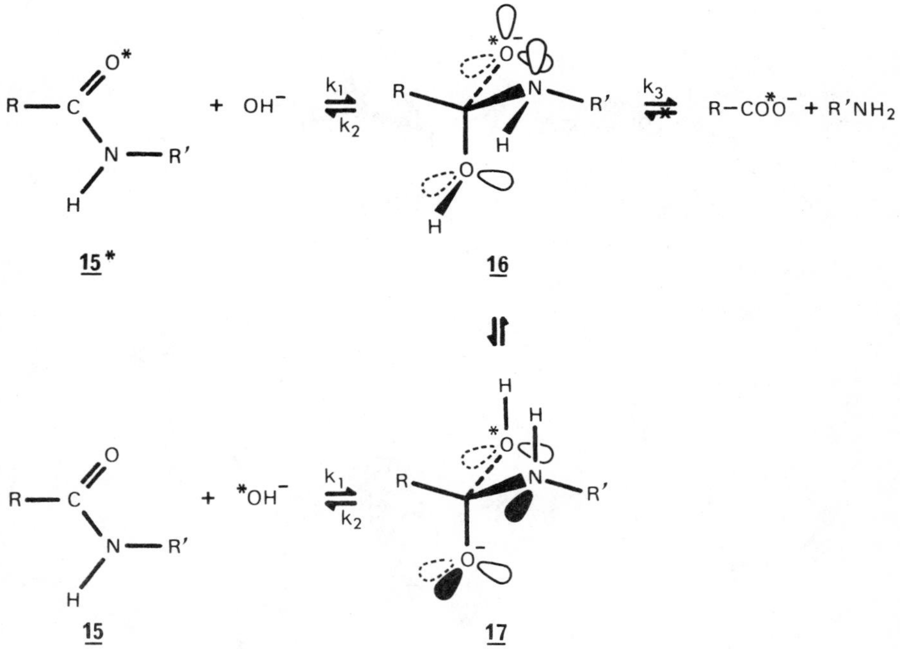

Fig. 2

A similar conclusion can be reached with secondary amides. In the most stable conformation for secondary amide, the alkyl group on the nitrogen atom is <u>syn</u> to the carbonyl-oxygen as in **15*** (25, 26). The reaction of hydroxide should give **16** which can also be converted into **17**. Thus, secondary amides should undergo carbonyl-oxygen exchange with the solvent during basic hydrolysis. This is also in accord with the experimental results (23, 24).

The fact that there is carbonyl-oxygen exchange in primary and secondary amides is in accord with the principle of stereoelectronic control but it does not constitute a proof since these experimental results can be explained without the use of this principle.

The stereoelectronically controlled reaction of hydroxide ion with an ^{18}O-labeled tertiary amide (**18***) (Fig. 3) should give the intermediate **19** which can fragment in only two ways, yielding the starting labeled amide **18*** or the hydrolysis products; direct cleavage of **19** to give unlabeled amide **18** cannot take place with the help of the primary electronic effect. In order to form the unlabeled amide **18** with stereoelectronic control, intermediate **19** must first be converted into another conformer such as **20**. Oxygen

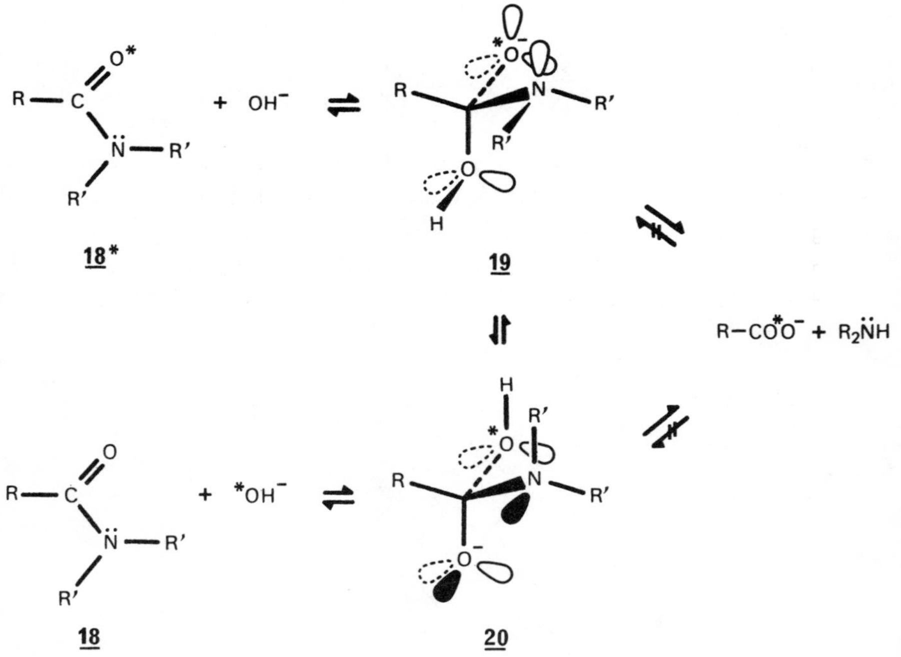

$$R-C\overset{*}{O}O^- + R_2\ddot{N}H$$

Fig. 3

exchange in tertiary amides depends therefore on the relative ease with which intermediate **19** can give either intermediate **20** or the hydrolysis products by direct fragmentation. Thus, the main difference between primary, secondary and tertiary amides, is that the first two can undergo ^{18}O-exchange without invoking a conformational change at the nitrogen in the corresponding tetrahedral intermediate, whereas in the case of tertiary amide, ^{18}O-exchange will take place only if conformational change at the nitrogen is allowed.

The change of **19** into **20** can be obtained either by a 120° rotation around the C — N bond, or by inversion of the nitrogen atom followed by a 60° rotation. Appropriate proton transfer on the two oxygens must also take place. For conformational change to occur at rates competitive with fragmentation, it is necessary that the energy barrier for that process be less than or comparable to that for fragmentation. Thus, there may or may not be oxygen exchange concurrent with the basic hydrolysis of tertiary amides, and that depends on the relative energy barriers for conformational change and fragmentation.

Bunton, Nayak, and O'Connor (27) have studied carbonyl-oxygen exchange during the hydrolysis of a primary, a secondary and a tertiary amide. They have observed that the alkaline hydrolysis of benzamide and N-methylbenzamide but not of N',N-dimethylbenzamide, is accompanied by extensive oxygen exchange between water and the amide. Thus, the tetrahedral intermediate (**19**, $R'=CH_3$ and $R=C_6H_5$) derived from N,N-dimethylbenzamide fragments more easily than it can undergo conformational change. The fact that there is no carbonyl-oxygen exchange in N,N-dimethylbenzamide constitutes a strong support for the principle of stereoelectronic control because this result can be rationalyzed only if that principle is taken into consideration.

The rates of hydrolysis and carbonyl-oxygen exchange carried out at 27°C with potassium hydroxide (1.5 N) on labeled N-benzyl-N-methyl derivatives of formamide (**21**), acetamide (**22**), and propionamide (**23**) have been reported (14).

$$R - \overset{\overset{\displaystyle O}{\|}}{C} - \overset{\displaystyle \ddot{N}}{\underset{\underset{\displaystyle CH_2C_6H_5}{|}}{}} - CH_3$$

21 R = H

22 R = CH_3

23 R = C_2H_5

It was found that there is significant carbonyl-oxygen exchange in the formamide, very little in the acetamide and apparently none in the propionamide. Thus, as the R group increases in size (R=H, CH$_3$, C$_2$H$_5$), carbonyl-oxygen exchange is less favored. This observation can be readily explained. In intermediate **19,** the barrier for internal rotation or inversion of the amino group should be lower when R is small and higher when R is large. At the same time, the energy barrier for the breakdown of **19** should be higher when R is small and lower when R is large. When R is a large group, it should favor the breakdown of the intermediate due to steric decompression. The reverse of this steric decompression effect is the classical steric hindrance caused by the size of the R group in esters (R—COOR') and amides (R — CONR'$_2$) which influences the rate of hydrolysis. For instance, formamides are hydrolyzed more rapidly than acetamides. Thus, as the R group in intermediate **19** increases in size, there should be less conformational change and that should result in decreasing carbonyl-oxygen exchange, and this is in accord with the experimental results.

The rates of carbonyl-oxygen exchange and hydrolysis of these three tertiary amides were also measured at higher temperatures (45, 65, 90°C) and it was observed that the increase in the rate of oxygen exchange is greater than that of hydrolysis with an increase of temperature (14). For instance, at 27°C, N-benzyl-N-methylformamide is hydrolyzed at a faster rate than it undergoes cabonyl-oxygen exchange, but at 90°C, the reverse is true; oxygen exchange occurs at a faster rate than hydrolysis. In principle, an increase in temperature should enhance the rate of breakdown of the intermediate as well as that of conformational change. The large increase in rate of one process over that of the other may appear surprising as both are essentially of the same nature involving internal molecular motion. There must therefore be an entropy factor which should either disfavor the breakdown of the tetrahedral intermediate or favor the conformational change with an increase of temperature.

When a hemi-orthoamide tetrahedral intermediate exists in the T$^-$ ionic form, the amide ion is not ejected previous to protonation by the solvent, to give the secondary amine. The formation of an amide ion **24** → **25** is a process so high in energy, that both the protonation and the ejection processes must be synchronized **24** → **26** → **27** (28). This means that in aqueous solution, the nitrogen electron pair must first be hydrogen bonded with the solvent, so that the group can leave as a secondary amine.

$$R-COOH + R_2\ddot{N}:^- \rightleftharpoons\!\!\!\!\!/ \quad R-\underset{\underset{OH}{|}}{\overset{\overset{O^-}{|}}{C}}-\ddot{N}R_2 \rightarrow R-\underset{\underset{OH}{|}}{\overset{\overset{O^-}{|}}{C}}-\overset{H-O-H}{\underset{}{\ddot{N}R_2}} \rightarrow R-COOH + R_2\ddot{N}H + OH^-$$

$$\underline{25} \qquad\qquad \underline{24} \qquad\qquad \underline{26} \qquad\qquad\qquad \underline{27}$$

Thus the tetrahedral intermediate must first be hydrogen bonded with the solvent before the breakdown can occur. The hydrogen bond to the nitrogen will be clearly favored at low temperature and disfavored at higher temperatures. On that basis, the breakdown of the intermediate is favored at low temperature and disfavored at higher temperature. At the same time, conformational changes should be easier at higher temperature (less hydrogen bonds) than at low temperature. Thus, by taking into account the importance of hydrogen bonding to the nitrogen in the tetrahedral intermediate, the increase of the rate of oxygen exchange over the rate of hydrolysis with an increase of temperature is readily explained.

A clear demonstration of the importance of a hydrogen bond to the nitrogen was obtained by studying N-2,6-dimethylphenyl-N-methylformamide (<u>28</u>) (14) and N-methyl-N-phenylformamide (<u>29</u>). The essential difference between these two formamides is believed to be that in <u>28</u>, contrasting to <u>29</u>, the benzene ring is not conjugated with the amide function. The benzene ring in <u>28</u> is perpendicular to the plane of the amide function. X-Ray analysis of an imidate salt derived from <u>28</u> supports this assignment (<u>vide infra</u>, p. 121). Interestingly, formamide <u>28</u> does not hydrolyze (0.15 N, KOH, 90°C, 70 h) but undergoes considerable carbonyl-oxygen exchange (>90%). This is in contrast with N-methyl-N-phenylformamide (<u>29</u>) where the hydrolysis as well as the carbonyl-oxygen exchange proceeded with ease. Formamide <u>28</u> must form the tetrahedral intermediate <u>30</u> as it undergoes carbonyl-

$$\underline{28} \quad R = CH_3$$
$$\underline{29} \quad R = H$$

$$\underline{30}$$

oxygen exchange, but since the hydrogen bonding to the nitrogen in **30** is prevented by the presence of the two methyl groups on the phenyl ring (which is parallel with the nitrogen electron pair) cleavage of the C — N bond does not take place.

The hydrolyses of labeled N-methyl-N-phenyl derivatives of acetamide (**31**) and of propionamide (**32**) were also carried out (14). It was found that there is appreciable oxygen exchange at 27°C in the N-methyl-N-phenylamides **31** and **32** which was not the case for the N-benzyl-N-methylamides **22** and **23**. This difference in behavior between these two types of amides can be explained by the phenyl ring in **31** and **32** which can be conjugated with the amino group in the tetrahedral intermediate. This conjugation effect must lower the barrier for nitrogen inversion. As a consequence, one of the two possible modes for conformational change at the nitrogen atom, i.e. nitrogen atom inversion plus a rotation of 60°, will be a lower energy process. At the same time, the amino group is less basic, and should therefore be weakly hydrogen-bonded with the solvent so the inversion-rotation process will again be favored.

31 R = CH$_3$
32 R = C$_2$H$_5$ **33**

It was also found that N,N-diisopropylformamide **33** is hydrolyzed slowly but undergoes extensive oxygen exchange (14). In **33**, the hindrance caused by the presence of the two isopropyl groups must slow down the rate of formation of the tetrahedral intermediate. The isopropyl groups must also create steric hindrance for hydrogen bonding in the tetrahedral intermediate. The cleavage of the tetrahedral intermediate becomes unusually difficult and conformational change is thus favored over that of hydrolysis in this case.

The rates of hydrolysis and carbonyl-oxygen exchange of ^{18}O-labeled N-benzyl-N-methylamides **34** (R=H, CD$_3$ and CD$_2$CH$_3$) were carefully measured at several temperatures in D$_2$^{16}O (18). The activation parameters, found by plotting

34

$\ell n(k/T)$ __versus__ $1/T$ where k is the second order rate constant for hydrolysis (k_h) or exchange (k_{ex}), are summarized in Table 2. The activation parameters for hydrolysis ($\Delta H_h^{\neq}$, $\Delta S_h^{\neq}$ and $\Delta G_h^{\neq}$) are in accord with values obtained previously on similar amides (12, 29). The large negative entropy of activation indicates that the nitrogen in the tetrahedral intermediate must form a hydrogen bond with the solvent in order for hydrolysis to occur. As expected, the formamide is more reactive towards hydrolysis than the other amides.

TABLE 2   Activation Parameters for Hydrolysis and ^{18}O Exchange of Amides **34**

R	$k_h^{25°a}$ (ℓ mol^{-1} s^{-1})	$\Delta H_h^{\neq}$ (kcal mol^{-1})	$\Delta S_h^{\neq}$ (eu)	$\Delta G_{h-25°}^{\neq}$ (kcal mol^{-1})	$k_{ex}^{25°a}$ (ℓ mol^{-1} s^{-1})	$\Delta H_{ex}^{\neq}$ (kcal mol^{-1})	$\Delta S_{ex}^{\neq}$ (eu)	$\Delta G_{ex-25°}^{\neq}$ (kcal mol^{-1})
H	1.15×10^{-4}	12.4 ± 0.3	-34.9 ± 0.5	22.8	3.56×10^{-5}	15.3 ± 0.6	-27.5 ± 1.9	23.5
CD$_3$	2.75×10^{-6}	13.3 ± 0.2	-39.3 ± 0.5	25.0	1.54×10^{-7}	19.0 ± 0.3	-25.9 ± 0.9	26.7
CH$_3$CD$_2$	1.27×10^{-6}	13.4 ± 0.5	-40.5 ± 1.3	25.5	6.85×10^{-8}	20.7 ± 1.0	-21.8 ± 2.8	27.2

^{a}Extrapolated.

Comparison of the activation parameters for carbonyl-oxygen exchange and for hydrolysis shows that, for the formamide, the rate of exchange is only slightly lower than that of hydrolysis, whereas in the case of acetamide and propionamide, the exchange occurs at a significantly lower rate.

Using the equilibrium constants estimated by Guthrie (12) for hemi-orthoamide tetrahedral intermediates (derived from N,N-dimethylformamide and N,N-dimethylacetamide) and the activation parameters described in Table 2, it was possible to obtain the free energy of activation for the breakdown ($\Delta G_{cleav}^{\neq}$) and for conformational change ($\Delta G_{conf}^{\neq}$) of the tetrahedral intermediates derived from the N-benzyl-N-methyl derivatives of formamide, acetamide and propionamide. These values are the following.

	$\Delta G_{cleav}^{\neq}$ (kcal/mol)	$\Delta G_{conf}^{\neq}$ (kcal/mol)
formamide	5.2	5.8
acetamide	6.2	8.0
propionamide	6.5	8.2

The free energy of activation for the breakdown ($\Delta G^{\neq}_{cleav}$) does not vary considerably with the amide structure, while that for the conformational change at the nitrogen ($\Delta G^{\neq}_{conf}$) changes appreciably from the formamide to the acetamide (or propionamide). This confirms that the conformational step is influenced appreciably by the steric interaction of the R group in these N,N-dimethylamides.

It is interesting to compare tert-butylbenzylmethylamine (**35**) with the tetrahedral intermediate **36** derived from N-benzyl-N-methylacetamide which has a similar degree of substitution. The rotation barrier for the $(CH_3)_3C-N$ bond and the nitrogen inversion barrier in **35** have been found identical and estimated at 6.2 kcal/mol (30). The higher value of 8.0 kcal/mol for the intermediate **36** must be a consequence of the double-bond character of the $C-N$ bond (nitrogen atom has one secondary electronic effect (n-σ*)).

35   **36**

Lactams have also been studied (31). ^{18}O-Labeled N-alkyllactams **37*** (Fig. 4) must react with hydroxide ion to give the tetrahedral intermediate **38** which can either revert to labeled lactam **37*** or produce the aminocarboxylic acid salt **39**; intermediate **38** cannot give unlabeled lactam **37** with primary stereoelectronic control. Unlabeled lactam **37** can be obtained only if the tetrahedral conformer **38** can undergo a chair inversion and proton exchange to give conformer **40**. Indeed, conformer **40** can give unlabeled lactam **37** with stereoelectronic control. Thus, carbonyl-oxygen exchange should be observed in N-alkyllactams only if the energy barrier for the chair inversion (**38** ⇌ **40**) can compete with the breakdown (**38** → **39**).

A similar conclusion can be reached with $N-H$ lactams (**37***, R=H). Such lactams should give intermediate **38** (R=H). Again, **38** can only yield either **37*** (R=H) or **39** (R=H). Unlabeled lactam **37** (R=H) can be obtained only via conformer **40** (R=H). Thus, as in the case of N-alkyllactams, $N-H$ lactams will undergo carbonyl-oxygen exchange only if the conformational change

38 ⇌ 40 and proton exchange can compete with the cleavage of intermediate 38 to give the hydrolysis product 39 (R=H).

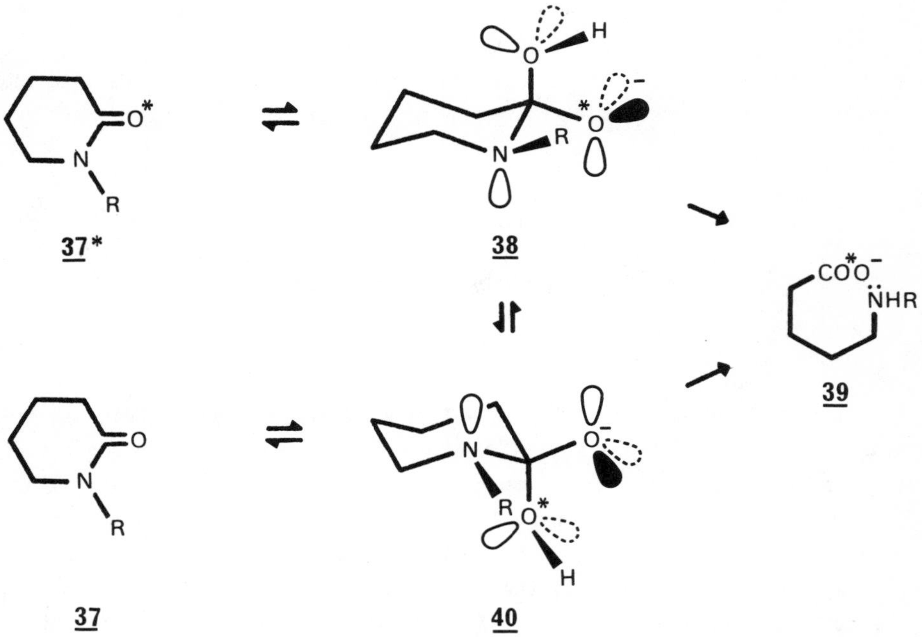

Fig. 4

^{18}O-Labeled N-methylpiperidone (37*, R=CH$_3$) and ^{18}O-labeled piperidone (37*, R=H) have been studied (31) and it was found that basic hydrolysis (1 N, NaOH) at room temperature occurs readily but no carbonyl-oxygen exchange was observed. These results show clearly that the conformational change 38 ⇌ 40 (R=H or CH$_3$) cannot compete with the breakdown of 38 to yield the hydrolysis product 38 (R=H or CH$_3$). Again, these results are consistent only if the principle of stereoelectronic control is taken into consideration; indeed, if it is neglected, 37* should give directly 40 as well as 38.

It is interesting to point out that N—H lactams which are secondary amides react differently (no ^{18}O-exchange) from acyclic secondary amides. This difference in behavior can however be readily explained because acyclic secondary amides exist in a different conformation (Z) from N—H lactams which are locked in the E conformation. Indeed, oxygen exchange can occur without conformational change, i.e. via appropriate oxygen and nitrogen proton transfers with the solvent, only in tetrahedral intermediates derived from secondary amides having the Z conformation.

β-Lactams **41*** and **42*** have also been studied (17). Concurrent carbonyl-oxygen exchange upon hydrolysis was not observed with β-lactam **41***. The β-lactam **42*** is hydrolyzed at a considerably slower rate than the β-lactam **41*** and contrary to **41***, **42*** does undergo concurrent carbonyl-oxygen exchange.

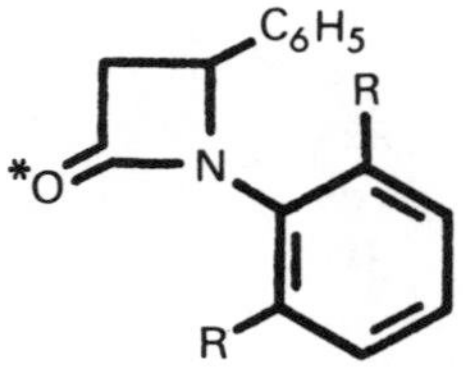

41* R = H

42* R = CH$_3$

The result obtained with β-lactam **41*** shows that the corresponding tetrahedral intermediate cleaves more readily to the hydrolysis product rather than undergo a conformational change. The result obtained with β-lactam **42*** can also be readily explained: **42*** can form a tetrahedral intermediate. However, the barrier to ring opening to give the hydrolysis product is raised in this case because of the steric hindrance to hydrogen bonding with the solvent (caused by the two methyl groups on the phenyl group). Conformational change can therefore compete with hydrolysis, with the result that carbonyl-oxygen exchange is observed.

Hydrolysis of imidate salts

Imidate salts are O-alkyl derivatives of tertiary amides. Being activated tertiary amides, they are extremely reactive towards nucleophiles. There is instantaneous reaction with hydroxide ion; they also react rapidly at room temperature with water under acidic conditions. When an imidate fluoroborate salt such as **43** reacts with sodium hydroxide, it gives sodium fluoroborate and the tetrahedral intermediate **44** which breaks down in an irrevers-

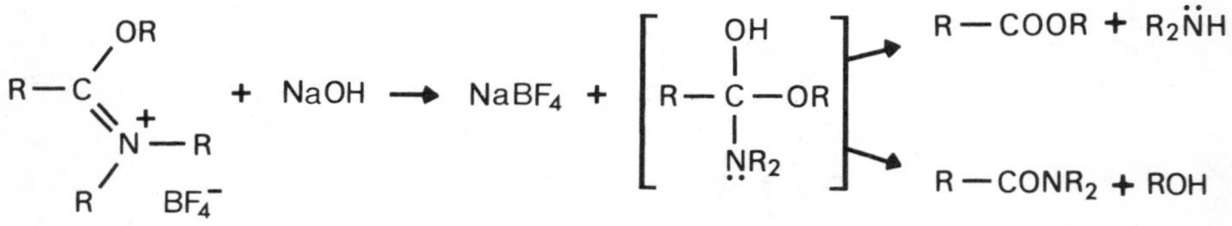

43   **44**

ible manner to yield the products of the reaction which can be either the corresponding ester and amine or amide and alcohol. The formation of **44** has been verified with ^{18}O-labeling experiments.

The hydrolysis of imidate salts is a technique to generate in situ hemi-orthoamide tetrahedral intermediates (**44**), and to observe their breakdown to yield the reaction products under kinetically controlled conditions. Such conditions can be ascertained by verifying that the reaction products are not interconverted (amide + alcohol $\rightleftharpoons$ ester + amine) during the reaction. This technique can therefore be used to test the principle of stereoelectronic control in the cleavage of tetrahedral intermediates derived from amides.

A hemi-orthoamide tetrahedral intermediate can take several ionic forms, T^+, $T^\pm$, T°, and T^-, depending on the pH of the reaction medium. In acidic medium, it will exist in the T^+ form, in slightly basic medium (near the pKa of the intermediate, pH $\approx$ 10), it will exist as $T^\pm$ and in basic medium (pH > 11), as T^-. In systems where the nitrogen can be readily protonated, T° is neglected since it is rapidly converted into the $T^\pm$ form which has a low energy barrier for fragmentation.

We have already discussed (p. 106) that T^+ and $T^\pm$ ionic forms can give the ester and amine products only. Thus, in acidic and neutral media which favor the formation of T^+ and $T^\pm$, imidate salts should always give the ester and amine products. In basic medium, which favors the formation of T^-, there is the possibility for the formation of both types of products, i.e., ester and amine or amide and alcohol. The cleavage of the C $-$ N bond in the T^- tetrahedral intermediate will take place only if the nitrogen electron pair can form a hydrogen bond with a solvent molecule. Thus, experimental evidence in favor of the principle of stereoelectronic control can be obtained with imidate salts, only when the hydrolysis is carried out under basic conditions.

As in tertiary amides, the primary electronic effect (n-π*) in imidates corresponds to the delocalization of two electron pairs, one from the nitrogen and one from the oxygen atom, which is normally represented by resonance structures **45**, **46**, and **47**. The central atoms (C, N, and O) of the imidate function are therefore sp^2 hybridized and this is confirmed by X-ray analysis (32) which shows that this function is planar. As a consequence, an imidate function can exist in two different conformations, the anti or the syn form.

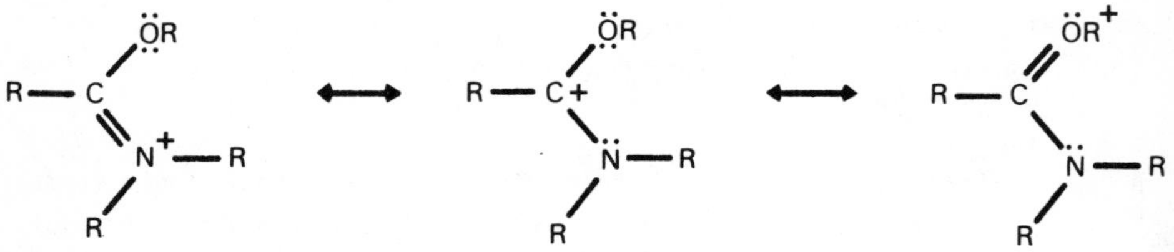

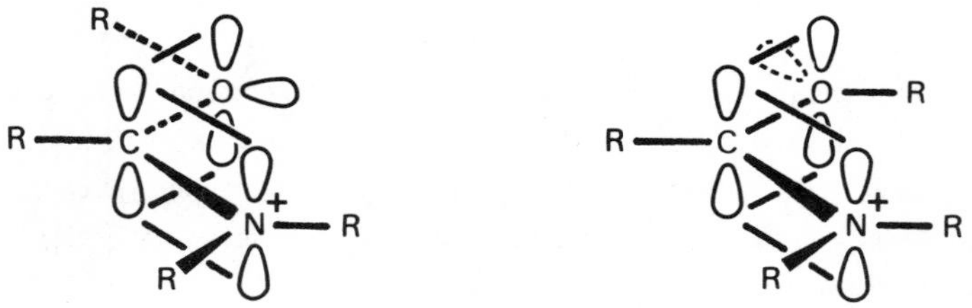

The syn and the anti forms can be represented in two dimensions by structures **48** and **49**, and in three dimensions by structures **50** and **51** respectively. In the syn conformation, the O — R bond is syn to the C — R bond whereas in the <u>anti</u> conformation, the O — R bond is antiperiplanar to the C — R bond.

In the <u>anti</u> conformation, the second electron pair on the oxygen atom (<u>cf.</u> <u>49</u> and <u>51</u>) is oriented antiperiplanar to the polar C — N bond, so this electron pair orbital can overlap with the antibonding orbital ($\sigma *$) of the C — N sigma bond. Thus, contrary to the <u>syn</u> isomer, <u>anti</u> imidates have in addition to the primary electronic effect, one secondary electronic effect (n-$\sigma*$). This additional electronic delocalization should stabilize the <u>anti</u> form relative to the <u>syn</u> form.

Steric effects must also be taken into consideration. In the <u>anti</u> form, there is a severe steric interaction between the R group on the oxygen and one of the R groups on the nitrogen atom. In the <u>syn</u> form, there is a steric interaction between the R group of the oxygen and the R group on the carbon atom. When the R group on the carbon atom is small (R=H), the steric interaction in the <u>syn</u> form is minimized and this form predominates. X-Ray analysis (32) of imidate salt <u>52</u> of N-2,6-dimethylphenyl-N-methyl-formamide confirms this conclusion and further shows that the 2,6-dimethyl-phenyl group is orthogonal to the imidate function. It was also shown (11, 15, 33) by a nuclear Overhauser effect study that the formamide imidate salt <u>53</u> exists in the <u>syn</u> form in solution.

52   **53**

When the R group linked to the carbon atom is a large group (such as a <u>t</u>-butyl or a phenyl group conjugated with the imidate function), it is assumed that the <u>anti</u> form predominates. When that R group is of an interme-diate size (R=CH$_3$ or cyclohexyl), it is assumed that there is a mixture of the <u>syn</u> and the <u>anti</u> forms. These assumptions are supported by the re-sults obtained from the hydrolysis of imidate salts <u>54</u>-<u>57</u> (11).

$$R - C \overset{\displaystyle OC_2H_5}{\underset{\displaystyle \overset{|}{N} - CH_3}{\big\|}}$$
$$\underset{\displaystyle CH_3}{\overset{+}{}}$$

54	$R = CH_3$
55	$R = C_6H_{11}$
56	$R = C_6H_5$
57	$R = (CH_3)_3C$

The results of hydrolysis of these imidate salts as a function of pH are the following: at pH 8.5 or lower, the imidate salts **54** and **55** yield the ester and amine products exclusively. At pH greater than 8.5, they start to produce the amide and alcohol products which reach a maximum yield at pH 11 (20% for **54** and 25% for **55**), and this yield remains unchanged at higher pH. The imidate salts **56** and **57** behaved completely differently as they give exclusively the ester and amine products over the entire range of pH values.

These results confirm that under acidic or neutral conditions, the hydrolysis of imidate salts yield only the ester and amine products via the T^+ and $T^{\pm}$ ionic form. They also show that under basic conditions some imidate salts (**56** and **57**) yield only the ester and amine product whereas others (**54** and **55**) give a mixture of ester and amine plus amide and alcohol products. This difference in behavior of imidate salts can be readily explained by taking into account the principle of stereoelectronic control and by assuming that imidate salts **56** and **57** exist in the anti conformation whereas imidate salts **54** and **55** exist either in the syn conformation or as a mixture of the syn and anti conformations.

Application of the principle of stereoelectronic control to the hydrolysis of syn and anti imidate salts leads to the following analysis. Syn imidate salts are first considered.

The stereoelectronically controlled reaction of the syn imidate salt **58** (Fig. 5) with hydroxide ion must give specifically conformer **59**, in which the nitrogen and the oxygen of the OR group have each an electron pair antiperiplanar to the C — OH bond; also, the R groups on the central carbon and on the oxygen atom which were syn in **58** are gauche in **59**.

Intermediate **59** cannot eject the OR or the NR₂ group with stereoelectronic control. It is therefore assumed that the energy barrier for the fragmentation of **59** is too high and this process cannot compete with internal mole-

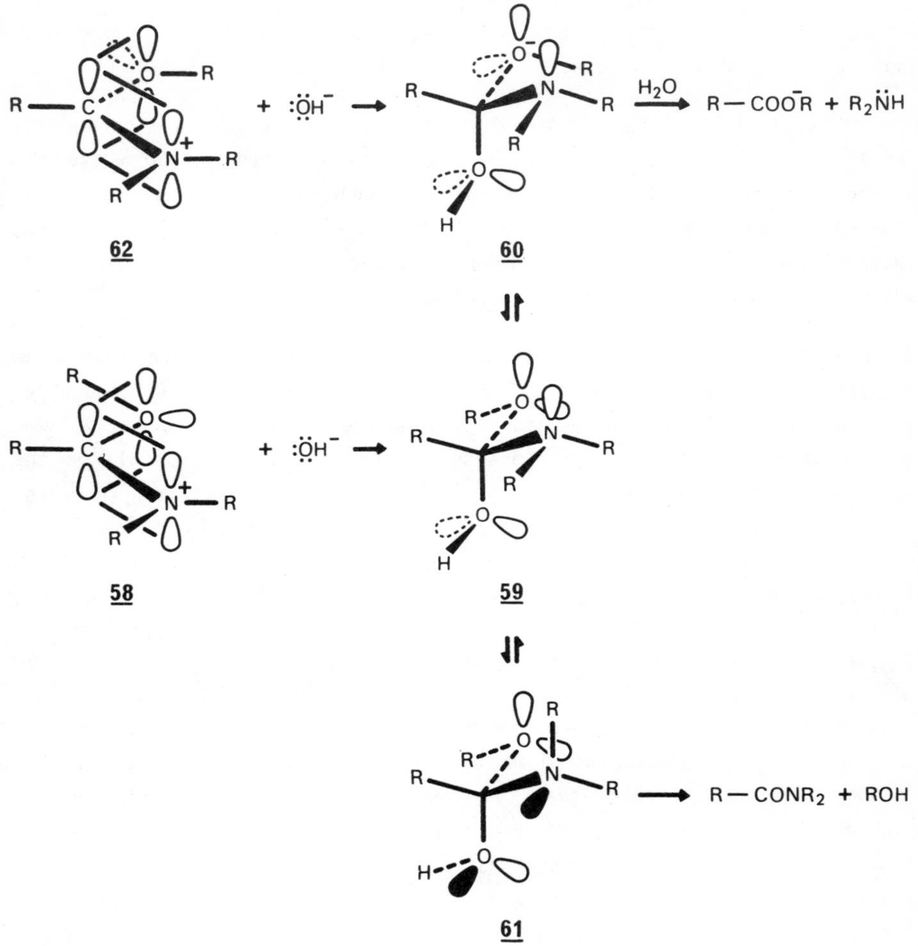

Fig. 5

cular rotation. Thus, **59** would undergo conformational changes either at the OR or NR$_2$ groups yielding in principle a mixture of the nine conformers described in Fig. 1 (p. 104, where **59** corresponds to conformer **G**). Thus, a syn imidate salt would first form intermediate **59** which is then converted into a mixture of several conformers some of which give the ester and amine, others the amide and alcohol products. For example, intermediate **59** would give intermediate **60** by rotation of the OR group and intermediate **61** by rotation of the NR$_2$ group. A stereoelectronically controlled fragmentation of the T$^-$ ionic form of **60** can only give the ester and amine products whereas that of **61** can only yield the amide and alcohol products. Thus, the basic hydrolysis of <u>syn</u> imidate salts should give ester and amine plus amide and alcohol as products.

The stereoelectronically controlled reaction of hydroxide ion with an <u>anti</u> imidate salt (**62**) must give the hemi-orthoamide conformer **60** where the nitrogen and the oxygen of the OR group have each an electron pair antiperiplanar to the C — OH bond; also, the O — R bond and the N — R bond which were antiperiplanar to the C — R bond in <u>anti</u> imidate salt **62** remain in the same relative orientation in intermediate **60**.

We have just seen that intermediate **60** can only give the ester and amine products under stereoelectronically controlled conditions. Consequently, in cases where the energy barrier for cleavage of the C — N bond is lower than that for conformational changes, the ester and amine should be the exclusive products of the reaction of hydroxide ion with <u>anti</u> imidate salts. In cases where the energy barrier for conformational changes is lower than that for fragmentation, intermediate **60** would then give in principle a mixture of the nine different conformers of Fig. 2 (where **60** corresponds to conformer <u>I</u>). Under such conditions, <u>anti</u> imidate salts also would give a mixture of ester and amine plus amide and alcohol products.

Thus, the results obtained with imidate salts **56** (R = C_6H_5) and **57** (R = $(CH_3)_3C$), <u>i.e.</u> exclusive formation of the ester and amine products, can be readily explained provided that they exist in the <u>anti</u> conformation, and that the energy barrier for fragmentation of the corresponding intermediate **60** is lower than that for conformational changes. This analysis is completely consistent with the study on the ^{18}O-carbonyl-oxygen exchange which showed that when the R group (R — $CONR_2'$) of a tertiary amide is large, conformational change cannot compete with fragmentation at room temperature. Also, this is in agreement with the fact that when the R group is large, imidate salts should exist in the <u>anti</u> conformation.

The mixture of ester and amine plus amide and alcohol products obtained from the imidate ion **54** (R=CH_3) and **55** (R=C_6H_{11}) can also be explained. These ions could exist in the <u>anti</u> form only. However, this would be surprising on the basis of the steric argument discussed previously. Also, the results obtained from the carbonyl-oxygen exchange in tertiary amides definitely show that conformational change can easily compete with the breakdown at room temperature only when R=H in tertiary amide ($RCONR_2'$). Thus, this possibility must be eliminated. These ions can exist either exclusively in the syn or as a mixture of syn and anti isomers. These two possibilities are likely because both predict the same results which correspond to the experimental observation. Indeed, if these ions are syn, it means that the intermediate 59 will first be produced and then, it would give different conformers which would yield the mixture of ester and amine plus amide and alcohol products. If these ions are a mixture of anti and syn, the former will give the ester and amine products whereas the latter would give a mixture of ester and amine plus amide and alcohol products.

This is also consistent with the results obtained from the basic hydrolysis of formamide imidate salt 63 (33) for which there is evidence (cf. 53) that this compound exists in the syn conformation. This salt gave a ≈1:1 mixture of ethyl formate and N,N-dimethylformamide.

$$C_2H_5$$
$$O$$
$$H—C$$
$$N^+—CH_3$$
$$CH_3 \qquad BF_4^-$$

63

The basic hydrolysis of imidate salt 64 was carried out (33), and it gave a mixture of aminoester 65 (83%) and N-methylpiperidone (66) (17%). This result can be explained in the following way. Assuming that this salt exists as a mixture of the syn and anti forms 67 and 68 (Fig. 6), these two isomeric forms would give the tetrahedral conformers 69 and 70 respectively. Conformer 70 can yield the aminoester 65 with stereoelectronic control whereas conformer 69 cannot break down. Thus, 69 would either be converted into 70 and 71 by rotation of the ethoxy group or undergo a chair inversion to conformer 72. Interestingly, 71 as well as 70 which come from the rota-

$$CH_3—N^+ \quad BF_4^- \quad + \quad :\ddot{O}H^- \quad \longrightarrow \quad CH_3—NH \quad COOC_2H_5 \quad + \quad N—O$$
$$OC_2H_5 \qquad\qquad\qquad\qquad\qquad\qquad\qquad\qquad\qquad\qquad CH_3$$

64   **65**   **66**

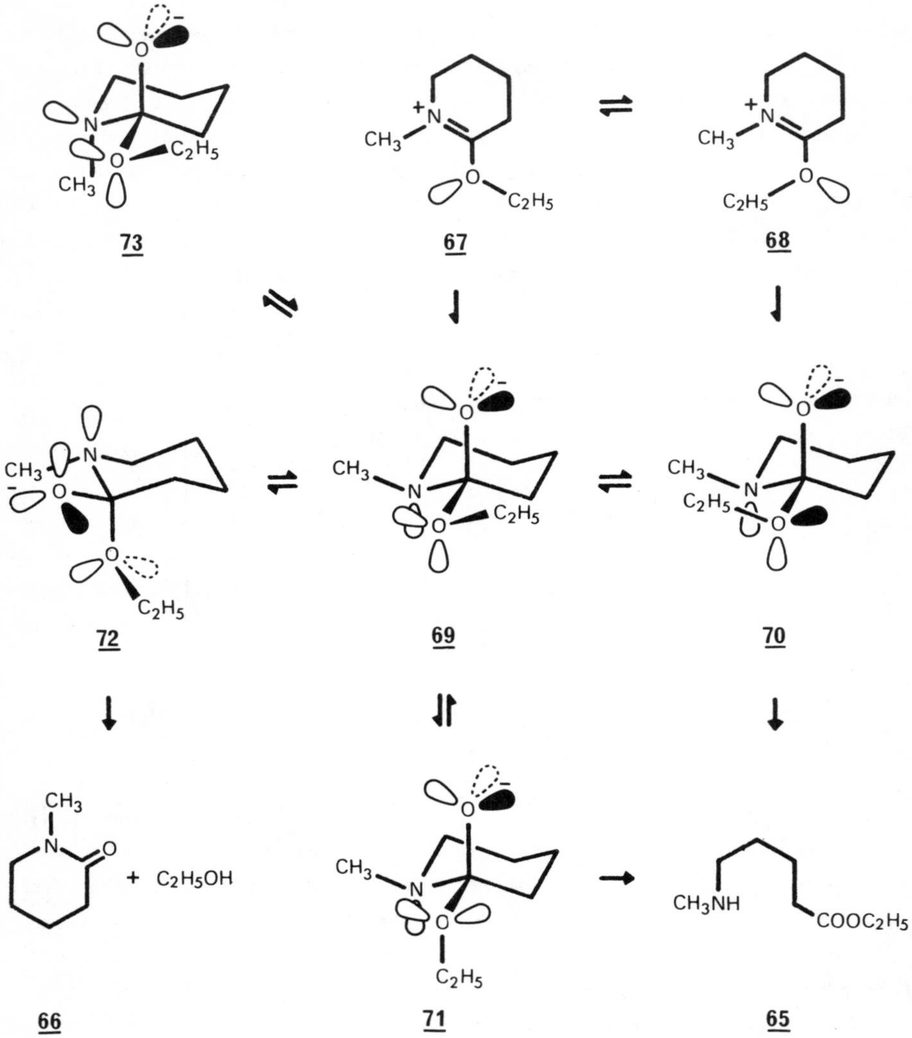

Fig. 6

tion of the ethoxy group can only give the aminoester **65**, whereas conformer **72** which comes from the chair inversion can give N-methylpiperidone (**66**). The chair inversion should be a higher energy process than the ethoxy group rotation and on that basis a large percentage of aminoester is expected. Note that a simple nitrogen inversion in **69** yields the intermediate **73** which cannot break down with stereoelectronic control.

The above rationalization is confirmed by the study of imidate salt **74** which can be considered a bicyclic analog of imidate salt **64**. Contrary to **64**, bicyclic imidate salt **74** gave first only the aminoester **75** (33). The bicyclic lactam **76** appeared only after a certain time in the reaction mixture, indicating that the aminoester **75** is clearly the exclusive kinetic product of the rotation. As in the case of imidate salt **64**, the bicyclic imidate salt must exist as a mixture of the syn and anti isomeric forms **77** and **78** (Fig. 7). The reaction of hydroxide ion with **77** and **78** must give the intermediates **79** and **80** respectively. Intermediate **80** can yield the aminoester **75**. Intermediate **79** cannot break down with stereoelectronic control; it will therefore be converted into **80** or **81** which can also fragment to give the aminoester **75**. Intermediate **79** can also undergo a nitrogen inversion by inverting ring A or ring B giving respectively intermediate **82** or **83**. Intermediate **82** cannot undergo a C—N bond cleavage with stereoelectronic control. Intermediate **83** can yield the bicyclic lactam **76** with stereoelectronic control, but this intermediate has a severe steric interaction between the axial methoxy group and ring A. The formation of **80** or **81** from **79** should be a much easier process than that of **83**, and on that basis, imidate salt **74** should give exclusively the aminoester **75**, in agreement with the experimental result.

The formation of bicyclic lactam **76** from the aminoester **75** must also take place with stereoelectronic control. It presumably occurs from the intermediate **84** which comes from the cyclization of the aminoester **75** where the ester function is in the Z conformation (cf. **85**). Note that the direct formation of **84** from the imidate salt **74** is impossible unless the principle of stereoelectronic control is violated.

The basic hydrolysis of imidate salt **86** was also reported (33). This salt gives a mixture of aminoester **87** (65%) and lactam **88** (35%). This result can be interpreted in the same manner as that of imidate salt **64**.

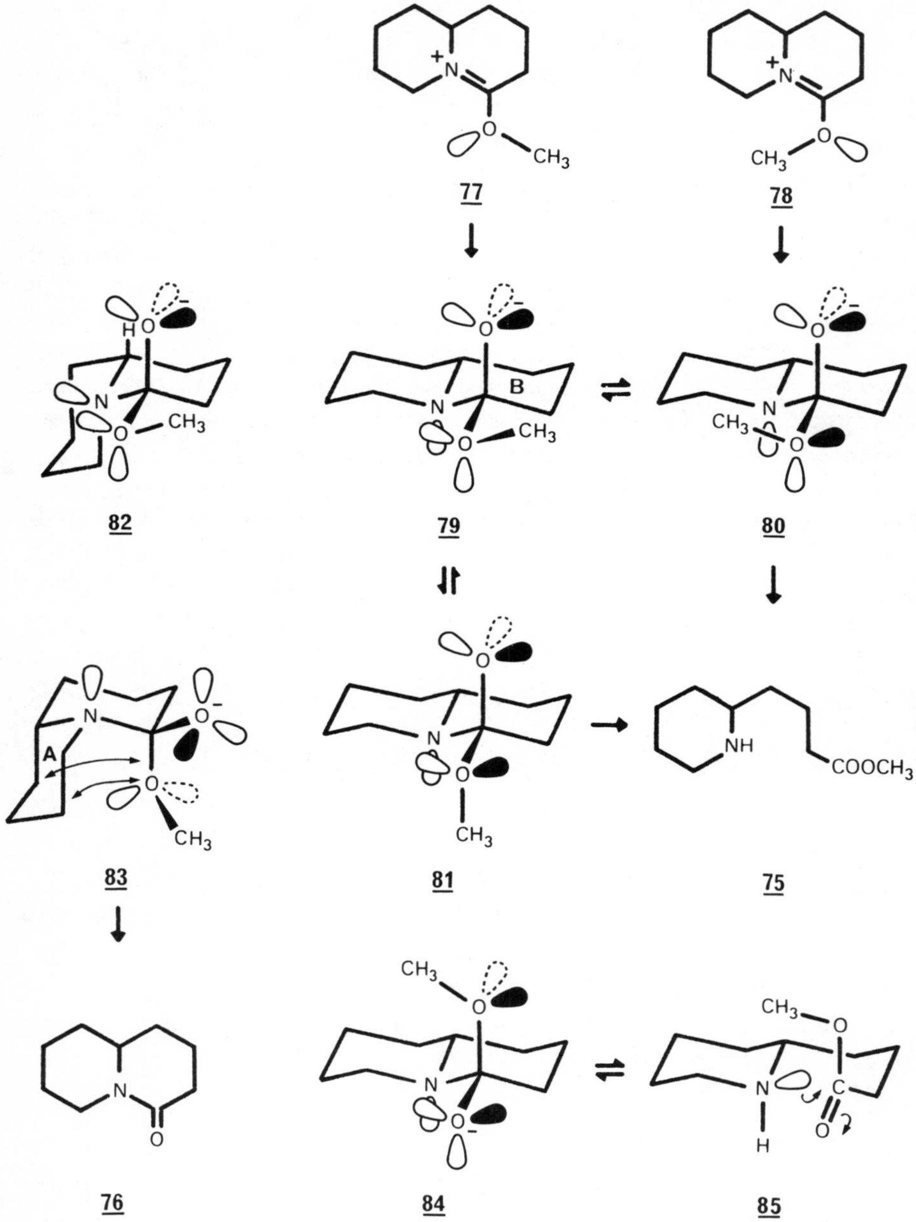

Fig. 7

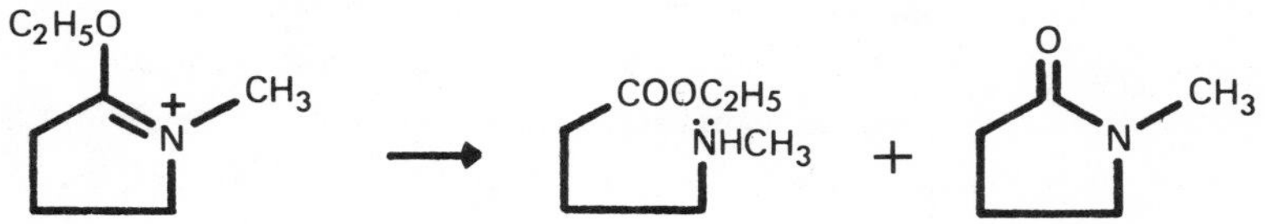

$$86 \longrightarrow 87 \; + \; 88$$

The preceding results obtained from the various imidate salts can be cor-
rectly interpreted by taking into account the principle of stereoelectronic
control. They constitute good evidence of its all important role, but they
cannot be considered as absolute proof because the conformation of the
imidate salts has not been thoroughly established.

The results which follow have been obtained on imidate salts where there
is absolutely no doubt about the syn or anti conformation, and they are
in complete accord with the preceding rationalization.

The basic hydrolysis of a series of cyclic anti imidate salts has been
investigated (1, 33). For instance, the six-membered imidate salt **89** where
the anti conformation is assured by its cyclic structure, gave first under
basic conditions, only the aminoester **90**. The aminoester **90** was then slowly
converted into the thermodynamic product of the reaction, i.e. the benzamidoalcohol **91**. The reaction of imidate salt **89** with hydroxide ion must first
give intermediate **92** following the principle of stereoelectronic control.
It can also be seen that **92** can only give the aminoester **90** by following

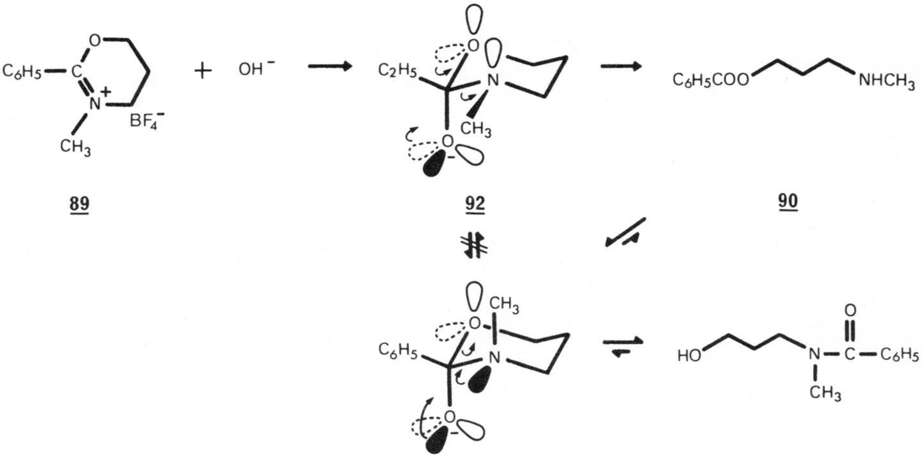

the same principle. Thus, the nitrogen inversion process to give **93** which can then yield the benzamidoalcohol **91** cannot compete with the breakdown of **92**. The slow appearance of benzamidoalcohol **91** would be due to the slow formation of intermediate **93** from aminoester **90**.

Similar studies were carried out (33) with cyclic imidate salts **94** ($R=C_6H_5$ or CH_3). They behaved like imidate salt **89**, yielding first the aminoester **95** followed by the slow formation of the thermodynamic product, i.e. the corresponding amidoalcohol **96**.

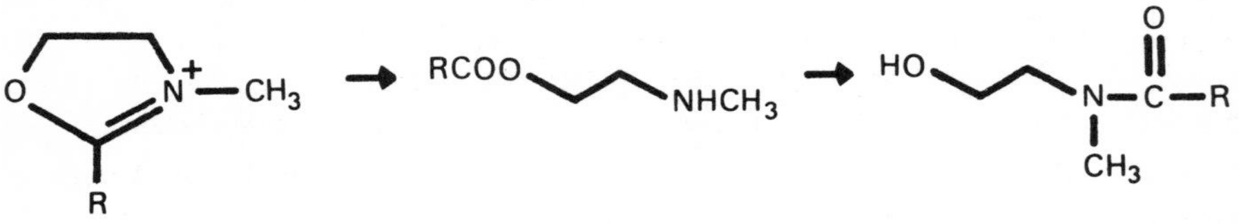

94   **95**   **96**

Imidate salt **97** also gave the aminoester **99** (33). Allen and Ginos (34) have reported that the basic hydrolysis of imidate salts **100** ($R=CH_3$, C_2H_5 or $(CH_3)_3C$) yielded only the corresponding aminoester **102**.

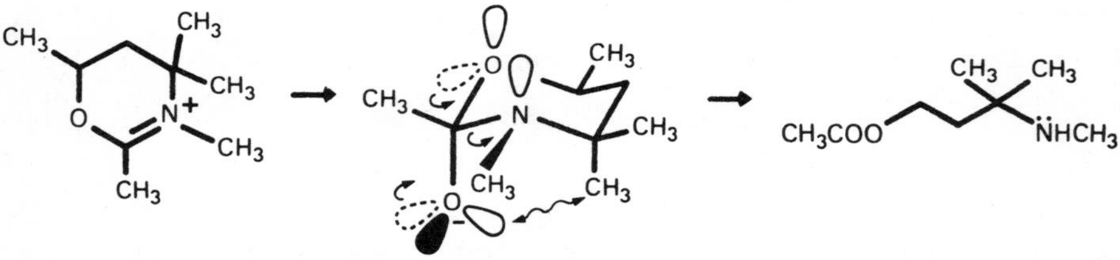

97   **98**   **99**

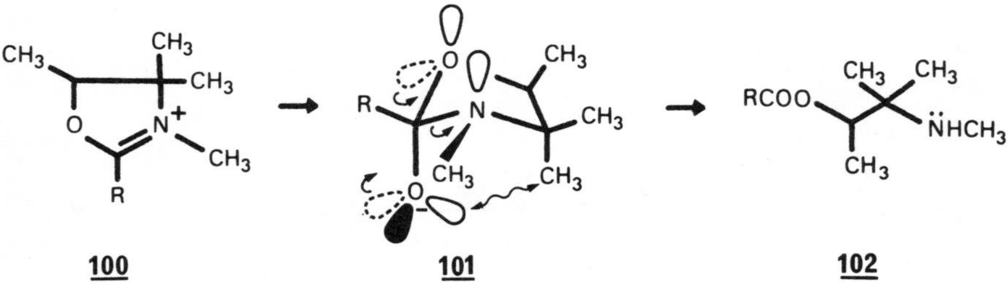

100   **101**   **102**

There are two factors which help the cleavage of the C — N bond in the hydrolysis of imidate salts **97** and **100**. Imidate salt **97** should form the intermediate **98** which has proper electron pair alignment to yield the aminoester **99**. Also, in **98** there is a 1,3-diaxial steric interaction between the OH group and one of the methyl groups which should promote the cleavage of the carbon-nitrogen bond. Similarly, compound **100** should give intermediate **101** in which there is again a strong steric interaction. This, combined with the stereoelectronic effect, favors the carbon-nitrogen bond cleavage.

The importance of this steric effect was verified by carrying out the ozonolysis of acetal **103**. Oxidation of **103** by ozone gave the ester **105** exclusively (33). In this reaction, the hydrotrioxide intermediate **104** has proper electron pair orientation to favor the opening in both directions, but cleavage occurs in only one, yielding **105** because of this steric decompression factor.

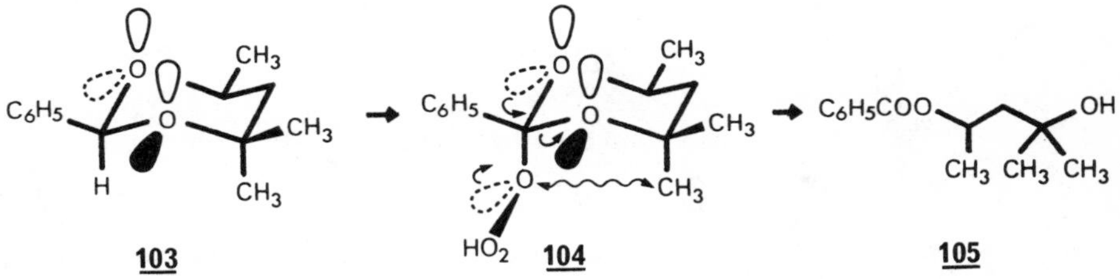

The reaction of hydroxide ion with imidate **106** should give the intermediate **107** in which stereoelectronic control promotes the cleavage of the C — N bond, while the 1,3-diaxial methyl-hydroxyl steric interaction favors the cleavage of the C — O bond. Hydrolysis of **106** gave exclusively the aminoester **108** (33); thus, stereoelectronic effects still control this reaction despite an important steric effect which favors the C — O bond cleavage. This result was also confirmed by the basic hydrolysis of imidate salt **109** which gave first the aminoester **110** which was then rapidly converted under the experimental conditions into the benzamidoalcohol **111**.

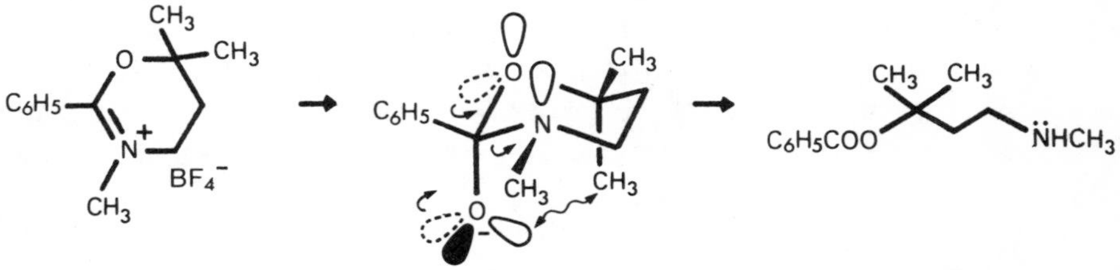

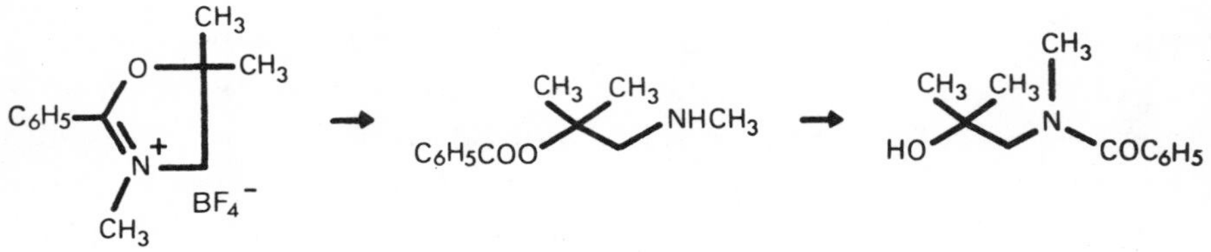

109   **110**   **111**

The acidic and basic hydrolysis of the cyclic imidate salt **112** was investigated (16). Under acidic conditions, imidate salt **112** was slowly hydrolyzed to yield the ester ammonium salt **113** exclusively. This is the expected result for any imidate salt.

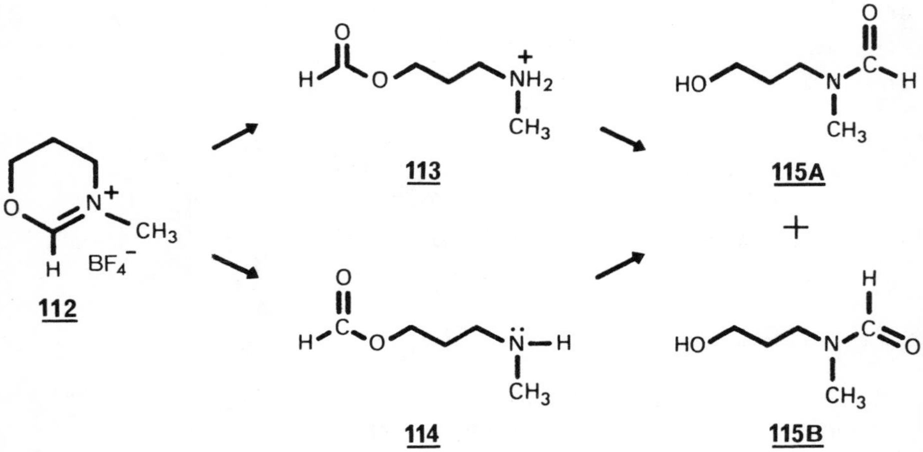

112

113   **115A**

+

114   **115B**

Under basic conditions, the hydrolysis of imidate salt **112** at 0°C gave a mixture (2:8) of the amidoalcohol rotamers **115A** and **115B** as the kinetic products of the reaction. Isomerization followed to yield the equilibrium ratio (4:6) of **115A** and **115B**. Imidate **112** is the first _anti_ imidate salt which does not give the anticipated product, _i.e._ the aminoester **114**. However, being a formate, thus a reactive ester, it is possible that under the reaction conditions, **114** recyclizes rapidly to give new tetrahedral intermediates which then yield a 2:8 mixture of amide rotamers **115A** and **115B**. This was proven by showing that the treatment of ester ammonium salt **113** under the same basic conditions at 0°C led directly to a 2:8 mixture of **115A** and **115B**.

Being a formate ester, aminoester **114** must exist as a mixture of the Z and the E forms **114A** and **114B** where the Z form predominates. Cyclization of **114A** and **114B** with stereoelectronic control should give intermediates **116** and **117** which can then give the amide rotamers **115A** and **115B** respectively. The experimental results show that the formation of **115B** is favored. The peculiar reactivity of imidate salt **112** is thus readily explained.

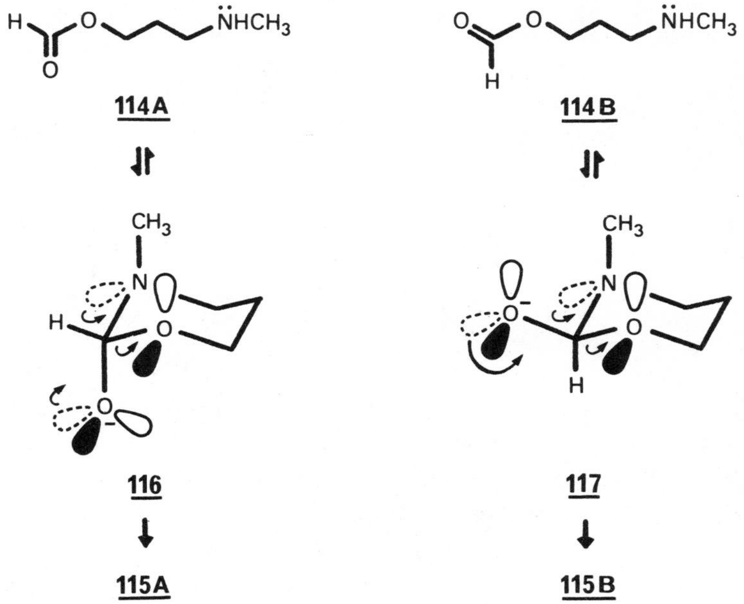

114A   **114B**

116   **117**

↓   ↓

115A   **115B**

The hydrolysis of the cyclic imidate salt **118** was also studied (16). The imidate salt **118** (R=$(CH_3)_2C_6H_3-$) behaved in a completely different manner from the salt **112**. Under acidic conditions, it yielded a ≈1:1 mixture of ester ammonium salt **119** and the amidoalcohol **120**. Again, the hydrolysis is a slow process, and it could be observed (at the beginning of the reaction) that the amidoalcohol was first formed as the least stable rotamer **120B** only. Rotamer **120B** was then slowly isomerized to give an equilibrium mixture of **120A** (67%) and **120B** (33%).

The behavior of imidate **118** under acidic conditions can be readily explained by the presence of the two methyl groups on the phenyl ring which create an important steric hindrance to protonation of the nitrogen atom in the resulting tetrahedral intermediate. The salt **118** reacts with water to give first a tetrahedral intermediate in the neutral T° form. However, the conversion

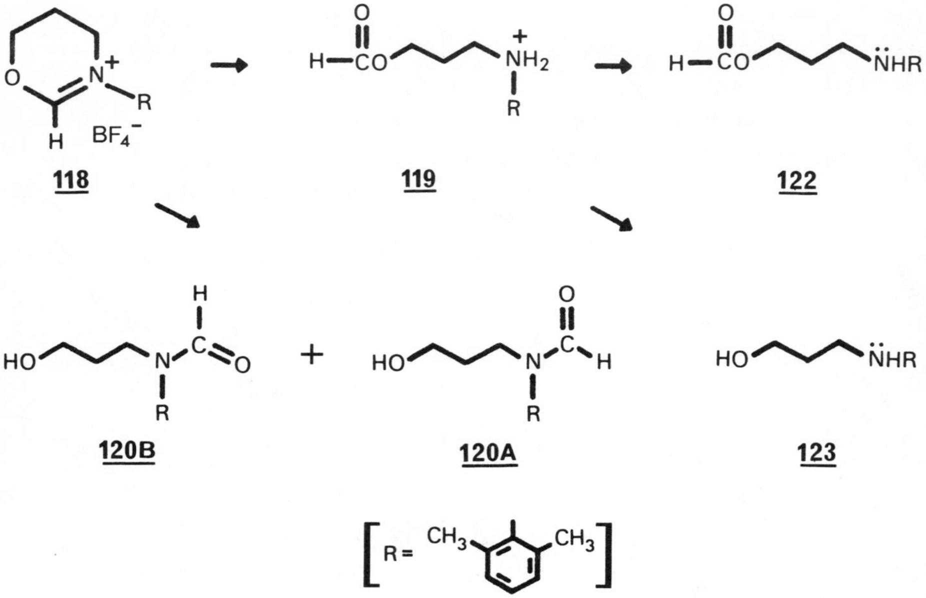

118   **119**   **122**

120B + **120A**   **123**

$$\left[R = \text{CH}_3\text{—}\bigcirc\text{—CH}_3 \right]$$

of T° into the $T^{\pm}$ or T^{+} ionic form does not occur readily. So, the intermediate gives in part the ester ammonium salt **119** via T^{+} or $T^{\pm}$ and in part the
amidoalcohol **120** via T° or more likely via T° protonated on the OR group as
in **121**. The specific production of rotamer **120B** is discussed below.

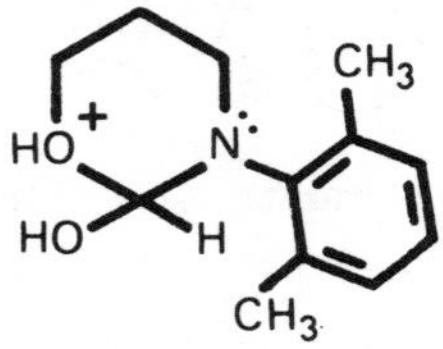

121

The basic hydrolysis of imidate salt **118** takes a different course from
that of imidate salt **112**, yielding first only the amide rotamer **120B** which
is then slowly isomerized to the equilibrium mixture (ratio 3:1) of **120A** and
120B. Treatment of the ester ammonium salt **119** under the same basic conditions gave directly the aminoalcohol **123**. This result shows that the aminoester **122** is not an intermediate in the basic hydrolysis of imidate **118**. The
formation of the amide rotamer **120B** is therefore the result of the direct
fragmentation of a tetrahedral intermediate which is formed from **118**.

The two methyl groups on the phenyl ring of imidate salt 118 are responsible for its different reactivity by comparison with the other anti imidate salts. These two groups create enough steric hindrance in the resulting tetrahedral intermediate that the tertiary nitrogen cannot be hydrogen-bonded with the solvent and the cleavage of the C—N bond is prohibited. Thus, the reaction of hydroxide ion on imidate 118 must give intermediate 124 (Fig. 8). Intermediate 124 cannot break down with stereoelectronic control to yield the amidoalcohol 120, and it cannot give the aminoester 122 because the nitrogen cannot form a hydrogen-bond with the solvent.

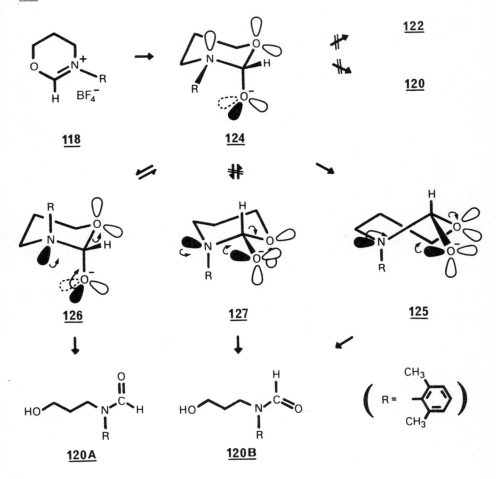

Fig. 8

Intermediate **124** will have to undergo a conformational change to cleave with stereoelectronic control. In **124**, there is a very severe steric interaction between the hydroxyl and the 2,6-dimethylphenyl groups which can be lessened on going from **124** to the half boat **125**. This steric interaction would be the main driving force for the specific conversion of **124** into **125** which has proper electron pair orientation to cleave the C − O bond and to produce exclusively the amide rotamer **120B**. The half boat **125** would be formed in preference to intermediate **126** (via nitrogen inversion) or **127** (via a chair inversion) because these intermediates have the bulky R group on the nitrogen axially oriented. Note also that **126** and **127** lead to amide rotamers **120A** and **120B** respectively. The chair inversion process which leads to **127** with an axial R group cannot be a lower energy process than the nitrogen inversion which gives **126** also with an axial R group. Thus, because the amide rotamer **120B** is the only product observed experimentally, intermediates **126** and **127** must be eliminated.

Imidate salts having a syn conformation were also studied (33). Imidate salt **128** which has a syn conformation due to its cyclic structure, gave on basic hydrolysis a mixture of amidoalcohol **129** (66%), δ-valerolactone (**130**, 33%) and dimethylamine (33%). Likewise, the hydrolysis of imidate salt **131** gave a one to one mixture of the corresponding amidoalcohol **132** and γ-butyrolactone **133** plus dimethylamine.

These syn salts give a mixture of products as predicted. Thus, for instance, the imidate salt **128** (Fig. 9) must react to produce first the intermediate **134** which cannot break down with stereoelectronic control. However, intermediate **134** can undergo a conformational change to give either **135** by carbon-nitrogen rotation (or inversion) or **136** by chair inversion. Intermediate **135** can only yield the amidoalcohol **129** whereas intermediate **136** can only give δ-valerolactone (**130**) and dimethylamine.

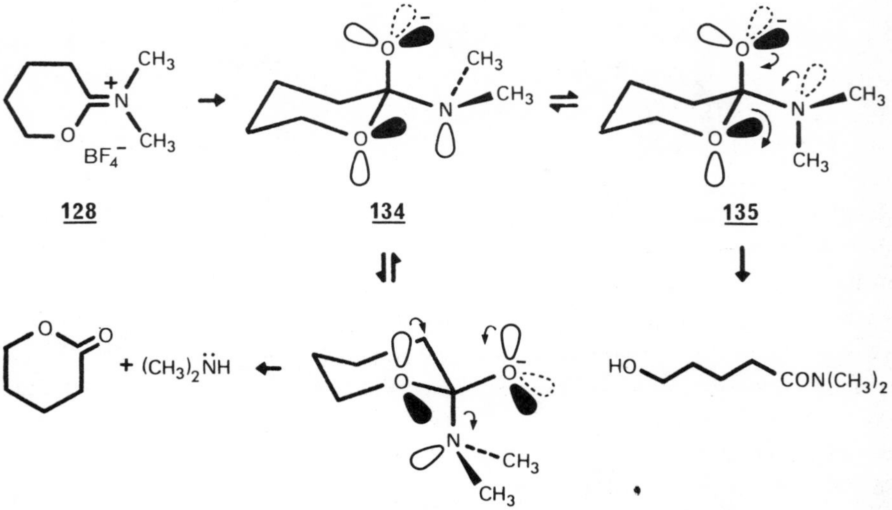

Fig. 9

An interesting study (15) has been carried out with formamide derived imidate salts which have two different alkyl groups on the nitrogen atom. The unsymmetrical imidate 137 was obtained as a 3:1 mixture of isomers 137A and 137B and the imidate 138 as isomers 138A and 138B which were obtained pure on separation. It has already been mentioned that imidate salts derived from formamide exist in the syn conformation and that salt 138B has been analyzed by X-ray (cf. 52).

Hydrolysis of the mixture of the two isomers 137A and B under acidic conditions gave only the ester and amine products (methyl formate and the salt of N-benzyl-N-methylamine). The imidate salt 138A behaved differently yielding 70% of the amide and alcohol products (2,6-dimethylphenyl-N-methyl-formamide and methanol) with only 30% of the ester and amine products (methyl formate and the salt of N-methyl-2,6-dimethylaniline). The hydrolysis of 138A was repeated with ^{18}O enriched water and the resulting amide showed a complete incorporation of the labeled oxygen. This shows that the hydrolysis of 138A proceeds via the formation of a tetrahedral intermediate.

Consequently, the isomers 137A and 137B in acidic conditions gave the expected result. The difference in behavior of salt 138A can again be explained

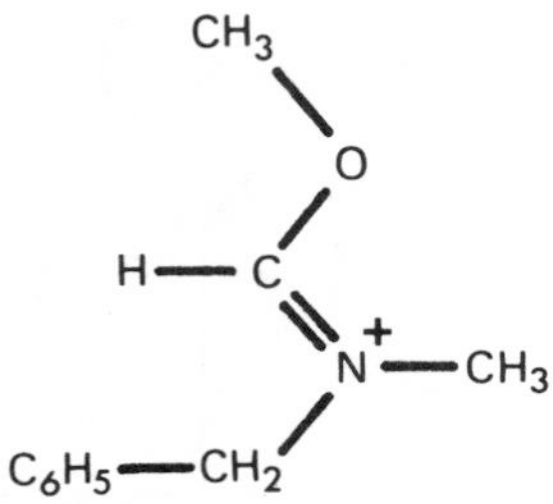

137A

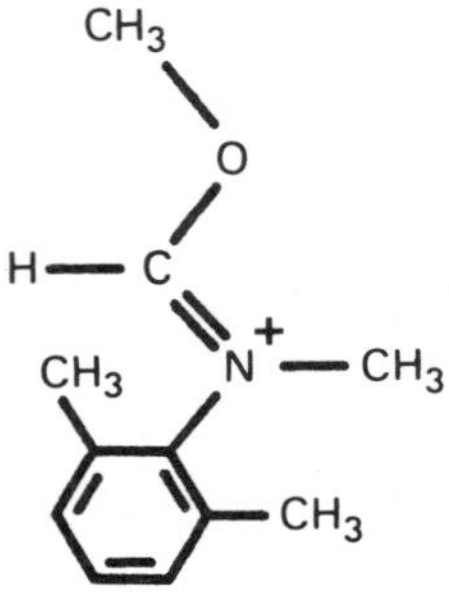

138A

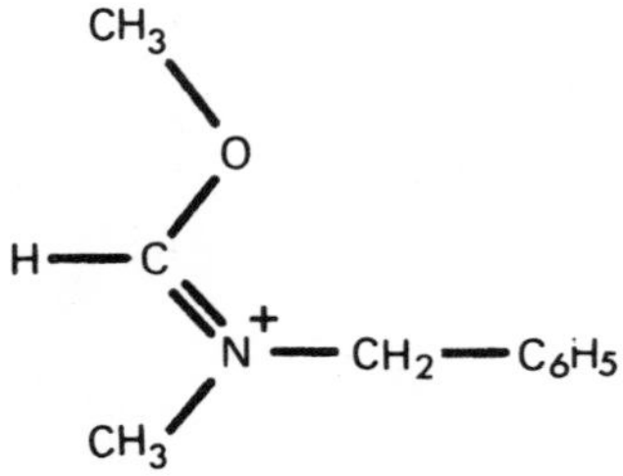

137B

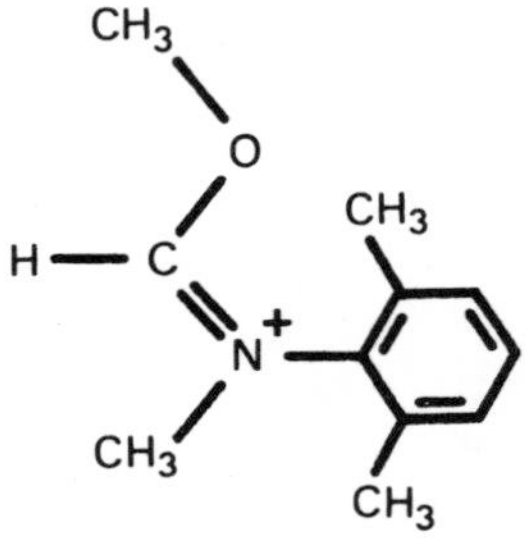

138B

by the presence of the two methyl groups on the phenyl ring which caused a severe steric hindrance to the protonation of the nitrogen in the tetrahedral intermediate. Thus salt **138A** forms first a tetrahedral intermediate present as neutral $T°$ which is then partly converted into the T^+ or $T^±$ ionic form to give the ester and amine products. The $T°$ form is also partly protonated at the OR oxygen to give the amide and alcohol products.

The results obtained under basic conditions are more interesting. Hydrolysis of the 3:1 mixture of **137A** and **B** under basic conditions gave 25% of the ester and amine and 75% of the amide and alcohol products. Under these conditions the imidate salt **138A** behaved differently giving 100% of the amide and alcohol products. An identical result was obtained with imidate salt **138B**. It was further observed that in the case of **137A** and **137B**, the tertiary amide produced was an equilibrium mixture of the corresponding rotamers **139A** (56%) and **139B** (44%). In the hydrolysis of **138A** or **138B**, an equilibrium mixture of the amide rotamers **140A** (77%) and **140B** (23%) was also observed.

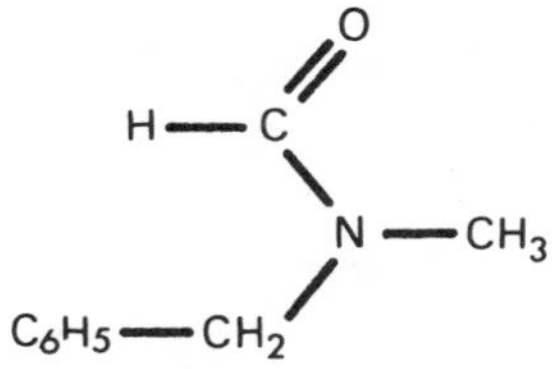

139A

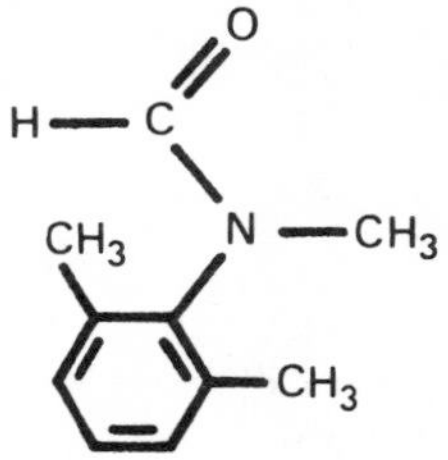

140A

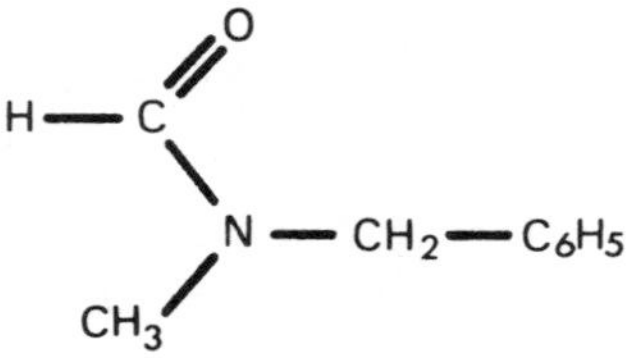

139B

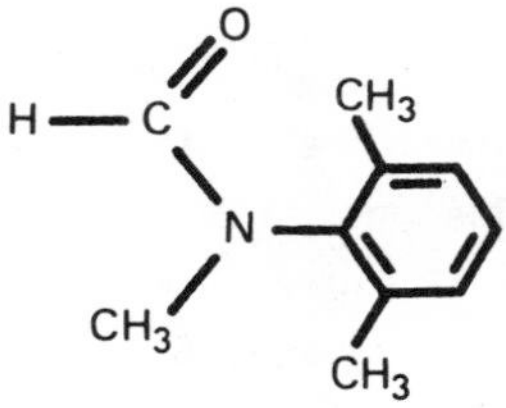

140B

The difference in products observed from imidate salts **137A** and **137B** by comparison with imidate salts **138A** and **138B** can again be readily explained. Imidate salts **137A** and **137B** behave like typical syn imidate salts as they yield a mixture of ester and amine plus amide and alcohol products. In the case of imidate salts **138A** and **138B**, the formation of the ester and amine products is prohibited because the nitrogen atom in the tetrahedral intermediates cannot form a hydrogen bond with the solvent. The next task is to explain the formation of the equilibrium mixture of the amide rotamers **139A-B** and **140A-B**.

It should be pointed out (a) that the half-life of each rotamer in formamides **139** and **140** is several minutes at room temperature, (b) that the basic hydrolysis of the imidate salts **137** and **138** is completed in less than one minute at that temperature. Consequently, the production of the equilibrium mixture of the amides rotamers comes from the direct fragmentation of tetrahedral intermediates. It cannot come from a subsequent equilibration of the amide rotamers.

These results can be explained by postulating that the tetrahedral intermediates can freely rotate prior to the cleavage even if they can break down with stereoelectronic control. This postulate is supported by the important carbonyl-oxygen exchange observed in the course of the basic hydrolysis of tertiary formamides. Indeed, this study showed clearly that in formamide derived tetrahedral intermediates, conformational changes can readily occur prior to the breakdown, even if the cleavages are allowed stereoelectronically.

Consequently, each unsymmetrical imidate salt must first produce a tetrahedral intermediate which has a specific conformation, then equilibration occurs to give a mixture of different tetrahedral conformers which can then break down to give the reaction products.

Strong evidence has been obtained that there is conformational change in these tetrahedral intermediates by carrying out the hydrolysis of imidate salts **138A** and **138B** in water (15). Under these conditions, the hydrolysis reaction is slow. It was observed that imidate **138A** (or **138B**) undergoes an isomerization to **138B** (or to **138A**), with which it reaches an equilibrium before the hydrolysis reaction is complete. This result shows that **138A** reacts with water to form intermediate **141**. Intermediate **141** can then equilibrate with intermediate **142** (by nitrogen inversion followed by a 180° rotation of the C — N bond). Intermediates **141** and **142** then give back **138A** and **138B** at a faster rate than their breakdown into the hydrolysis products.

The basic hydrolysis of these unsymmetrical formamide-derived imidate salts can therefore be rationalized in the following way. The unsymmetrical imidate **138A** reacts with hydroxide ion to give first the intermediate **143** (Fig. 10). Similarly, its isomer **138B** gives first the intermediate **144**. Intermediates **143** and **144** cannot break down to the hydrolysis products with stereoelectronic control. Intermediate **143** can be converted into intermediate **145** via a C — N bond rotation and into **146** via a nitrogen inversion process. Intermediate **144** can also give intermediates **145** and **146**. The intermediates **143**-**146** are therefore easily interconvertible and the cleavages occur through

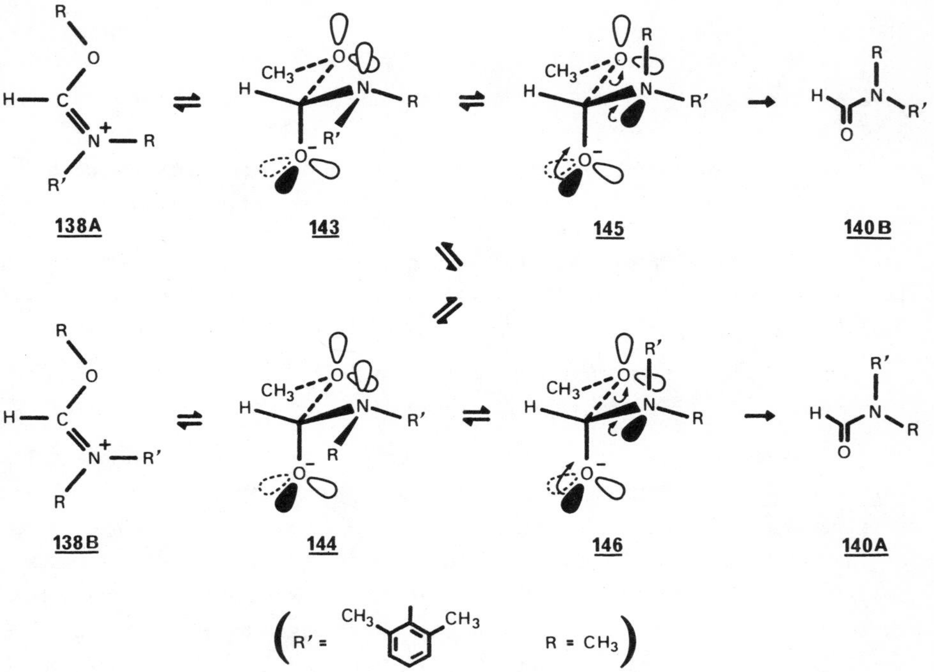

Fig. 10

145 and 146, yielding the amide rotamers 140B and 140A respectively. As a consequence, the conversions 145 → 140B and 146 → 140A are the determining steps of this hydrolysis reaction. Since an equilibrium mixture of the amide rotamers is produced, it means that the difference in energy between the two transition states involved in 145 → 140B and 146 → 140A are similar to that of the amide rotamers.

The relative stability of the rotamers 140A to 140B depends upon their difference in resonance energy and the relative steric repulsions which the oxygen and the hydrogen are experiencing with the two different R and R' groups. Thus, the relative populations of the two rotamers are directly related to their relative steric repulsions (35, 36).

The two electron pairs, one on the nitrogen and one on the oxygen (O^-), which are oriented antiperiplanar to the C — OR bond constitute the most important driving force which governs the cleavages 145 → 140B and 146 → 140A. However, in both cleavages, on going from the tetrahedral intermediate to

the amide, steric interactions increase. Therefore the discriminating factors in these cleavages are the steric effects of the R and R' groups relative to the oxygen and the hydrogen. On that basis, the energy difference
between the two transition states **145** → **140B** and **146** → **140A** will be similar to
the energy difference of the two rotamers. This also means that the geometry
of the transition states must be close to that of the amide rotamers.

The direct formation of the equilibrium mixture of amide rotamers from
the fragmentation of tetrahedral intermediates for imidate salts **138A** and
138B is thus rationalized. The same explanation is also valid for the formation of the equilibrium mixture of the amide rotamers **139A** and **139B** from
imidate salts **137A** and **137B**.

It is however interesting to point out that in the case of imidate salts
137A and **137B**, the ester and amine products were also observed. Thus, in
this particular case, C − N bond cleavage is able to compete with cleavage
of the C − O bond (even though conformational change is allowed). It has already been mentioned that in tetrahedral conformers which can give either
an amide or an ester in the **Z** conformation, the ejection of the amino group
is favored by one additional secondary electronic effect. This might therefore be the explanation for the observed results with formamide derived
imidate salts.

Imidate salt **138A** was also found to be isomerized with **138B** in the presence
of methanol, but no further reactions took place (15). This is in contrast
with the closely related imidate salt **147** (Fig. 11) which reacts with methanol to yield trimethylorthoformate and N-methyl aniline fluoroboric acid
salt. The isomerization of imidate salts **138A** and **138B** takes place because
a tetrahedral intermediate is formed by the reaction of methanol. Contrary
to **147**, the salts **138A** and **138B** do not undergo further reaction because
the nitrogen in the resulting tetrahedral intermediate cannot be protonated.

The hydrolysis of the β-lactam derived imidates **148** and **151** was also reported (17). The results are described in Table 3. The results obtained with imidate **148** are those expected for an imidate existing in the _syn_ or as a
mixture of the _syn_ and _anti_ conformation. The difference in behavior of
imidate **151** can again be explained by the presence of the bulky 2,6-di-
methylphenyl group. In basic conditions, this group prohibits hydrogen
bonding of the tertiary nitrogen and, consequently, the β-lactam **152** is
formed exclusively. Under mild acid conditions, this bulky group prevents

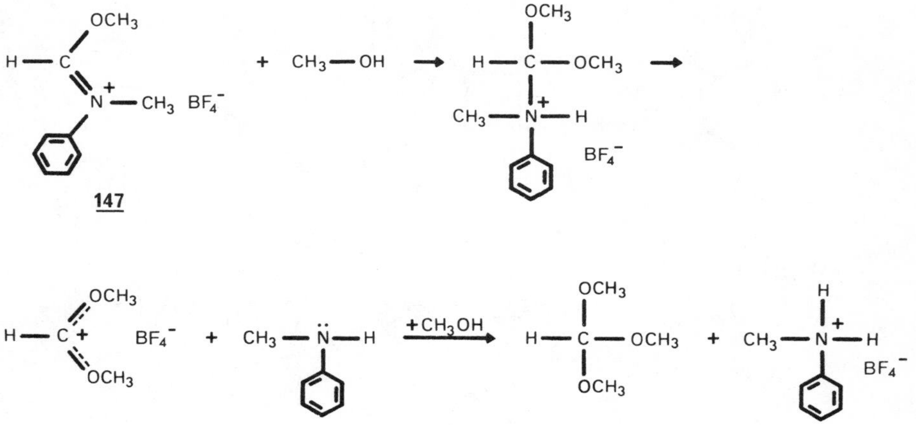

Fig. 11

TABLE 3   Hydrolysis of Imidate Salts **148** and **151** as a
Function of pH at Room Temperature

medium	imidate salt **148**		imidate salt **151**	
	β-lactam **149** (%)	ester-amine **150** (%)	β-lactam **152** (%)	ester-amine **153** (%)
1.0 N NaOH	20	80	100	0
pH 12.7	12	88	100	0
pH 9	0	100	100	0
pH 7	0	100	100	0
pH 4	0	100	68	32
0.1 N HCl	0	100	56	44
0.5 N HCl	0	100	65	35
1.5 N HCl	0	100	87	13
3.0 N HCl	0	100	100	0
6.0 N HCl	0	100	100	0

in part the protonation of the tertiary nitrogen, so a mixture of β-lactam
152 and aminoester **153** is produced. Interestingly, under increasingly acidic
conditions, the formation of β-lactam **152** becomes the exclusive process.
It was however demonstrated by using ^{18}O-labeling that the formation of
β-lactam **152** under these strong acid conditions occurs by an SN_2 type displacement reaction on the methoxy group by chloride ion (cf. **154**), and
the formation of methyl chloride was observed experimentally.

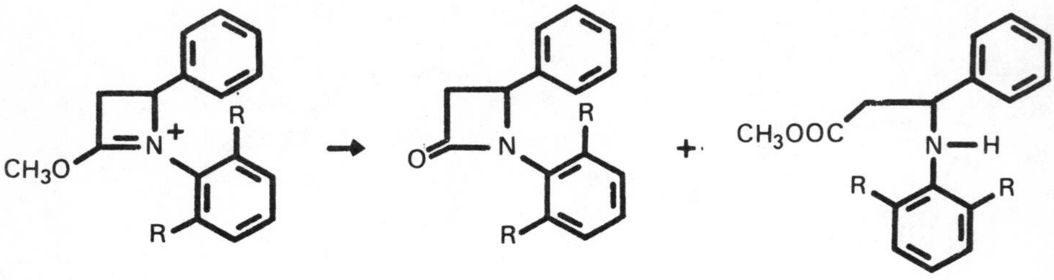

148 R = H
151 R = CH$_3$

149 R = H
152 R = CH$_3$

150 R = H
153 R = CH$_3$

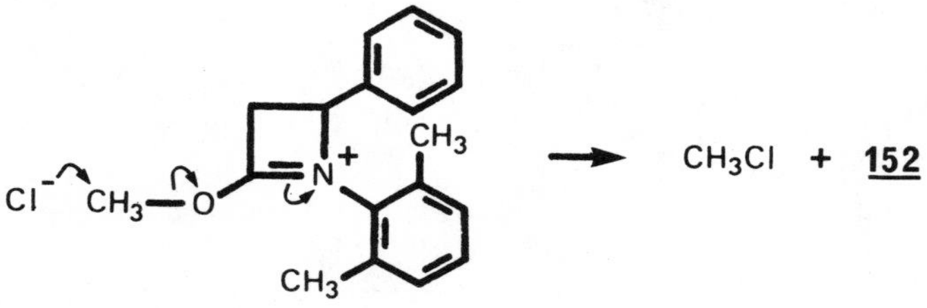

CH$_3$Cl + **152**

154

Cleavage of hemi-orthothioamide tetrahedral intermediates

Kaloustian and co-workers (37) have reported that the kinetic breakdown of hemi-orthothioamide tetrahedral intermediates (**156**) was found to involve the preferential cleavage of the C — N bond (→**158**) rather than the C — O bond (→**157**). The intermediate **156** were produced in situ by the reaction of sodium hydrosulfide on an imidate salt (**155**) in acetone.

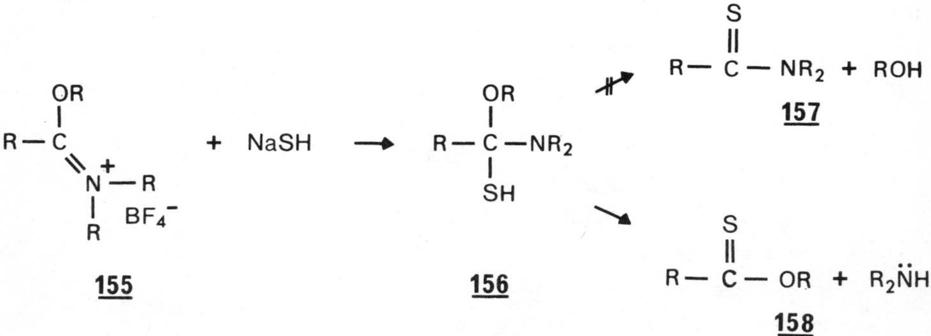

$$R-\overset{S}{\underset{\|}{C}}-NR_2 + ROH$$

157

155

156

$$R-\overset{S}{\underset{\|}{C}}-OR + R_2\ddot{N}H$$

158

Thin layer chromatography analysis (at room temperature) of the reaction mixture obtained from equimolar amounts of 2,N-dimethyl-1,3-oxazolinium fluoroborate (**159**) and sodium hydrosulfide at -78° gave the thioamidoalcohol **161A** only. However, tlc analysis of a similar reaction mixture <u>after</u> low temperature trapping (AcCl/pyridine, -78°C) showed the thioester **160B** as the major product along with the thioamide **161B** (ratio ≈9:1).

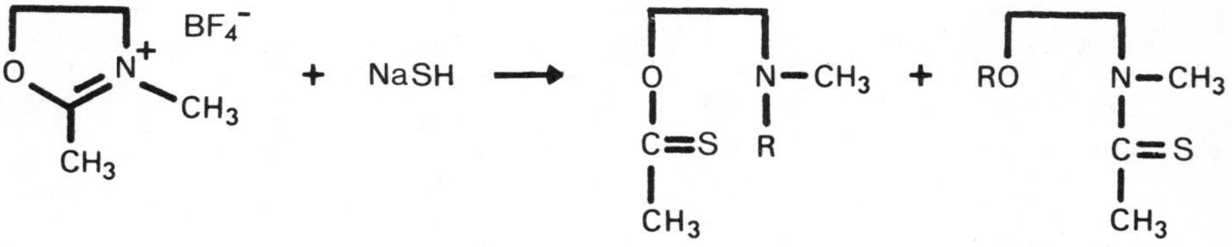

159

160A R = H
160B R = COCH$_3$

161A R = H
161B R = COCH$_3$

The reaction of hydrosulfide anion with the imidate salt **159** must give the tetrahedral conformer **162** which has an electron pair each on the nitrogen and the oxygen atoms oriented antiperiplanar to the C — SH bond. In **162**, cleavage of the C — O bond is not allowed whereas the cleavage of the C — N bond can take place with primary and secondary (two secondary electronic effects) stereoelectronic control yielding **160A** in the **Z** conformation **163**.

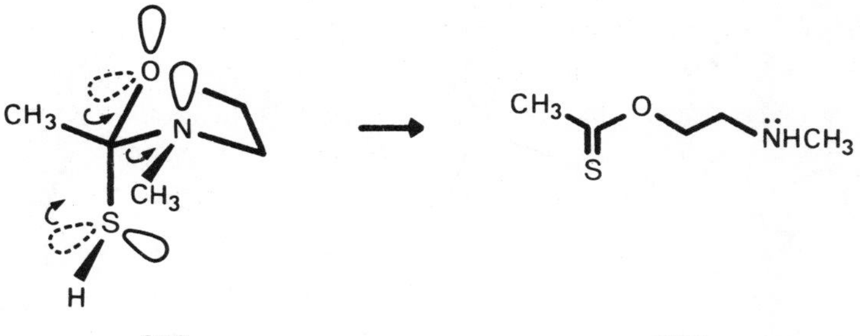

162   **163**

Analysis of the reaction mixture from equimolar amounts of anhydrous sodium hydrosulfide and N,N-dimethyliminobutyrolactonium fluoroborate (**164**) in acetone at room temperature revealed the thioamidoalcohol **165A** only (32). However, when the reaction was run at -78°C, and the mixture acetylated (-78°C), the thionolactone **166** and N,N-dimethylacetamide (**167**) were the major detectable products.

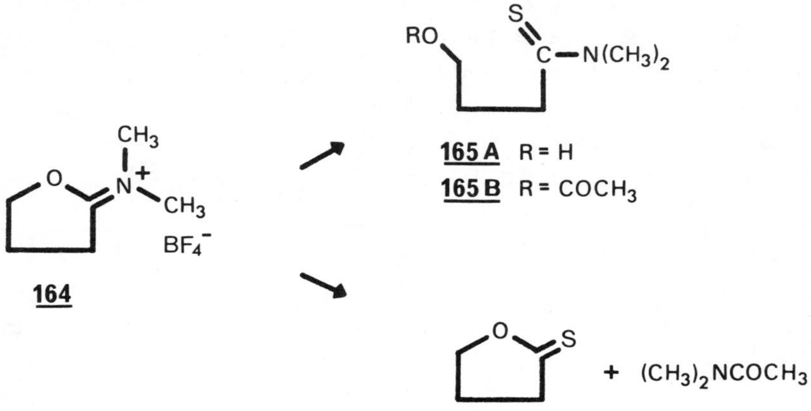

164

165 A R = H
165 B R = COCH₃

166   + (CH₃)₂NCOCH₃

166   **167**

The stereoelectronically controlled reaction of hydrosulfide anion with imidate salt **164** must first yield the tetrahedral conformer **168**. This intermediate cannot break down to yield either **165** or **166** and dimethylamine, but it can undergo a conformational change either at the nitrogen atom (→**169**) or at the ring (→**170**). Conformer **169** can only give the thioamidoalcohol **165A** whereas conformer **170** can only produce the thionolactone **166** and dimethylamine. The experimental results indicate that the conformational change at the ring (→**170**) is the preferred pathway at low temperature.

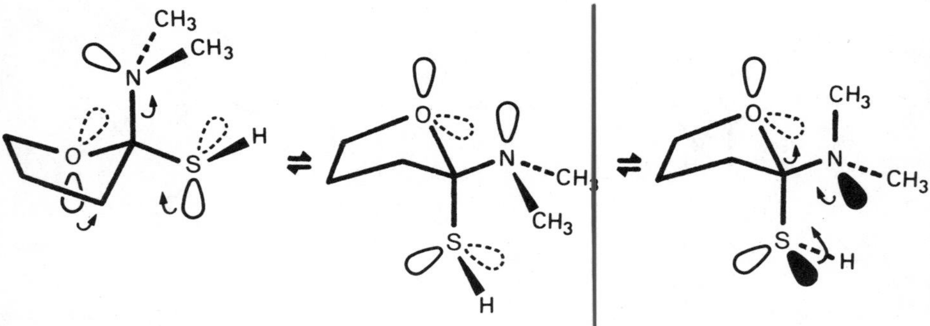

170   **168**   **169**

Khouri and Kaloustian (38) have recently observed that the hydrolysis of 2-phenyl-N-methyl-1,3-thiazolinium (**171**) and 2-phenyl-N-methyl-5,6-dihydro-1,3-thiazinium tetrafluoroborates (**172**) under kinetically controlled conditions (NaOH, 15-crown-5, anhydrous n-C₃H₇CN, Ac₂O, -78°C) proceed by preferential cleavage of the C─N bond (giving **173** and **174** respectively).

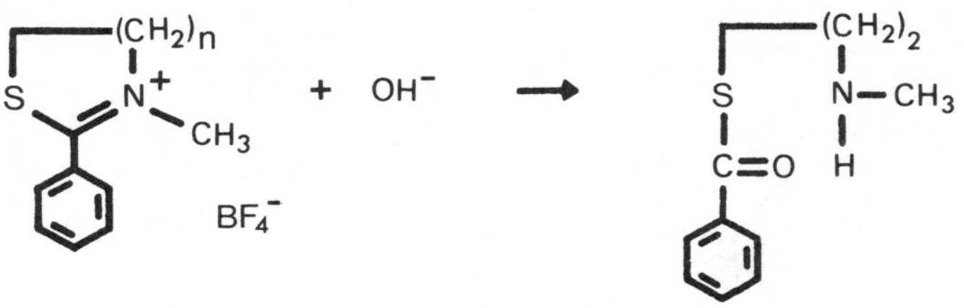

171 n = 1   **173** n = 1
172 n = 2   **174** n = 2

They have also observed that the sulfhydrolytic cleavage of O,N-dimethyl-butyrolactonium and O,N-dimethyl-valerolactonium tetrafluoroborates (**175** and **176**) involves C—N cleavage (giving **177** and **178** respectively), under kinetic control (61°C, NaSH, Ac$_2$O, CHCl$_3$) (39). These results can again be rationalized in terms of the principle of stereoelectronic control.

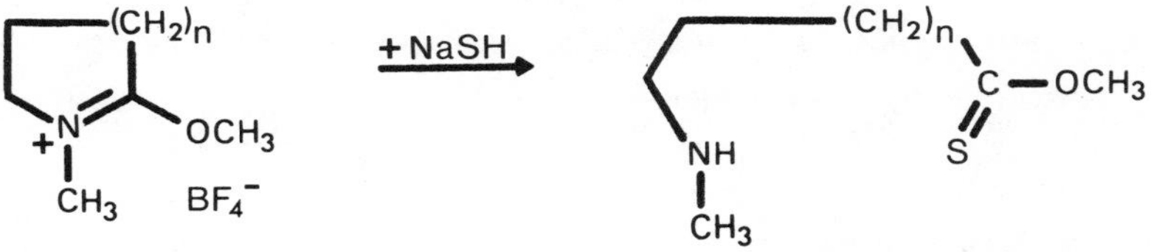

175 n = 1   **177** n = 1
176 n = 2   **178** n = 2

Imino-ethers, amidines, etc.

Imino-ethers can adopt theoretically the four conformations **179-182**. In addition to the primary electronic effect, conformer **179** possesses two secondary electronic effects, conformers **180** and **181** only one whereas conformer **182** has none. Conformer **179** has a strong steric interaction between the two R groups and must be eliminated. Imino-ethers must therefore exist either in conformation **180** and **181**. Meese, Walter, and Berger (40) have shown that the most stable conformation is **180**. In some compounds where there is an important steric effect between the R' and the RN group, conformation **181** is also observed. It was also found that in a polar solvent (CH$_3$OH), there is a mixture of a major (**180**) and a minor (**181**) conformer.

These results suggest that the secondary electronic effect caused by the electron pair of the oxygen atom in **180** is more important (cf. **183**) than that of the nitrogen electron pair in **181** (cf. **184**).

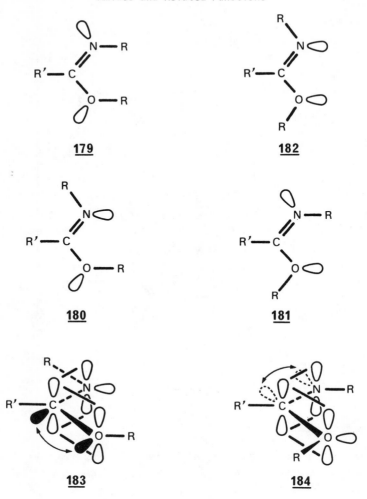

179   182

180   181

183   184

In a study on the aminolysis of 0-acetylethanolamine (**185**, R'=H) and 0-acetylserine (**185**, R'=COOH), it was observed (41) that the kinetics of acetyl-transfer reaction of these two aminoesters indicate that the break-down of **186** yields mainly the amidoalcohol **187**; only 1.5-3% of aminoester **185** was detected. The breakdown of **186** was compared with that of **188** obtained from imidate **94** (cf. p. 130) which gave only the aminoester **95** by C—N bond cleavage. Contrary to the conclusion reached by these authors (41), the difference in behavior between **186** and **188** can be readily understood. Imidate salt **94** reacts with hydroxide ion to give conformer **189** (R=CH₃) which can only give the aminoester **95** with stereoelectronic control; the amino-

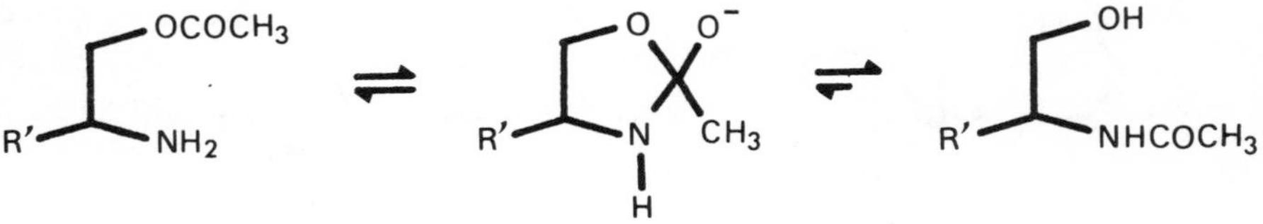

185   **186**   **187**

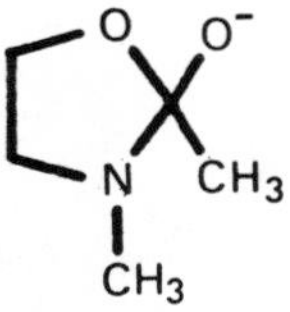

188 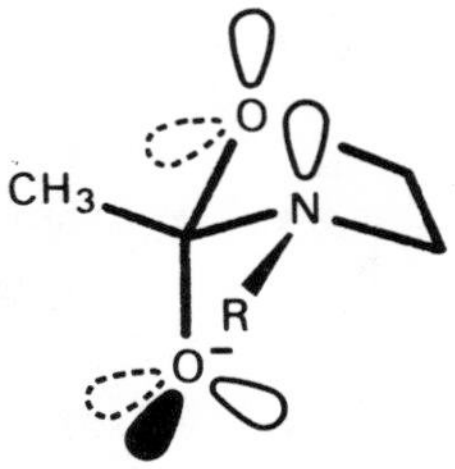   **189** 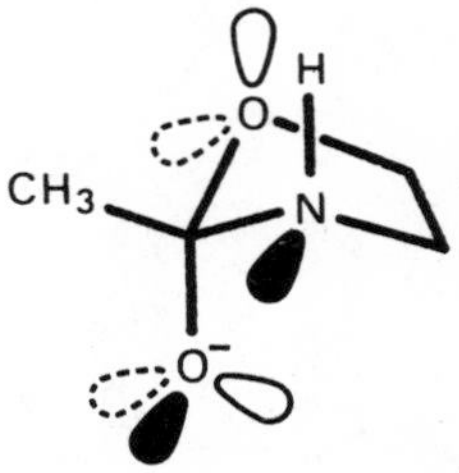   **190**

ester **185** can either form **189** (R=H) or **190**. The interconversion of **189** (R=H)
and **190** involves only a proton transfer which is a low energy process (cf.
p. 109); thus, **190** (or **189**, R=H) can give either the amidoalcohol or the
aminoester product with stereoelectronic control. The experimental results
show that the formation of the amidoalcohol product is favored.

The hydrolyses of a series of cyclic and acyclic amidines have been carried
out by Burdick, Benkovic, and Benkovic (42) in order to assess the possible
importance of stereoelectronic control of the ensuing tetrahedral intermediates.

The acyclic amidines **191** and **192** gave only the corresponding kinetic products, i.e., formamides **193** and **194** respectively. The basic hydrolysis of the
7-membered amidine **195** gave only the thermodynamic product **196**. However, on
partial hydrolysis, **195** gave a mixture of the thermodynamic product **196** and
the kinetic product **197**. The six-membered amidines **198** and **199** yielded only
the thermodynamic products **200** and **201**. Attempts to observe the kinetic
products **202** and **203** during partial hydrolysis failed. It was further observed that the basic hydrolysis of the trifluoroacetamide derivatives **204** and
205, two precursors of the kinetic products **202** and **203**, yielded directly
the thermodynamic products **200** and **201**.

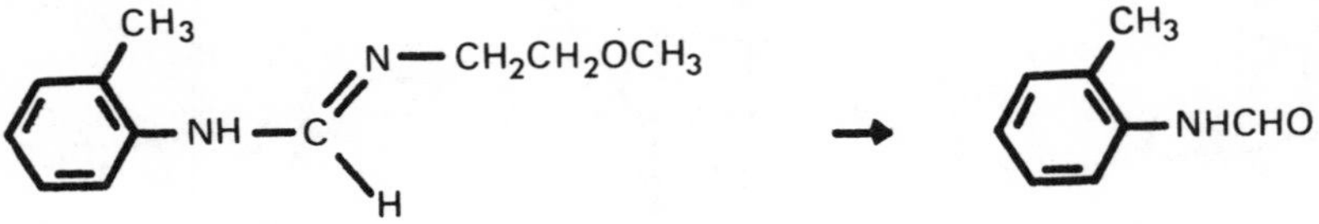

191 193

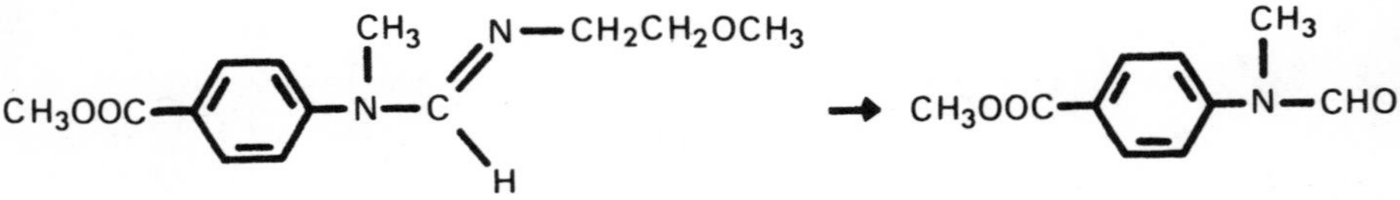

192 194

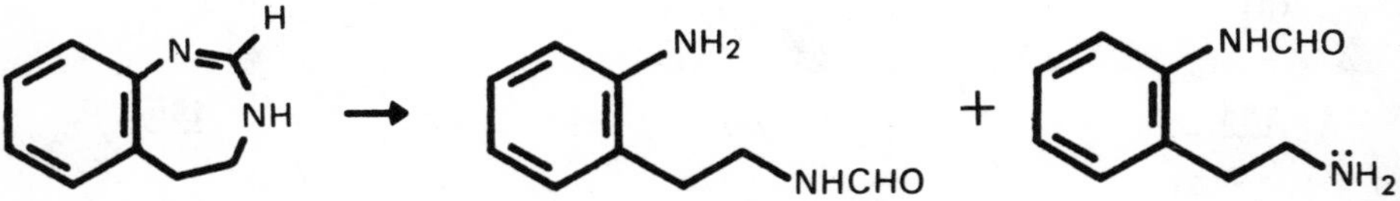

195 196 197

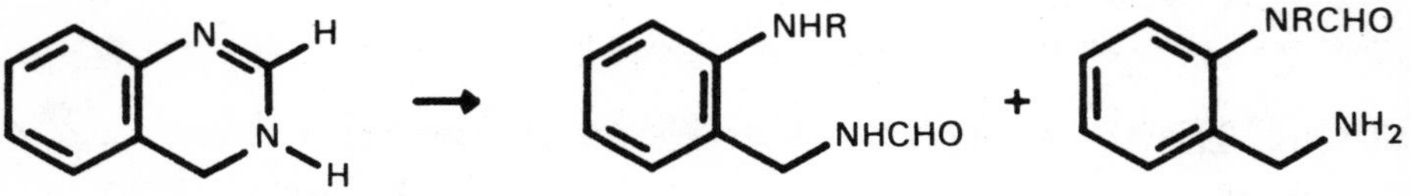

198

200 R = H
201 R = CH3

202 R = H
203 R = CH3

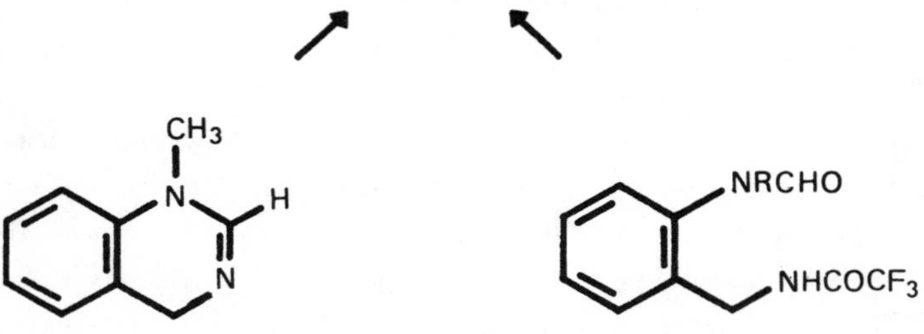

199

204 R = H
205 R = CH3

Thus, the hydrolysis of **191** and **192** takes place under kinetically controlled conditions. There is evidence that **195** gave first the kinetic product **197** which would then be rapidly converted into the thermodynamic product **196**. The hydrolysis of **198** and **199** occurs under thermodynamic control conditions to give **200** and **201** respectively. It is possible that **198** and **199** produced first the kinetic products **202** and **203** which would then be converted very rapidly into the thermodynamic products.

It does not seem possible to explain these results on the basis of the principle of stereoelectronic control alone. It is, however, interesting to notice that the ejection of the more basic nitrogen is favored in the kinetically controlled proces (43-45).

This may be due to the following: (a) the more basic nitrogen atom can form a stronger hydrogen bond with the solvent than the aniline nitrogen and (b) the hydrogen on the aniline nitrogen is acidic and can therefore be removed (or partly) under basic conditions. This would give transition state **206** which favors the cleavage of the more basic nitrogen.

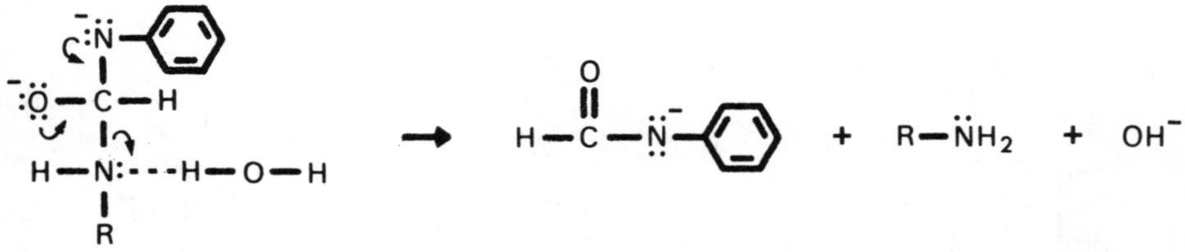

206

Halliday and Symons have reported (46) that the basic hydrolysis of N,N'-dimethylformamidine (**207**) yields the anti form **208** of N-methylformamide which is then isomerized into the most stable syn form **209**.

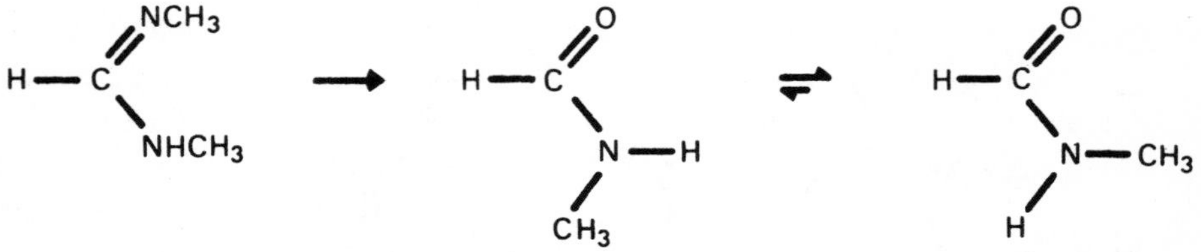

207   **208**   **209**

^{1}H nmr analysis showed that monoprotonated dimethylformamidine is present mostly as isomer **207B**, with about 5% of **207C** and no detectable **207A**. The reaction of hydroxide ion with **207B** should give conformer **210** which after appropriate proton transfer can only give syn-amide **209** with stereoelectronic control if conformational change at the nitrogen in **210** is not allowed. The same conclusion is reached with **207A** (cf. **211**). Furthermore, **207C** gives a tetrahedral intermediate (**212**) which cannot break down with stereoelectronic control.

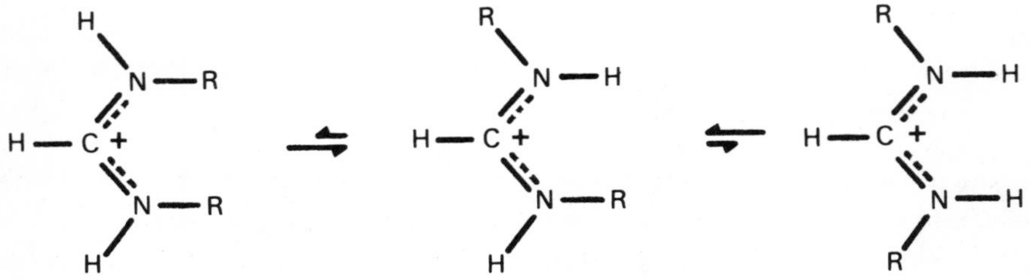

It is therefore not possible to form the anti-amide **208** unless conformational changes at the nitrogen are allowed. It is however reasonable to make such an assumption for formamide-derived tetrahedral intermediates (cf. p. 112). We can therefore analyze the cleavage of a tetrahedral conformer such as **213**. This conformer can either give the anti or the syn-amide isomers (**208** and **209**) with stereoelectronic control. It is however not clear on that basis, why the formation of the less stable anti form **208** is favored. There must be another parameter in this case which is not known yet.

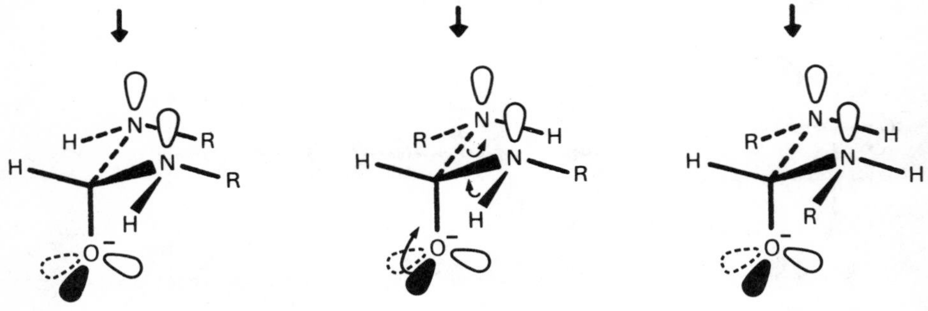

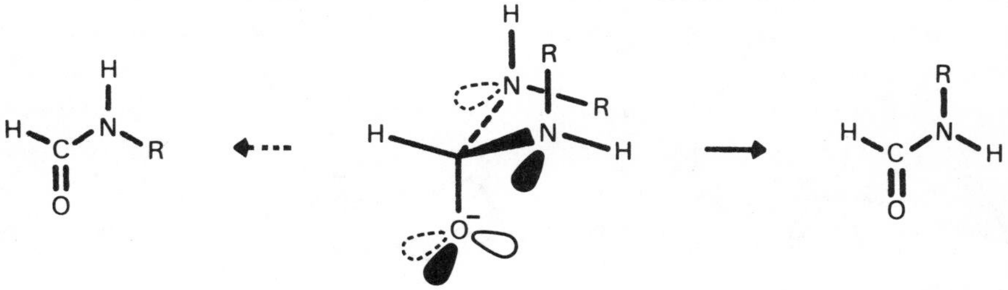

209   **213**   **208**

Recent developments

Perrin and Arrhenius (47) have studied the basic hydrolysis of cyclic amidines **214** and **215**. They found that the primary product of hydrolysis of **214**
(or **215**) is solely the aminoamide **216** (or **217**), and the lactam **218** (or **219**)
is a secondary product which is formed from the corresponding aminoamide.

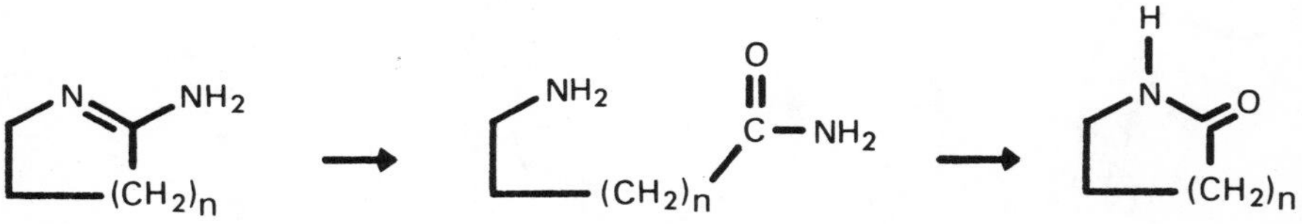

214 n = 2   **216**   **218**
215 n = 1   **217**   **219**

These results can be readily explained. For instance, the reaction of cyclic
amidine **214** with hydroxide ion in water must give first the tetrahedral intermediate **220**, which after appropriate proton transfer (→**221**) can only give
the aminoamide **216** with stereoelectronic control. In order to form the
lactam **218**, intermediate **220** must undergo a conformational change to intermediate **222** followed by an appropriate proton transfer (→**223**). Since lactam
218 is not observed under kinetically controlled conditions, the conformational change **220** ⇌ **222** cannot compete with the breakdown of **220** via **221**.
Lactam **218** is then slowly formed from aminoamide **216** via intermediate **222**.

Recently, Eschenmoser, Dunitz, and co-workers (48) have reported the hydrolysis of tricyclic ketene N,O-acetal **224** which yields only aminopropionic
acid ester **225**. The first step in this reaction must be the protonation of
224 which gives the anti imidate salt **226**. Since anti imidate salts always

yield the aminoester under kinetically controlled conditions (cf. p. 124), the above result is readily explained. A similar case was observed by Meyers and Nazarenko (49).

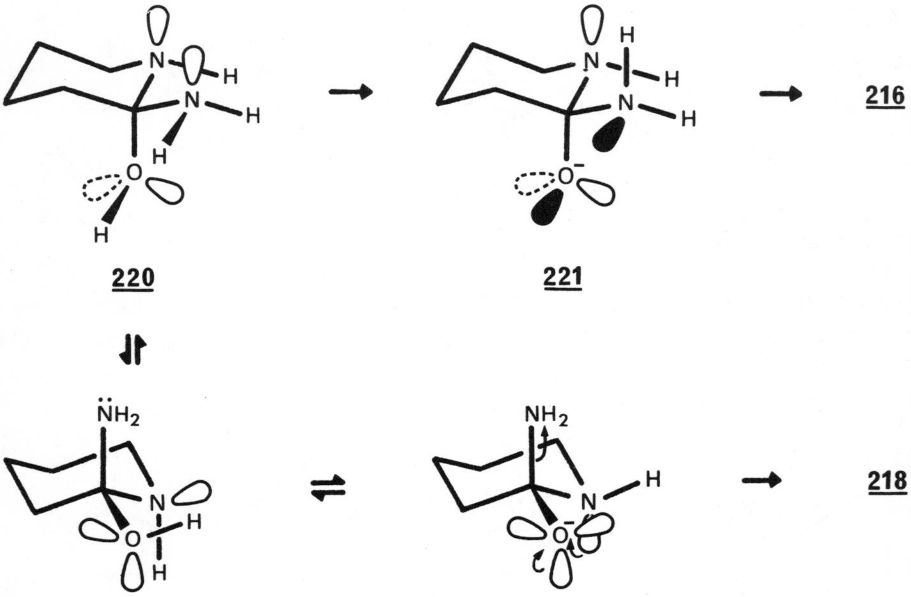

220   **221**   **216**

222   **223**   **218**

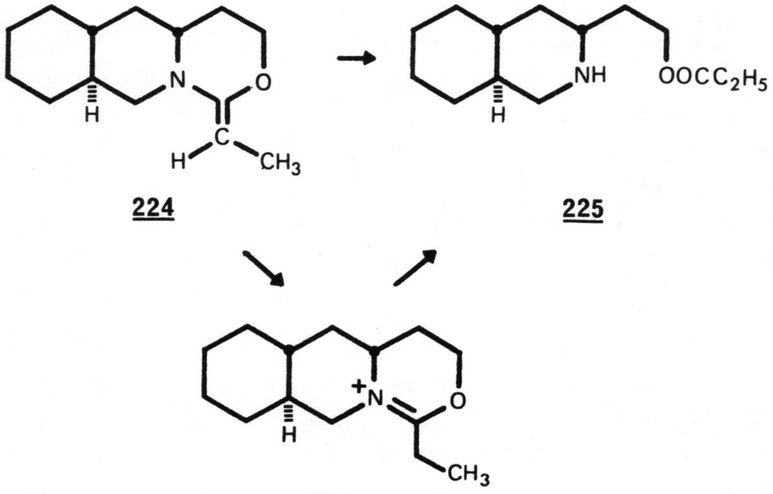

224   **225**

226

Johnson, Nalley, Weidig, and Arfan (50) have investigated the stereochemistry of the reaction of sodium methoxide with the isomeric O-alkylbenzo-
hydroximoyl chloride 227 and 230 in dimethylsulfoxide-methanol (9:1). They
found that 227 gave almost exclusively the Z-isomer 229 while 230 gave
a mixture of Z and E isomers 229 and 233 where the E isomer largely predominates.

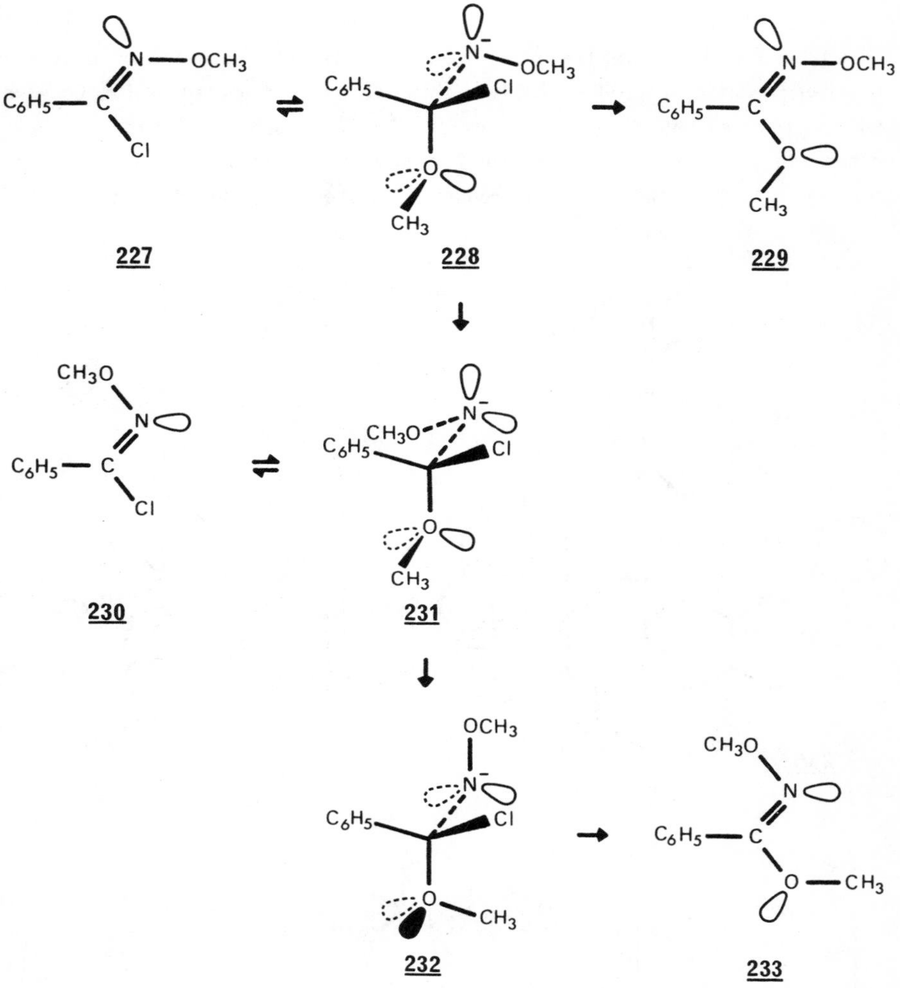

The authors have rationalized their results on the basis of the principle of
stereoelectronic control. The reaction of methoxide ion on 227 gives the
intermediate 228 which can eject chloride ion to give the Z isomer 229. The

reaction of methoxide ion on **230** gives the intermediate **231** which cannot undergo a stereoelectronically controlled ejection of chloride ion. Thus, intermediate **231** would undergo a stereomutation to intermediates **228** and **232**. Intermediate **232** can eject chloride ion to give the E isomer **233** and this process would predominate over the process **231** → **228** → **229**.

Stork, Jacobson, and Levitz (51) have recently reported that the reaction of the lithium carbanion **234** with benzaldehyde followed by reduction with sodium borohydride gave the phenylcarbinol **235**. The sequence of events in the transformation of **234** to **235** was shown to be as depicted below. Convincing spectral evidence was obtained for **236**, **238**, and **239**. Thus, the hemiorthoamide tetrahedral intermediate **237** which was generated *in situ* gave the aminobenzoate **238**, the expected product from stereoelectronic control.

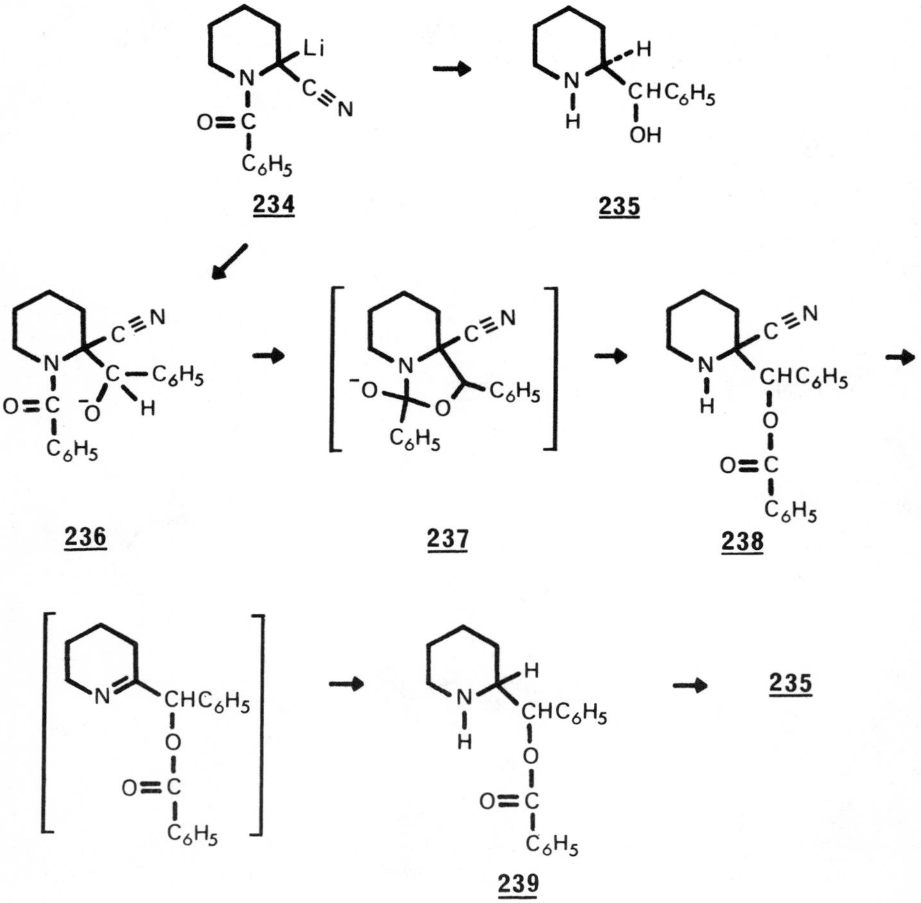

Lyapova, Pojarlieff, and Kurtev (52) have studied the N—O and O—N acyl migration in 1-amino-1,2,3-triphenylpropanols under acidic and basic conditions respectively. They found a great difference in the ease of migration which depends on the relative stereochemistry of the substrate and the degree of substitution of the nitrogen atom (NH or NCH$_3$).

Under acidic conditions (0.58 M HCl in dioxan:THF (1:1) at 50° for 24 h), the isomeric acetamidoalcohols **240** and **244** are converted into the ammonium salts of the isomeric aminoesters **243** and **247** respectively. These N—O acyl migrations take place via the intermediates **241**, **242**, and **245**, **246** respectively. Interestingly, they found that when R=H, the conversion **240** → **243** takes place readily but when R=CH$_3$, no reaction was observed. On the other hand, the conversions **224** → **247** (R=CH$_3$ or H) are both sluggish.

These results can be readily interpreted if the rate-determining step of these reactions is the formation of the tetrahedral intermediates **241** and **245**. Indeed, there is a direct relationship between the relative ease of formation of these two intermediates (which depends upon their respective steric hindrance) and that of the acetamidoalcohol products **243** and **247** (R=H or CH$_3$). In intermediate **241**, when R=CH$_3$, there is a strong 1,3-diaxial steric interaction between the N—CH$_3$ group and the axial phenyl group. Such steric interaction does not exist when R=H. Consequently, the formation of acetamidoalcohol **243** (via **241**) should occur with ease only when R=H. This conclusion agrees with the experimental results. In the case of intermediate **245**, when R=CH$_3$, there is no 1,3-diaxial steric interaction caused by the N-methyl group. On that basis, intermediate **245** (R=CH$_3$) should be readily formed. However in this intermediate, there is a strong 1,3-diaxial steric

interaction between the phenyl group at C-6 and the OH group. So, the formation of acetamidoalcohol **247** (via **245**) must be equally difficult when R=H
or CH$_3$. This is again in accord with the experimental results.

The authors have also observed that under basic conditions (0.02 M - Na$_2$CO$_3$
in aqueous acetonitrile) the process **243** → **240** is fast when R=H but slow
when R=CH$_3$. On the other hand, the process **247** → **244** is relatively slow
(15 min when R=H and 120 min when R=CH$_3$). Under basic conditions, these
migration reactions must take place via the T$^-$ tetrahedral intermediates
248 and **249** respectively. Again, if the formation of these two intermediates
is the slow step, these results can be readily interpreted. Indeed, there
is a strong 1,3-diaxial steric interaction in intermediates **248** (R=CH$_3$) and
249 (R=H or CH$_3$), so only intermediate **248** (R=H) should be readily formed.

248   **249**

Finally, it should be pointed out that the stereochemistry of the tetrahedral intermediates **241**, **245**, **248**, and **249** can only be obtained by application of the principle of stereoelectronic control in hydrolytic processes.
The above results must therefore be accepted as strong evidence in favor
of the validity of this principle.

Penicillin is a V-shaped molecule which has a fairly rigid structure due
to the fusion of the β-lactam and the thiazolidine rings. As a consequence,
the electron density of the β-lactam nitrogen lone pair should be concentra-
ted heavily on the α-face of the penicillin molecule (cf. **250**), and the β-
lactam nitrogen cannot adopt the sp^2 hybridization found in normal amides;
the conjugation between the nitrogen lone pair and the carbonyl group is
thus reduced. This effect (53) in addition to the strain of the β-lactam
ring (54), activates the carbonyl lactam function and the great susceptibil-
ity of penicillin to react with nucleophiles (**250** → **251**) is thus readily
understood.

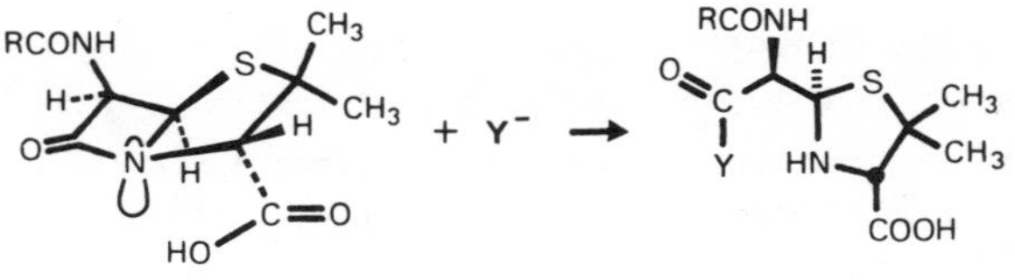

250   + Y⁻ →   **251**

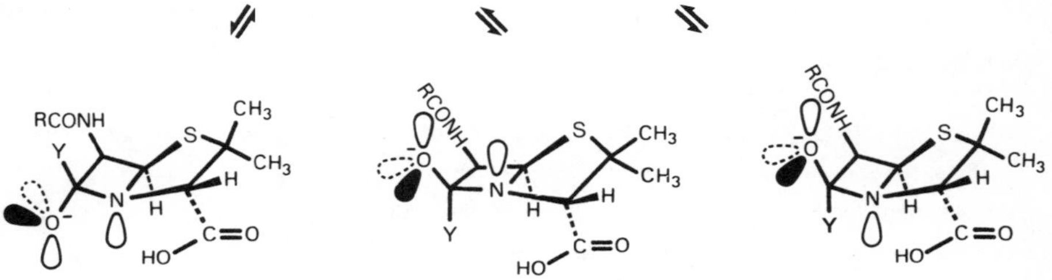

252   **253**   **254**

According to the theory of stereoelectronic control, the direction of nucleophilic attack on the lactam carbonyl carbon is such that the lone pairs
on the heteroatoms will be antiperiplanar to the attacking group. Nucleophilic attack on penicillin should therefore take place preferentially
from the β-side (cf. 252) rather than the α-side (cf. 253) as the former
gives a tetrahedral intermediate which has a cis rather than a trans ring
junction. Indeed, the cis intermediate 252 is much less strained than the
trans intermediate 253. However, the β-face of penicillin is sterically
hindered and Page and collaborators (55-59) have suggested that the nucleophilic attack would take place from the least hindered α-side yielding directly the tetrahedral conformer 254, in disagreement with the prediction
of the theory of stereoelectronic control. This suggestion was made because
in the course of the kinetic studies on the aminolysis and the hydrolysis
of penicillins (55-57), it was found that the reaction of 1,2-diaminoethane

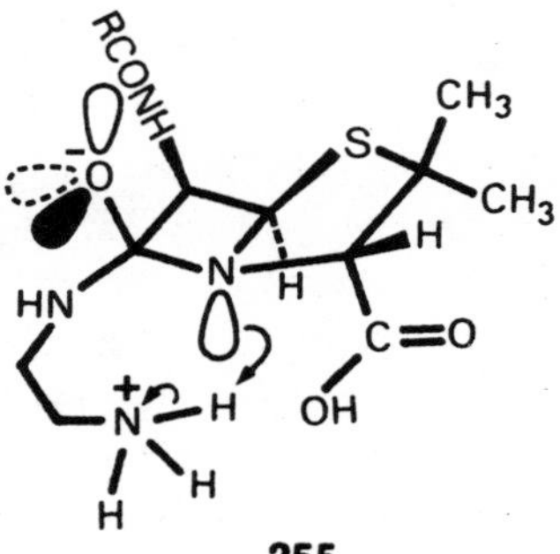

255

monocation with benzylpenicillin shows a rate enhancement of 100-fold compared with a monoamine of similar basicity. This rate enhancement was attributed to an internal proton transfer (intramolecular general acid catalysis) from the primary ammonium group to the tertiary nitrogen which can
occur from intermediate 255, implying that nucleophilic attack must take
place from the least hindered α-side of penicillin!

REFERENCES

(1) Deslongchamps, P.; Lebreux, C.; Taillefer, R.J. Can. J. Chem. 1973,
     51, 1665.
(2) Deslongchamps, P. Pure Appl. Chem. 1975, 43, 351.
(3) Deslongchamps, P. Tetrahedron 1975, 31, 2463.
(4) Deslongchamps, P. Heterocycles 1977, 7, 1271.
(5) Bürgi, H.B.; Dunitz, J.D.; Shefter, E. J. Am. Chem. Soc. 1973, 95,
     5065.
(6) Rice, F.O.; Teller, E. J. Chem. Phys. 1938, 6, 489; 1939, 7, 199.
(7) Hine, J. J. Org. Chem. 1966, 31, 1236.
(8) Altmann, J.A.; Tee, O.S.; Yates, K. J. Am. Chem. Soc. 1976, 98, 7132.
(9) Okuyama, T.; Pletcher, T.C.; Sahn, D.J.; Schmir, G.L. J. Am. Chem.
     Soc. 1973, 95, 1253 and references cited therein.
(10) Satterthwait, A.C.; Jencks, W.P. J. Am. Chem. Soc. 1974, 96, 7018,
     7031.
(11) Deslongchamps, P.; Taillefer, R.J. Can. J. Chem. 1975, 53, 3029.
(12) Guthrie, J.P. J. Am. Chem. Soc. 1974, 96, 3608.
(13) Eigen, M. Angew. Chem. Int. Ed. Engl. 1964, 3, 1.
(14) Deslongchamps, P.; Gerval, P.; Cheriyan, U.O.; Guida, A.; Taillefer,
     R.J. Nouv. J. Chim. 1978, 2, 631.
(15) Deslongchamps, P.; Cheriyan, U.O.; Pradère, J.-P.; Soucy, P.; Taille-
     fer, R.J. Nouv. J. Chim. 1979, 3, 343.
(16) Deslongchamps, P.; Cheriyan, U.O.; Taillefer, R.J. Can. J. Chem. 1979,
     57, 3262.
(17) Deslongchamps, P.; Caron, M. Can. J. Chem. 1980, 58, 2061.
(18) Deslongchamps, P.; Barlet, R.; Taillefer, R.J. Can. J. Chem. 1980,
     58, 2167.
(19) Lehn, J.M.; Wipff, G. J. Am. Chem. Soc. 1974, 96, 4048.
(20) Lehn, J.M.; Wipff, G. Helv. Chim. Acta 1978, 61, 1274.
(21) Lehn, J.M.; Wipff, G. J. Am. Chem. Soc. 1980, 102, 1347.

(22) Schweizer, W.B.; Procter, G.; Kaftory, M.; Dunitz, J.D. Helv. Chim. Acta 1978, 61, 2783.

(23) Bender, M.L. Chem. Rev. 1960, 60, 53.

(24) Johnson, S.L. "Advances in Physical Organic Chemistry" Vol. 5; Gold, V., Ed; Academic Press: London, 1967; pp. 237-331.

(25) Huisgen, R.; Walz, H. Chem. Ber. 1956, 89, 2616.

(26) LaPlanche, L.A.; Rogers, M.T. J. Am. Chem. Soc. 1964, 86, 337.

(27) Bunton, C.A.; Nayak, B.; O'Connor, C. J. Org. Chem. 1968, 33, 572.

(28) Jencks, W.P. Chem. Rev. 1972, 72, 705.

(29) Bolton, P.D.; Jackson, G.L. Aust. J. Chem. 1971, 24, 969.

(30) Bushweller, C.H.; O'Neil, J.W.; Bilofsky, H.S. J. Am. Chem. Soc. 1971, 93, 542.

(31) Deslongchamps, P.; Cheriyan, U.O.; Guida, A.; Taillefer, R.J. Nouv. J. Chim. 1977, 1, 235.

(32) Birnbaum, G.I. Unpublished result.

(33) Deslongchamps, P.; Dubé, S.; Lebreux, C.; Patterson, D.R.; Taillefer, R.J. Can. J. Chem. 1975, 53, 2791.

(34) Allen, Jr., P.; Ginos, J. J. Org. Chem. 1963, 28, 2759.

(35) Stewart, W.E.; Siddall, III, T.H. Chem. Rev. 1970, 70, 517.

(36) LaPlanche, L.A.; Rogers, M.T. J. Am. Chem. Soc. 1963, 85, 3728.

(37) Kaloustian, M.K.; Aguilar-Laurents de Gutierrez, M.I.; Nader, R.B. J. Org. Chem. 1979, 44, 666.

(38) Khouri, L.; Kaloustian, M.K. Personal communication.

(39) Kaloustian, M.K.; Nader, R.B. Personal communication.

(40) Meese, C.O.; Walter, W.; Berger, M. J. Am. Chem. Soc. 1974, 96, 2259.

(41) Caswell, M.; Chaturvedi, R.K.; Lane, S.M.; Zvilichovsky, B.; Schmir, G.L. J. Org. Chem. 1981, 46, 1585.

(42) Burdick, B.A.; Benkovic, P.A.; Benkovic, S.J. J. Am. Chem. Soc. 1977, 99, 5716.

(43) Benkovic, S.J.; Bullard, W.P.; Benkovic, P.A. J. Am. Chem. Soc. 1972, 94, 7542.

(44) Robinson, D.R.; Jencks, W.P. J. Am. Chem. Soc. 1967, 89, 7088.

(45) Robinson, D.R.; Jencks, W.P. J. Am. Chem. Soc. 1967, 89, 7098.

(46) Halliday, J.D.; Symons, E.A. Can. J. Chem. 1978, 56, 1463.

(47) Perrin, C.L.; Arrhenius, G.M.L. J. Am. Chem. Soc. 1982, 104, 2839.

(48) Kümin, A.; Maverick, E.; Seiler, P.; Vanier, N.; Damm, L.; Hobi, R.; Dunitz, J.D.; Eschenmoser, A. Helv. Chim. Acta 1980, 63, 1158.

(49) Meyers, A.I.; Nazarenko, N. J. Am. Chem. Soc. 1972, 94, 3243.

(50) Johnson, J.E.; Nalley, E.A.; Weidig, C.; Arfan, M. J. Org. Chem. 1981, 46, 3623.

(51) Stork, G.; Jacobson, R.M.; Levitz, R. Tetrahedron Lett. 1979, 771.

(52) Lyapova, M.J.; Pojarlieff, I.G.; Kurtev, B.J. J. Chem. Res. 1981, S, 351.

(53) Woodward, R.B. "The Chemistry of Penicillins"; Clarke, T.H.; Johnson, J.R.; Robinson, R., Eds; Princeton University Press, 1949, p. 441.

(54) Strominger, J.L. Antibiotics 1967, 1, 706.

(55) Martin, A.F.; Morris, J.J.; Page, M.I. J. Chem. Soc., Chem. Commun. 1979, 298.

(56) Morris, J.J.; Page, M.I. J. Chem. Soc., Perkin Trans 2 1980, 212.

(57) Gensmantel, N.P.; Page, M.I. J. Chem. Soc., Perkin Trans 2 1979, 137.

(58) Gensmantel, N.P.; Gowling, E.W.; Page, M.I. J. Chem. Soc., Perkin Trans 2 1978, 335.

(59) Gensmantel, N.P.; Proctor, P.; Page, M.I. J. Chem. Soc., Perkin Trans 2 1980, 1725.

CHAPTER 5

REACTIONS AT SATURATED CARBONS

SN$_2$ type reaction

The concerted displacement of a leaving group by a nucleophile in aliphatic and alicyclic compounds, i.e. the SN$_2$ reaction, is one of the first reactions which was found to take place with stereoelectronic control (1, 2). This reaction is a one step process without intermediate resulting in a Walden inversion. The nucleophile must approach the substrate from a 180° angle opposite to the leaving group. The stereochemistry of the resulting transition state corresponds 1 where the central carbon can be considered to be, in molecular orbital terms, sp^2 hybridized. The remaining p-orbital has one lobe overlapping with the nucleophile and the other with the leaving group (cf. 2). The mechanism of this reaction is therefore controlled by electronic effects which impose a specific stereochemistry at the transition state level.

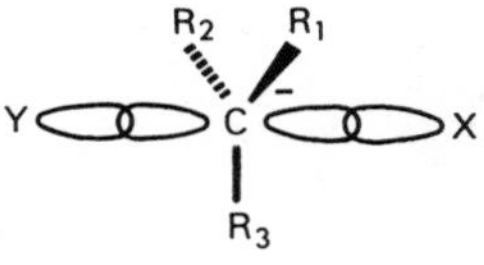

It is not easy to design an experiment to test the subtle stereochemical requirement of the transition state of a reaction. One experimental approach is to attach the nucleophile to the substrate, creating a situation where this nucleophile can undergo two different competing reactions. In many cases, the nucleophile can more easily fulfill the stereochemical require-ment of one process and only one reaction is observed. The experiment is almost perfectly designed when the process which takes place is "stereoelec-tronically allowed" and leads to the kinetic rather than the thermodynamic product, or when the process which does not take place is not "stereoelec-tronically allowed" but is otherwise favored on the basis of steric argu-ments or entropy consideration, especially when that would have led to the thermodynamic product.

Using this approach, Eschenmoser and co-workers (3) have provided strong evidence that, in the SN$_2$ transition state, the nucleophile is aligned colinearly. They found that the anion **3** gave the expected product **4** via an intermolecular process rather than the "formally appealing" intramolecular process. The logical conclusion of this experiment is that the intramole-cular attack (cf. **5**) although entropically the most favored path, does not occur because the proper alignment of the nucleophile with the substrate cannot be attained in the transition state. This work represents also a beautiful example of an experiment designed to investigate the importance of stereochemistry in the transition state of an organic reaction.

As a consequence of this stereoelectronic requirement, the opening of a symmetrical epoxide (**6**) by a nucleophile gives a product of defined stereo-chemistry (cf. **7**). The same requirement necessarily holds for the reverse process, i.e. **7** → **6**. Indeed, the formation of an epoxide from **7** can be regard-ed as an internal SN$_2$ reaction (footnote 20 in ref. 4, see also 5).

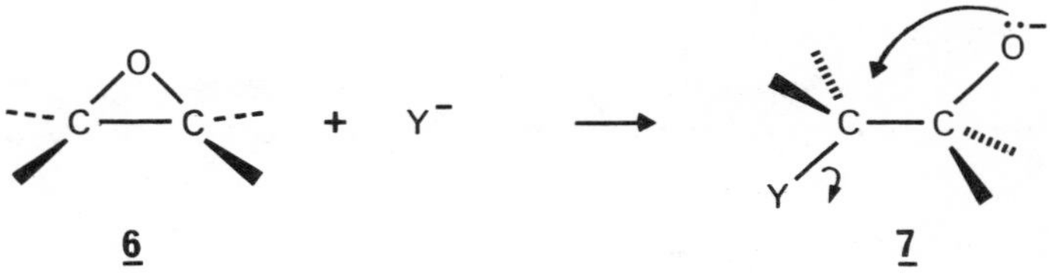

$$\underset{6}{} \quad + \quad Y^- \quad \longrightarrow \quad \underset{7}{}$$

In the case of an unsymmetrical epoxide which is conformationally rigid such as **8**, two different products are theoretically possible, the diequatorial product **10** resulting from the attack at C-2 or the diaxial product **12** from the attack at C-3.

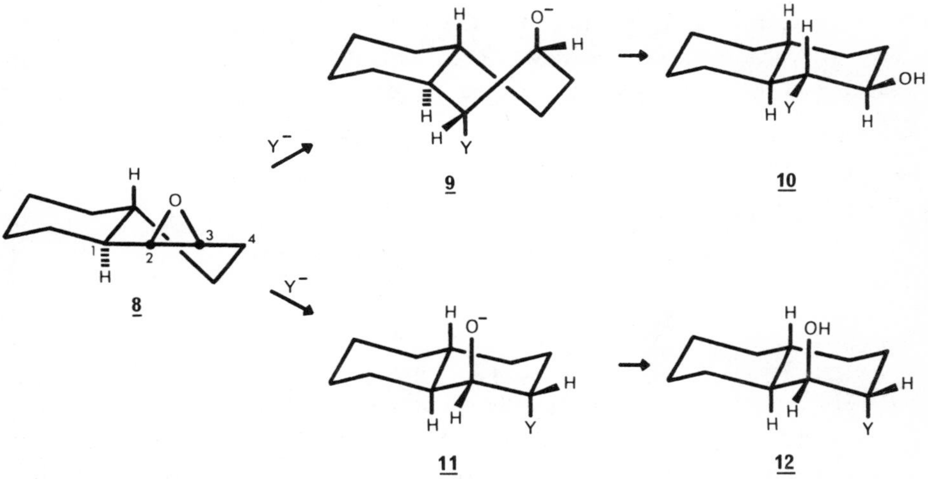

As discussed by Valls and Toromanoff in 1971 (6), the electronically controlled stereochemistry of the transition state in the SN$_2$ reaction requires that the reaction at C-2 must give the twist-boat intermediate **9** which then gives compound **10** after a conformational change whereas the reaction at C-3 gives the chair intermediate **11** and thence **12**. The formation of the transition state to produce **11** is a lower energy process than that leading to **9**. Product **12** is thus formed in preference to **10** under kinetically controlled conditions. Indeed, it is well known that the epoxide opening gives the trans-diaxial product.

Most addition reactions of electrophilic reagents to double-bonds which are known to proceed through an epoxide-like intermediate (i.e. the three-membered ring intermediate **14**), and, therefore, must follow the same principle, yielding the trans-diaxial product **15** (7).

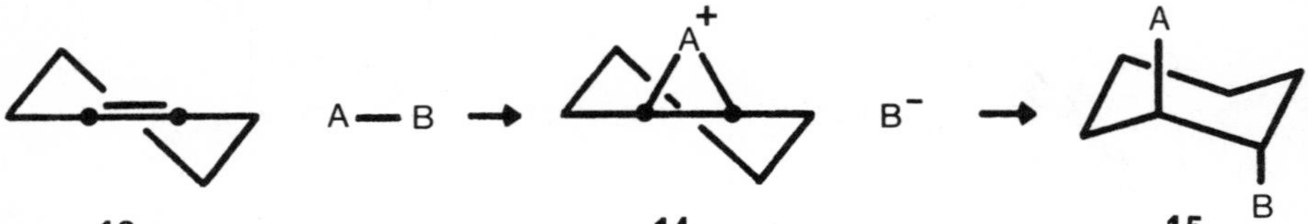

13   **14**   **15**

This is true for the addition of hydrogen halide (H—X), halogens (X—X) and other electrophilic reagents (NO — Cl, I — N_3, etc.) to cyclohexenes. The oxymercuration reaction ($Hg(OAc)_2$) in the presence of several nucleophiles (H_2O, AcOH, ROH) is another well known example.

The reverse process, i.e. **15** → **14** → **13**, which is a method to produce olefins, takes place also with the same stereochemical requirements. The familiar reaction of a diaxial diiodo compound **16** with zinc illustrates the principle.

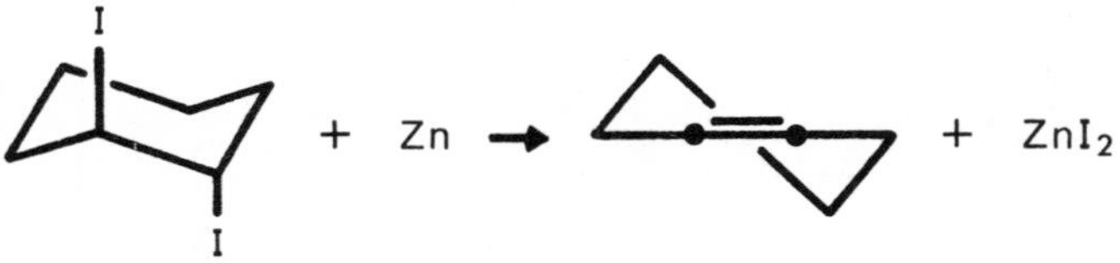

16

Following these general examples, we can now analyze the opening of epoxides in more detail. The size of the ring imposes geometric constraint and the SN_2 reaction demands colinearity. The approach of the nucleophile should then resemble **17** → **18**, which is different from that in displacement of a leaving group on an aliphatic chain, i.e. **19** → **20**.

17   **18**

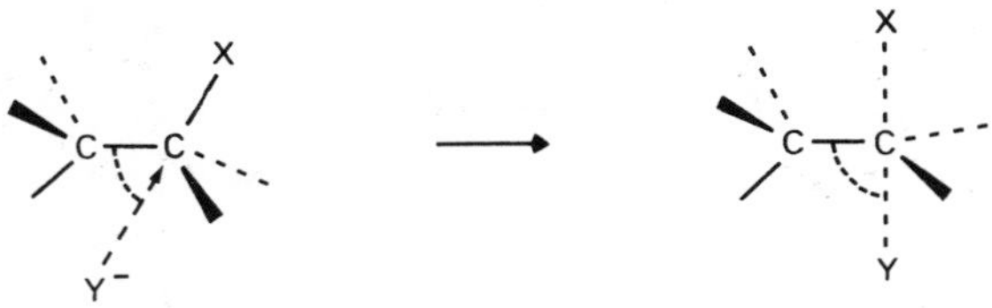

19   **20**

It is on the basis of this theoretical analysis that Stork deduced that the intramolecular opening of an epoxide yielding a six-membered ring should be more facile than that leading to a five-membered ring. Indeed, in the case of six-membered ring formation (21), the nucleophile is perfectly aligned for a backside colinear displacement whereas, in the five-membered ring formation (22), the positioning of the carbanion along the dotted line representing colinear approach requires considerable bond distortion. It should be pointed out that when the process leading to a five-membered ring involes a displacement as illustrated by 23, there is no particular constraint in locating the carbanion in the proper position for backside colinear displacement. Reactions of this type generally lead to faster formation of the five rather than the six-membered ring.

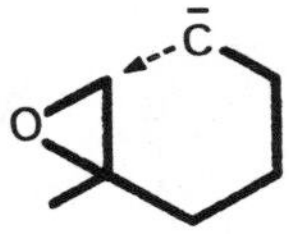

21 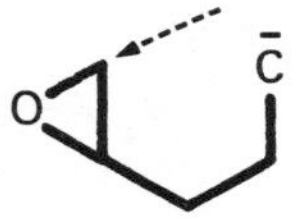   **22** 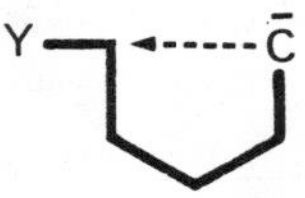   **23**

Experimental evidence in support of this was provided by Stork, Cama, and Coulson (8) who studied the cyclization under basic conditions (KNH$_2$, NH$_3$-DME) of epoxynitriles 24 and 26. They observed that the formation of the cyclohexane ring (24 → 25) was indeed easier than that of the cyclopentane ring (26 → 27).

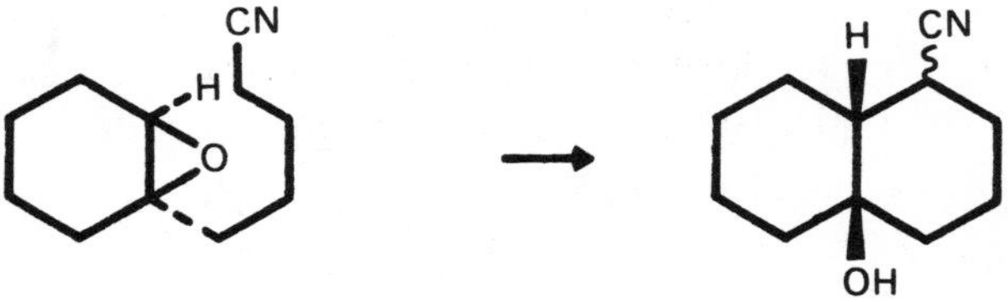

24   **25**

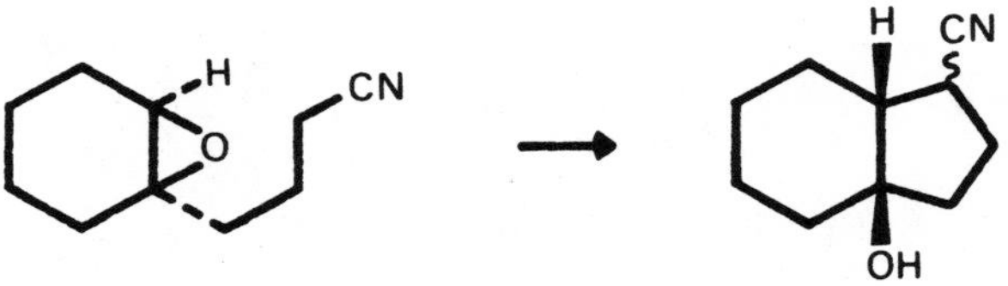

26   **27**

Stork then analyzed the transition state requirements for the formation of a four-membered ring (**28**) and realized that it seemed easier to attain collinearity in this case than in that of the five-membered ring (**22**), arriving at the unexpected conclusion that the former process could prevail over the latter.

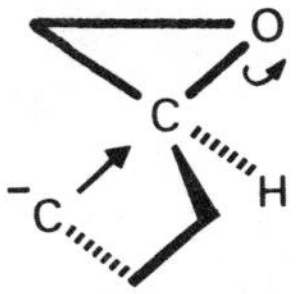

28

Stork and Cohen (9) have then found experimentally that indeed the cyclization of epoxynitrile **29** with lithium bis(trimethylsilyl)amide in benzene gave a 95:5 mixture of the isomeric cyano-hydroxy-cyclobutanes **30** and **31**. Thus, with equal substitution at both ends of the oxirane ring, a cyclobutane is formed in preference to a cyclopentane ring (**32**). Note also that the major isomer **30** is the less stable isomer (two larger groups cis to each other). This shows that formation of transition state **33** leading to **30** is of lower energy than that of transition state **34** which gives **31**. This new methodology of cyclobutane formation was used to synthesize (±)-grandisol starting from **30**.

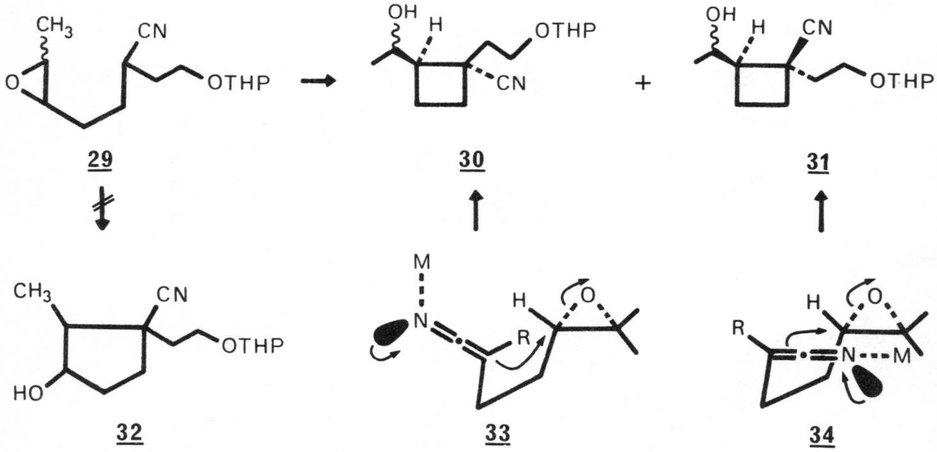

29   **30**   **31**

32   **33**   **34**

Obviously, when the oxirane carbon further from the nitrile is more substituted, the formation of a cyclobutane ring is even more favored since it involves a displacement at the less substituted end of the epoxide. Accord-

ingly, cyanoepoxides **35** (R=H or CH$_3$) gave the cyclobutanes **36** (R=H or CH$_3$)
(8). Note again that the specific formation of **36** (R=CH$_3$) shows that the
preferred transition state is **33**, not **34**. This indicates that the effective
steric hindrance of the cyano anion is in fact larger than that of a normal
alkyl group as a consequence of the allenic structure of the metal salt. It
is interesting to point out that the proposed geometrical requirement for
the nitrogen electron pair, i.e. antiperiplanar to the π system of the
double-bond which will eventually open the oxirane ring (see arrows in
33), is deduced from stereoelectronic principles (for more recent work
on this topic see refs. 10 and 11).

Barnett and Sohn (12, 13, see also 14) have discovered that the iodolactonization of β,γ-unsaturated carboxylic acid salts **37** yield, under kinetically
controlled conditions, the γ-iodo-β-lactones **39** in preference to the more
stable β-iodo-γ-lactones **41**. Similar results were obtained in the course of
the bromolactonization reaction. Thus, here again, the formation of a four-
membered ring is more facile than that of a five-membered ring. This can be
rationalized on the basis of Stork's analysis, i.e. the internal opening by
the carboxylate anion of the three-membered ring iodonium ion (or bromonium)
38 → **39** is preferred over the other mode of opening **40** → **41** for stereoelectronic reason.

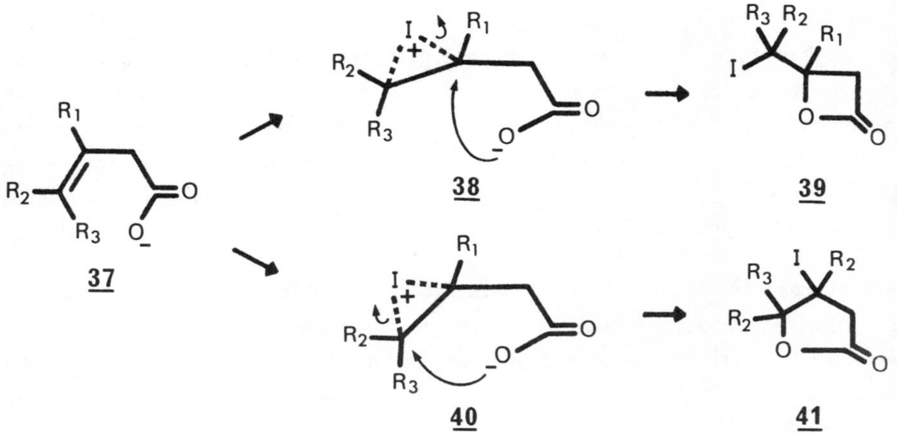

Returning to internal opening of cyanoepoxides, the rate of cyclopropane formation (**42**) is such that this ring is produced in preference to a cyclo-butane (**43**) regardless of the relative degree of substitution of the oxirane ring. For instance, cyclization of epoxynitrile **44** gave the cyanopropane **45** (9). It is well known that cyclopropanes are normally more readily formed than cyclobutanes; the internal opening of cyanoepoxides is therefore no exception. This suggests that, presumably through the use of "bent bonds", the nucleophilic alignment is better in **42** than in **43**. In that case, cyclo-propane would correspond to the bent-bond model of Coulson and Moffitt (15, see also 16).

42   **43**

44   **45**

It is pertinent to point out that in internal epoxide opening, six-membered ring formation (**21**), requires no bending of C — C bonds, while five- (**22**), four- (**28** and **43**) and three-membered ring (**42**) formation requires the simul-taneous bending of four, three, and two C — C bonds, in addition to the carbanion electron pair. The degree of bending is of course more pronounced in the formation of the three-membered ring than in the others; consequent-ly, the number of C — C bonds which must bend simultaneously would be more important than the degree of bond bending. In other words, this suggests the simultaneous bending of three C — C bonds (cyclobutane formation) would be less difficult than that of four C — C bonds (cyclopentane formation), but more difficult than that of two C — C bonds (cyclopropane formation).

In that respect, it is interesting to regard a double-bond as a "two-member-ed" ring, i.e. made of two bent bonds as proposed by Pauling in 1931 (17, see also 18). Its formation requires extensive bending of only one C — C bond in addition to the carbanion electron pair (**46** → **47** → **48**). So, the ready

formation of double-bonds could thus be explained. Note also that the anomeric effect (Chapter 1) is the equivalent of what is illustrated by **46** →
47.

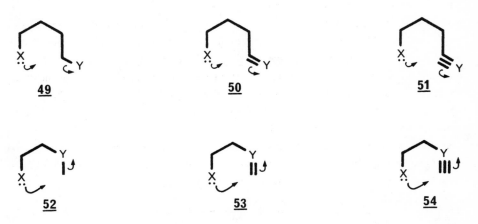

46   **47**   **48**

Baldwin (19) has proposed a set of pertinent rules for ring closure. Ring
forming reactions are designated by a numerical prefix which denotes the
ring size, which is then followed by either the term exo or endo depending
on whether the breaking bond is exocyclic or endocyclic (3) to the smallest
so-formed ring, and finally by one of the suffixes tet, trig or dig describ-
ing the hybridization of the carbon atom undergoing attack in the closure
reaction (tetrahedral, trigonal, and digonal, respectively).

For example, the 5-Exo-Tet, 5-Exo-Trig, and 5-Exo-Dig processes are represented by **49**, **50**, and **51** while the 5-Endo-Tet, 5-Endo-Trig, and 5-Endo-
Dig corresponds to **52**, **53**, and **54** respectively. In this Chapter, we will
consider the tetrahedral systems only. The trigonal systems are considered
in Chapter 6.

In tetrahedral systems, the possibilities are the following: 3-Exo-Tet
to 7-Exo-Tet (**55-59**) are all favored processes whereas 5-Endo-Tet (**60**)
and 6-Endo-Tet (**61**) are disfavored on stereoelectronic grounds.

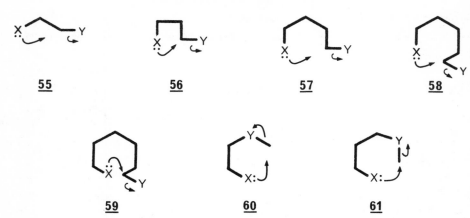

The previously described experiment of Eschenmoser (p.164) shows definitely that the 6-<u>Endo-Tet</u> process (<u>61</u>) is indeed not favored. These are numerous experiments in the literature which show that all the <u>Exo-Tet</u> processes are allowed and the relative ease with which these processes take place is the following: <u>55</u> > <u>56</u> < <u>57</u> > <u>58</u> > <u>59</u>. In other words, <u>55</u> is easier than <u>56</u>, and <u>57</u> is easier than <u>56</u> and <u>58</u> while <u>58</u> is better than <u>59</u>. However, Stork has clearly demonstrated that when Y is part of an oxirane ring, this order is changed to <u>55</u> > <u>56</u> > <u>57</u> < <u>58</u>. Thus, there is a change between <u>56</u>, <u>57</u> and <u>58</u>, and the fundamental reason is due to stereoelectronic effects.

Before completing this section on the SN_2 reaction, we must also consider the influence of the stereoelectronic effects caused by the electron pairs of atoms which are linked to the carbon bearing the leaving group.

$$R - O - CH_2 - Cl \quad + \quad Y^- \quad \longrightarrow \quad R - O - CH_2 - Y \quad + \quad Cl$$
$$\underline{62}$$

$$R - \overset{\overset{\textstyle O}{\|}}{C} - CH_2 - Cl \quad + \quad Y^- \quad \longrightarrow \quad R - \overset{\overset{\textstyle O}{\|}}{C} - CH_2 - Y \quad + \quad Cl^-$$
$$\underline{63}$$

For instance, the rate of SN_2 reaction is greatly enhanced in α-haloethers <u>62</u> and in α-haloketones <u>63</u>. This enhancement should however occur only when the oxygen atom in <u>62</u> has an electron pair antiperiplanar to the C—Cl bond (<u>cf.</u> <u>64</u>→<u>65</u>). Similarly, in an α-haloketone the π system of the carbonyl group must be parallel to the C—Cl bond (<u>cf.</u> <u>66</u>→<u>67</u>) (20).

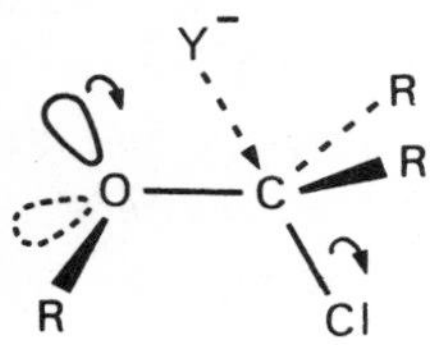

It is well known that bimolecular substitution in benzyl substrates ($\underline{68} \rightarrow \underline{69}$) takes place with great ease. This is due to an electronic stabilization of the transition state by conjugation with the aromatic ring (20, 21). The theory predicts that such conjugation will be greatest when the $C-X$ bond of the leaving group is parallel with the p-orbitals of the benzene ring, as in the arrangement $\underline{70}$, and smallest when the $C-X$ bond is orthogonal to the same orbital system as in the arrangement $\underline{71}$.

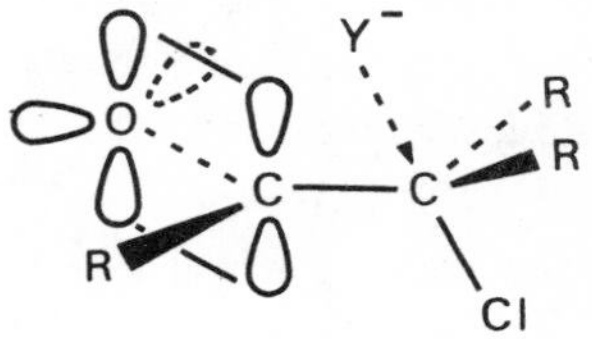

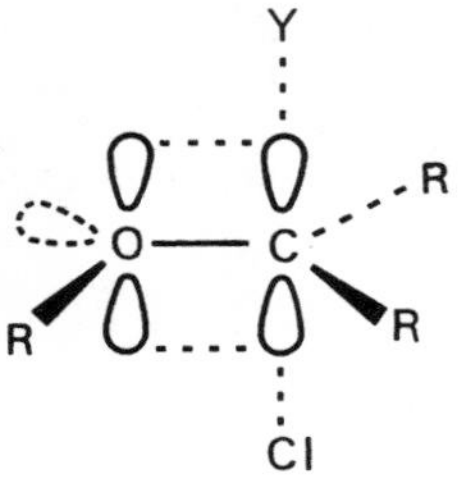

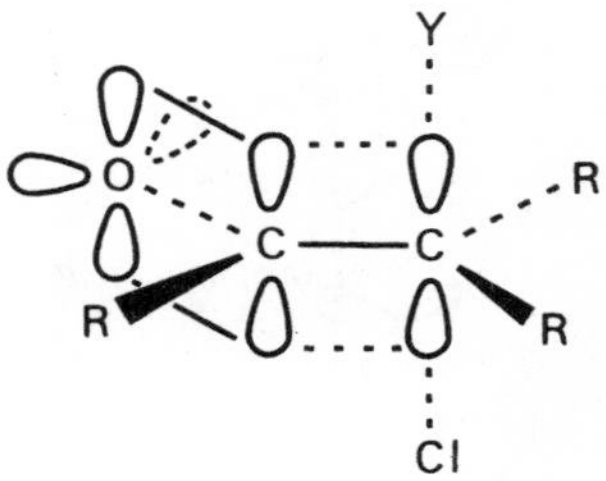

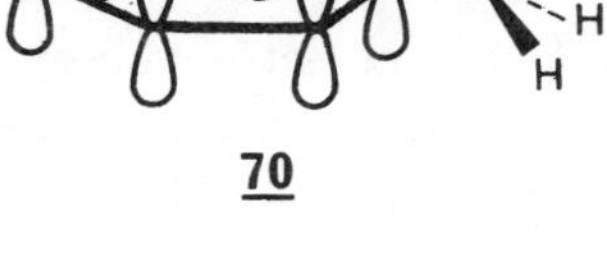

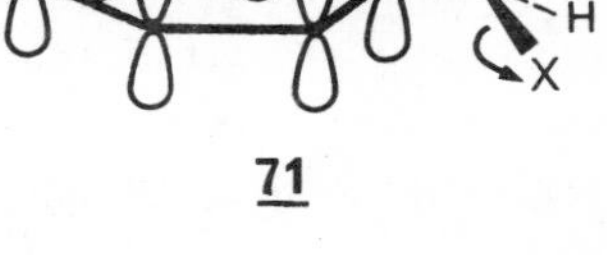

King and Tsang (22) have recently studied the relative rate of bimolecular substitution in dibenzylethylsulfonium salt **72** and the cyclic sulfonium fluoroborate salt **75**. They found that the reaction of thiocyanate anion with salt **72** which can easily take the arrangement **70** is 8000 times faster than the corresponding reaction with the cyclic sulfonium salt **75**, which is tied up in the arrangement **71**.

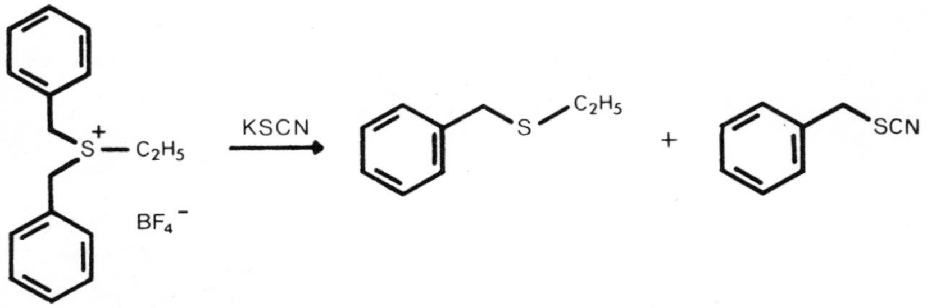

Moreover, product formation was not the same. Dibenzylethylsulfonium salt **72** gave only benzyl ethyl sulfide (**73**) and benzyl thiocyanate **74**, but no dibenzyl sulfide or ethyl thiocyanate. By contrast, the cyclic sulfonium salt **75** gave a mixture containing chiefly (80%) dihydroisothianaphthene (**76**) with an equivalent amount of ethyl thiocyanate (**77**) and a minor amount (20%) of the sulfide-thiocyanate **78**. This work represents the first direct experimental information on the rate of bimolecular nucleophilic substitution and relative orientation of the benzene ring and the leaving group and it demonstrates clearly the importance of stereoelectronic effects in these reactions.

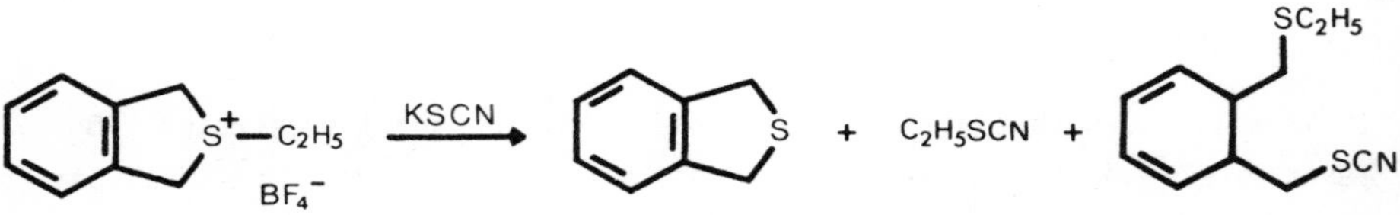

SN$_2$' type reaction

Stereoelectronic effects can also be considered in the SN$_2$' reaction. An excellent review on this topic has been published recently by Magid (23). Although there is still some discussion concerning the concertedness of this reaction, as pointed out by Magid, the stereochemistry of the process

can be analyzed, regardless of the precise timing of the bond-breaking
and bond-making steps.

The SN$_2$' reaction can theoretically take place by a nucleophilic attack on
the allylic system either syn or anti to the leaving group (cf. 79 and
80). Although he never fully elaborated on it in print, Winstein (24-26)
was the first to postulate that the nucleophilic attack should be syn. He
suggested that the approach of the nucleophile displaces the π electrons
in such a direction as to allow them to attack the C—X bond from the rear.
The syn attack would therefore be favored for stereoelectronic reasons.

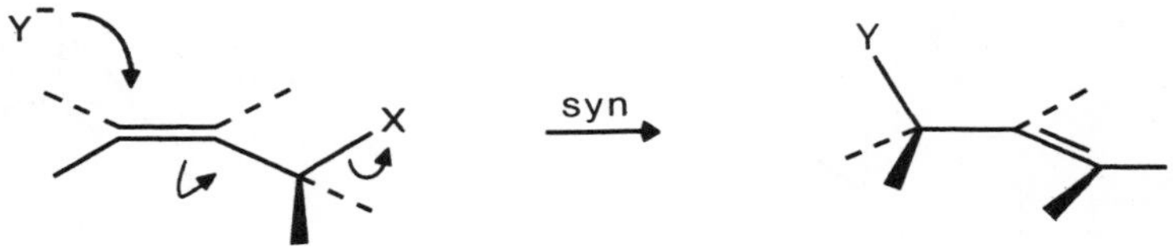

A cyclohexenyl system can exist in the two conformations 81 and 82. A syn
attack on 81 to give 84 should take place via a transition state having
the chair-like stereochemistry 83 where the electron pair is oriented antiperiplanar to the C — X bond. An anti attack on 81 should occur via the
twist-boat transition state 85 where the electron pair is syn to the C—X
bond, yielding the product in the twist-boat conformation 86. A syn attack
on conformer 82 should give 88 via the twist-boat 87 where the electron pair
is antiperiplanar to the C — X bond, whereas an anti attack should give
90 via the chair form transition state 89 where the electron pair is syn
to the C—X bond.

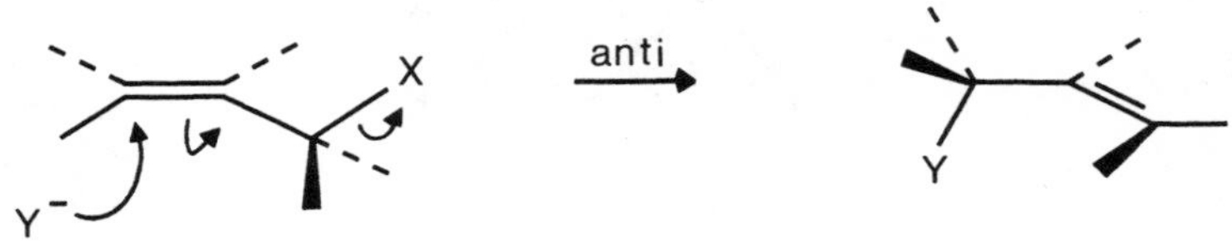

The syn attack on 81 and the anti attack on 82 both lead to the chair-like
transition states 83 and 89 so should be favored over the other two processes. Moreover, the transition state 83 which has an electron pair antiperi-

planar to the C — X bond should be favored electronically over transition state **89**. Thus, the syn process **81** → **83** → **84** should prevail.

A strong case can be made that all SN$_2$' reactions should proceed through conformation **81** where the C — X bond is nearly parallel to the p-orbitals of the double-bond. Support for this view comes from the cleavage of protonated cyclohexenols as shown by Goering and co-workers (27, 28). Also, theoretical studies have, for the most part, supported the syn attack notion.

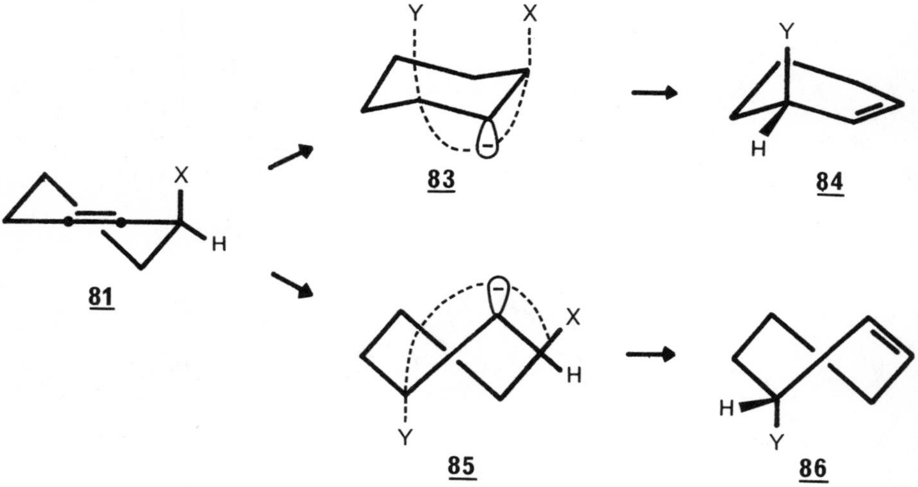

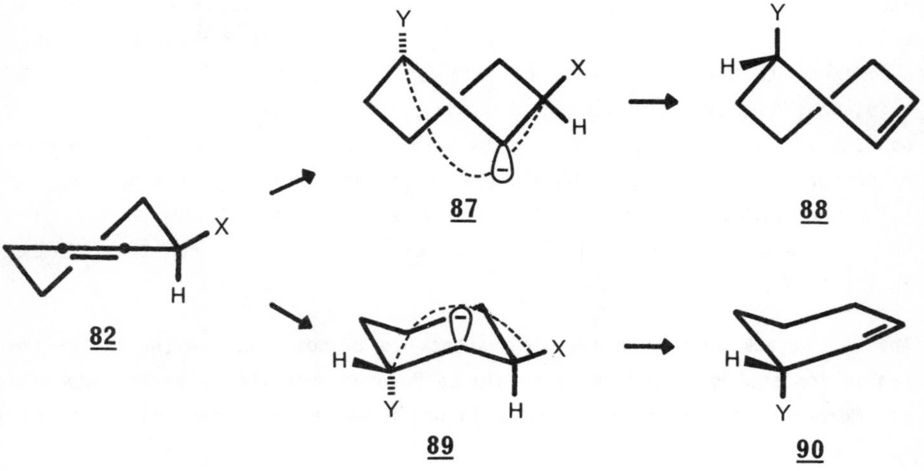

The first experimental study which showed that the SN_2' reaction takes place via a syn mode is due to Stork and White (29). They have observed that the cyclohexenyl dichlorobenzoates **91** (R=CH$_3$, CH(CH$_3$)$_2$ or C(CH$_3$)$_3$ and R'=Cl$_2$C$_6$H$_3$—) react with piperidine to give the syn SN_2' product **92**. This work was recently confirmed by Stork and Kreft (30) and Dobbie and Overton (31). Also, cis and trans mesitoates **91** and **93** (R'=C$_6$H$_2$(CH$_3$)$_3$, R=CH(CH$_3$)$_2$) yielded syn SN_2' products **92** and **94** (R=CH(CH$_3$)$_2$) respectively.

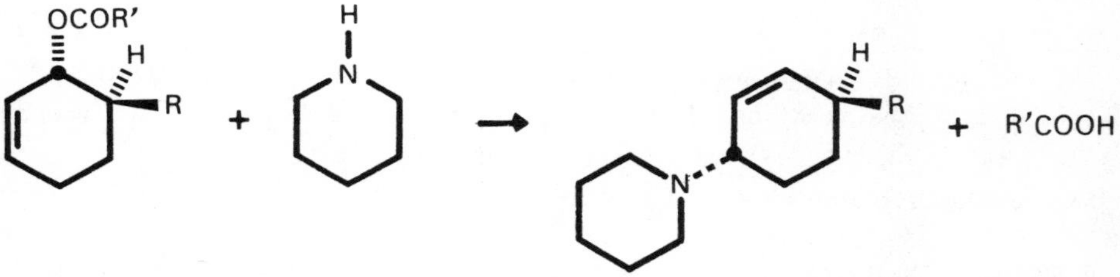

91   **92**

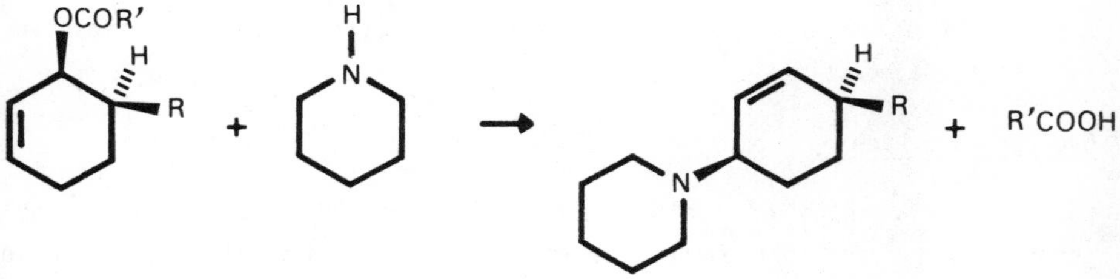

93   **94**

Other experiments confirm the syn SN_2' mode. Kirmse and co-workers (32) have found that cis-dichlorocyclobutene **95** undergoes two consecutive syn SN_2' displacements (→ **96** → **97**) with sodium methoxide.

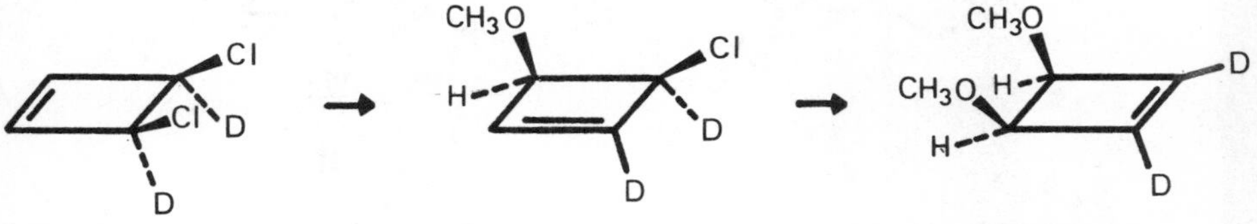

95   **96**   **97**

Magid and Fruchey (33) found that the optically active allylic chloride **98** (R=CH$_3$ and X=Cl) underwent an SN_2' reaction with at least 96% syn stereo-selectivity.

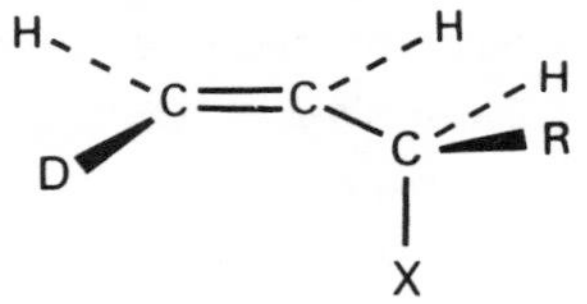

98

However, there are experiments which show that the anti mode can also take place. For instance, changing piperidine for sodium propanethiolate, the trans and cis mesitoates **91** and **93** (R'=$C_6H_2(CH_3)_2$, R=CH(CH$_3$)$_2$) gave a mixture of syn and anti SN$_2$' products (30). Also, Oritani and Overton (34) showed that the reaction of dichlorobenzoates **98** (R=CH$_3$ or C$_5$H$_{11}$, and X= Cl$_2$C$_6$H$_3$CO$_2$) with (S)-α-methylbenzylamine gave a mixture of syn and anti SN$_2$' products (ratio ≈6:4).

A complete survey of the SN$_2$' reaction can be found in the review of Magid (23). There are several experiments which clearly show that the syn mode is favored but the anti addition can also take place. The factors which favor the anti mode are not yet completely understood. Nevertheless, there appears to be no doubt that stereoelectronic effects play an important role in these reactions.

Cyclopropane opening

Nucleophilic opening of cyclopropanes doubly activated by alkoxycarbonyl groups is known to take place (see ref. 35 for a review). For instance, in an interesting study, Danishefsky, Dynak, and Yamamoto (36) showed the rapid base catalyzed (NaCH$_2$SOCH$_3$ in DMSO) interconversion of the two cyclopropanes **99** and **102** via their anions **100** and **101**. Danishefsky, Tsai, and

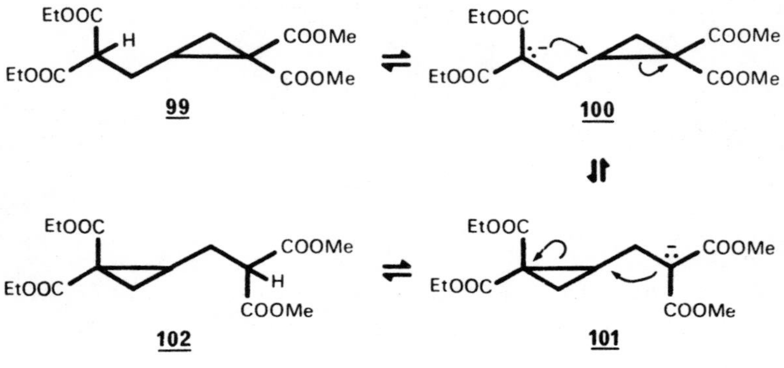

99   **100**

102   **101**

Dynak (37) have further reported the interconversion of **103** and **106** via their respective anions **104** and **105**.

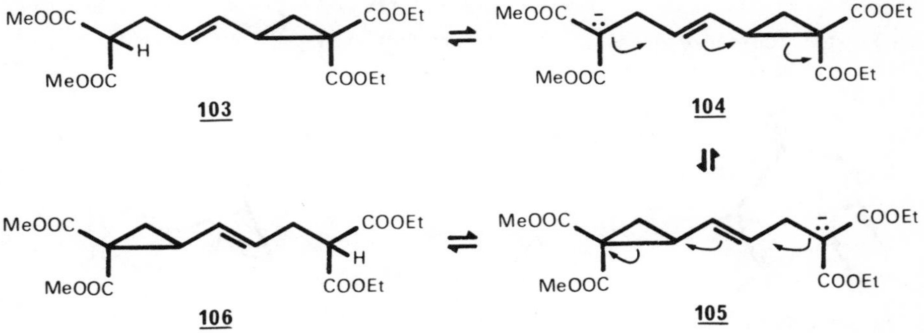

103   **104**

106   **105**

These reactions are essentially SN_2 type and take place because the cyclopropane $C_2 - C_1$ bond is parallel to the π system of one alkoxycarbonyl group (cf. **107**). As a consequence, the nucleophile attacks the polarized $C_2 - C_1$ bond and inversion at C-2 occurs to give the enolate ion **108**. Thus, in the reverse process, i.e. cyclopropane formation, the stereochemistry of the transition state should correspond to **108** → **107**. These reactions are therefore controlled by powerful stereoelectronic effects.

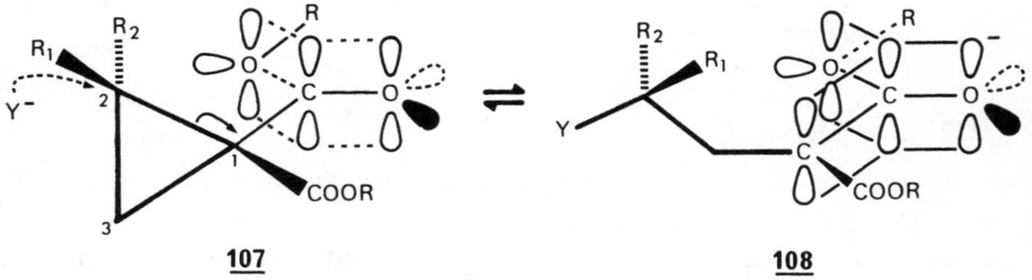

107   **108**

For instance, **109** produced in situ gave **111** via **110** while **112** gave **114** via **113** (38).

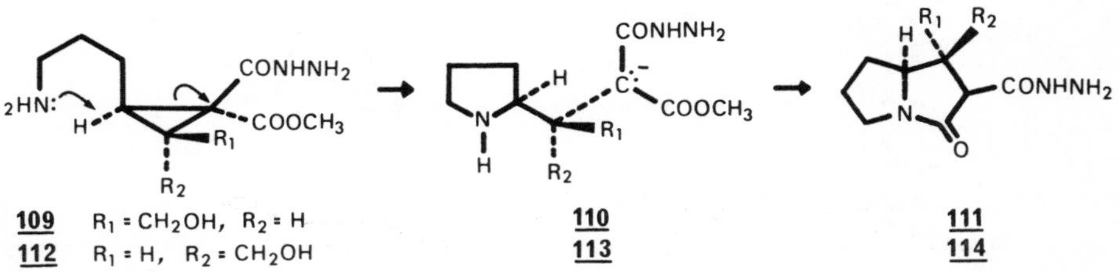

| **109** | R_1 = CH$_2$OH, R_2 = H |
| **112** | R_1 = H, R_2 = CH$_2$OH |

110   **111**
113   **114**

Corey and Fuchs (39) noted the vinyl cuprate addition on **115** to produce **116** stereospecifically. Clark and Heathcock (40) observed the lithium dimethylcopper addition on **117** which yielded **118** while Trost, Taber, and Alper (41) reported the conversion of **119** into **120** with the same reagent.

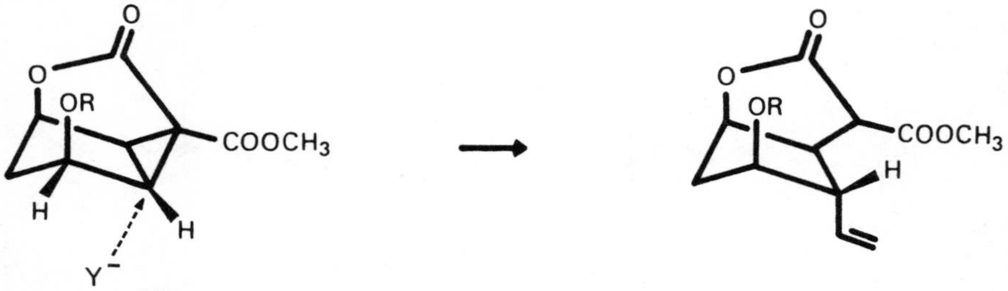

115

116

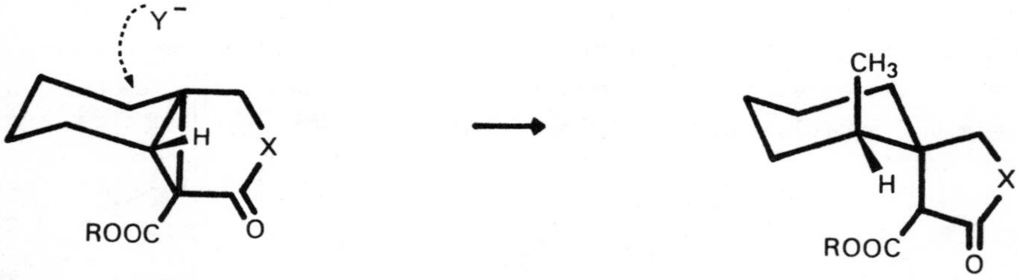

117   X = CH₂ or O

118

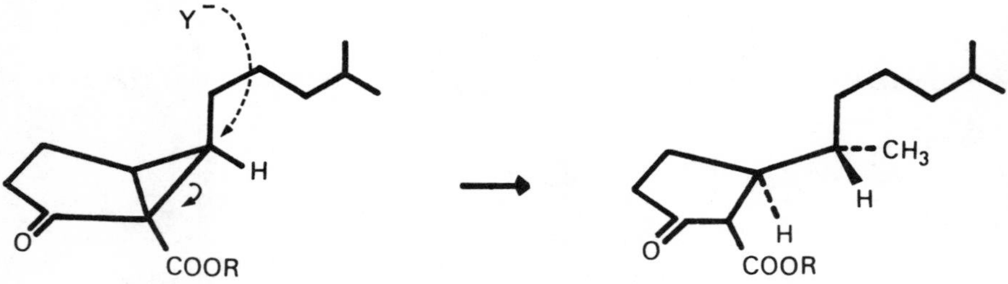

119

120

Taber (42) and Kondo and co-workers (43) have found the stereospecific opening of **121** → **122** with thiophenoxide. Using phenyl selenide anion as the reagent, Isobe and collaborators (44) discovered the conversion **123** → **124**.

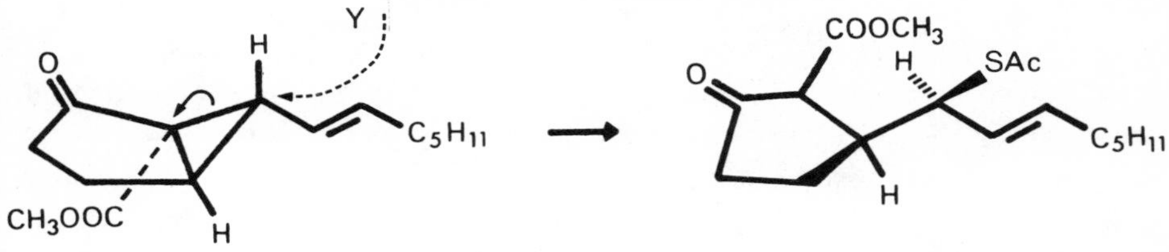

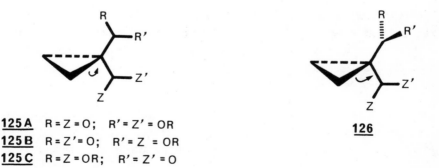

121 → **122**

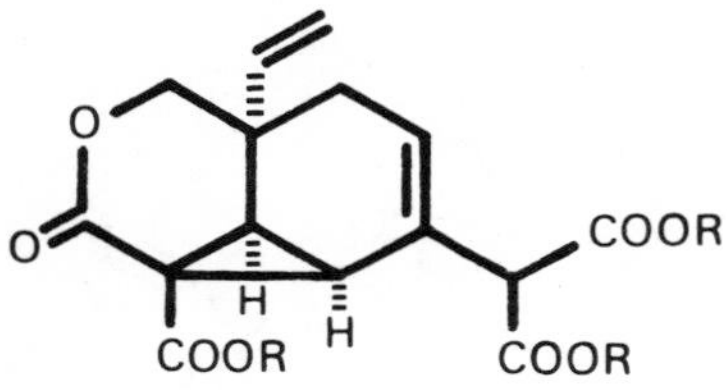

123

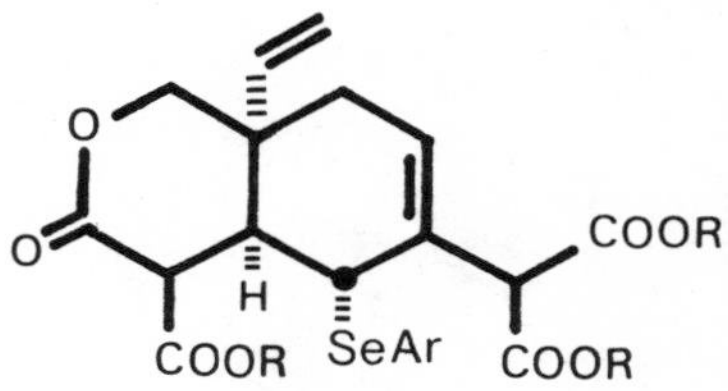

124

In the preceding examples, if both alkoxycarbonyl groups participate in the delocalization of the emerging carbanion, they must take either conformations exo-exo **125A**, exo-endo **125B** or endo-endo **125C**. These conformations are disfavored on dipolar and steric grounds and it is assumed that these compounds react through conformations of type **126** where only one alkoxy-carbonyl group (Z'—C—Z) is disposed to facilitate the reaction. Danishefsky and Singh (45) reasoned that if a compound could be designed to exist in one of the conformations of type **125**, its reactivity towards nucleophiles should be enhanced. They have verified their conclusion by studying the reactivity of cyclopropane acylals which exist in conformation **125A**.

125 A   R = Z = O;   R' = Z' = OR
125 B   R = Z' = O;   R' = Z = OR
125 C   R = Z = OR;   R' = Z' = O

126

They found that **127** reacted exothermically with piperidine in benzene at room temperature yielding adduct **128**. Other nucleophilic additions went smoothly (C_6H_5SH, $C_6H_5NH_2$ and $NaCH(COOCH_3)_2$), even the weakly nucleophile pyridine reacted to give the yellow betaine **129**. They have also proven that the nucleophilic attack takes place with inversion of configuration. Heating **130** with acetone-water at reflux gave after esterification the trans lactone **131** (46). This series of experiments represents a rather clear demonstration of the importance of stereoelectronic effects in relation to the reactivity of organic molecules.

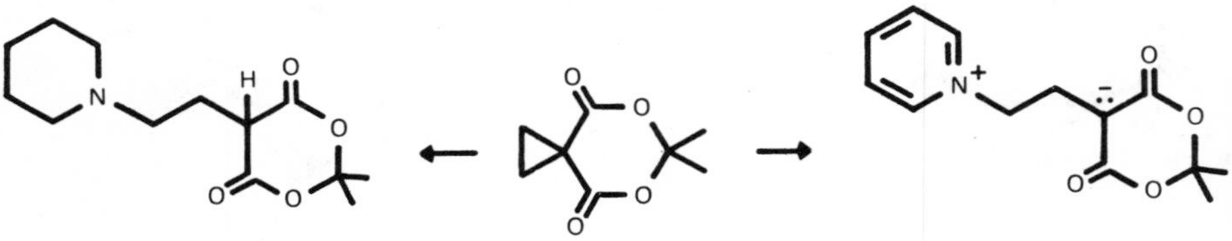

128   **127**   **129**

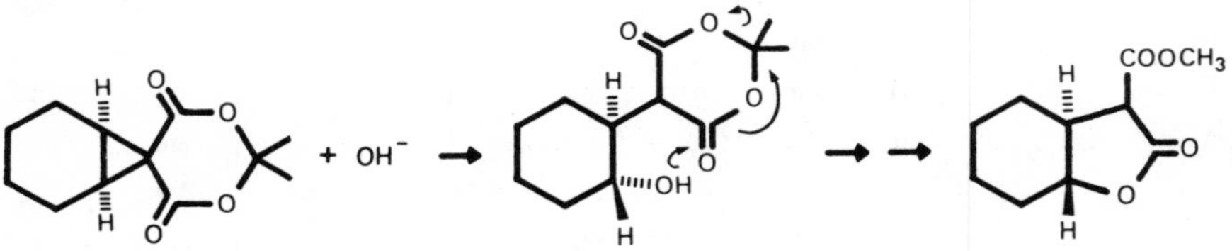

130   **131**

Molecular rearrangements

Cyclic systems with a tertiary hydroxyl group adjacent to a leaving group can undergo a molecular rearrangement in the presence of base yielding either a ring contraction (**132** → **133** → **134**) (47) or a ring expansion (**135** → **136** → **137**) (48) product.

In these rearrangements, there are two consecutive internal SN_2 type displacement processes: a) an electron pair of the oxygen atom displaces the electron pair of a C — C bond and b) the electron pair of a C — C bond displaces the leaving group. It is therefore pertinent to find out if these processes follow the stereoelectronic principle of the SN_2 reaction.

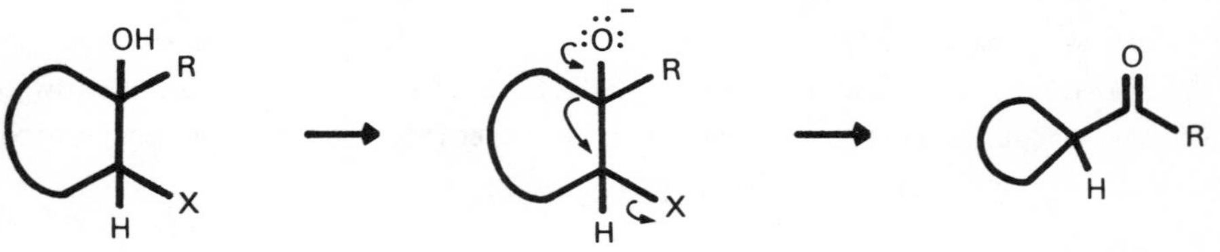

132   **133**   **134**

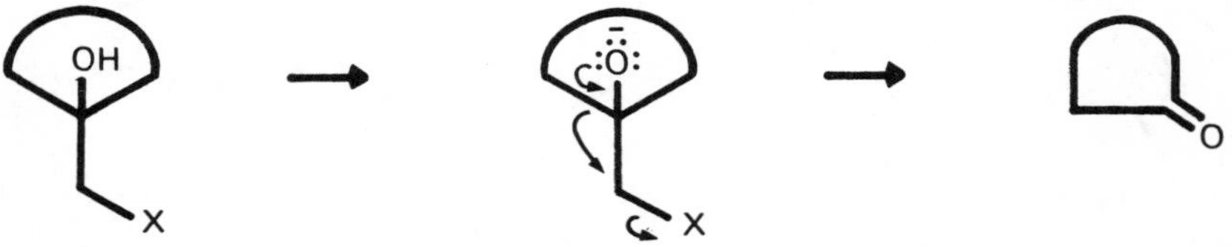

135   **136**   **137**

Numerous examples show that the migrating group is always oriented antiperiplanar to the leaving group. For instance, it is the R_1 group which migrates when a compound has a conformation equivalent to that of **138** or **139**. Note also that in **138** and **139**, the oxygen anion has an electron pair oriented antiperiplanar to the migrating R group. It is also pertinent to point out that when a molecule exists in the third alternative conformation **140**, no rearrangement occurs, but the epoxidation process (**141**) takes place readily. Thus, stereoelectronic effects play a dominant role in these reactions.

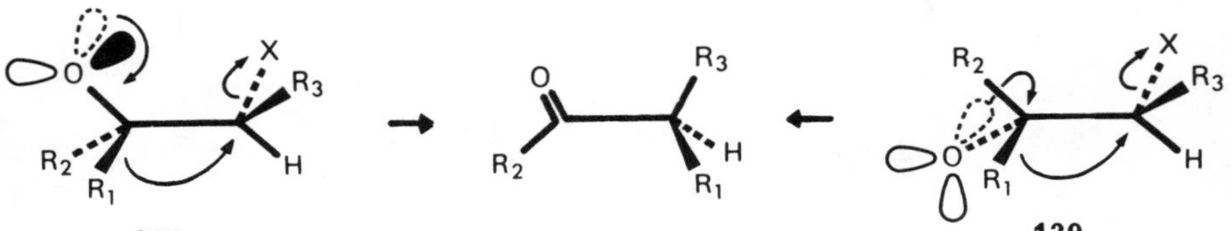

138   **139**

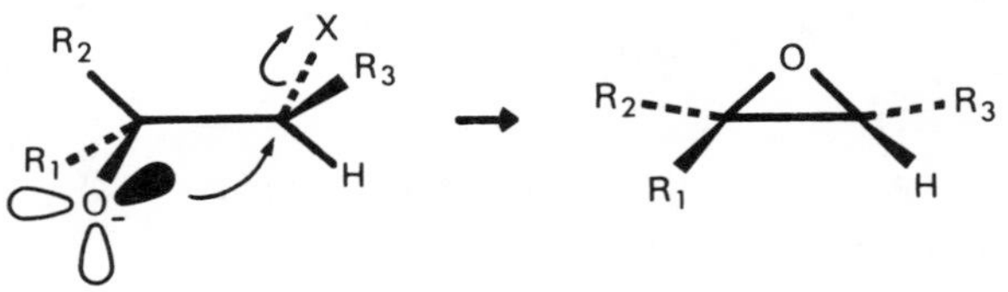

140   **141**

Only a few examples will be reported here (for a more complete survey, see refs. 47-53). Büchi and co-workers (54) have shown that lithium aluminium hydride reduction of the tosylate **142** gave specifically the ring contracted product **144** via the methyl ketone **143**.

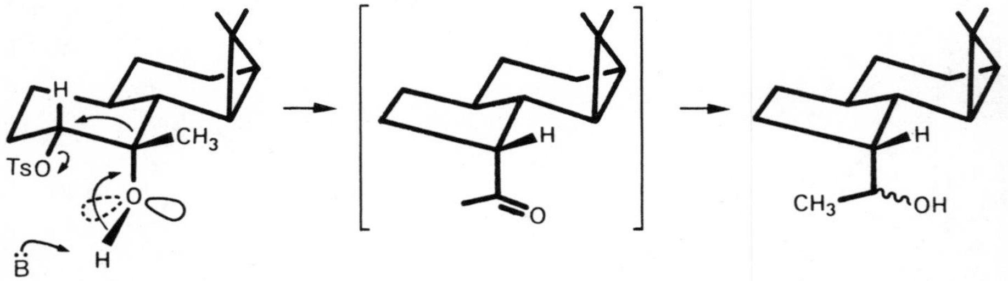

142   **143**   **144**

Nussim and Mazur (55) have observed that under mild basic conditions, the steroid derivatives **145**, **147**, and **149** yield specifically the rearranged ketones **146**, **148**, and **150** respectively, and Paukstelis and Macharia (56) found that treatment of **151** with sodium methoxide gave nopinone **152**. Compounds **142**, **145**, **147**, and **149** correspond to conformation **138** whereas that of **151** is similar to **139**.

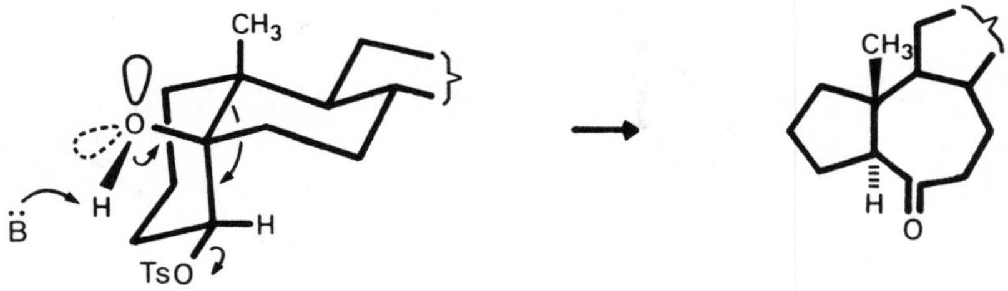

145   **146**

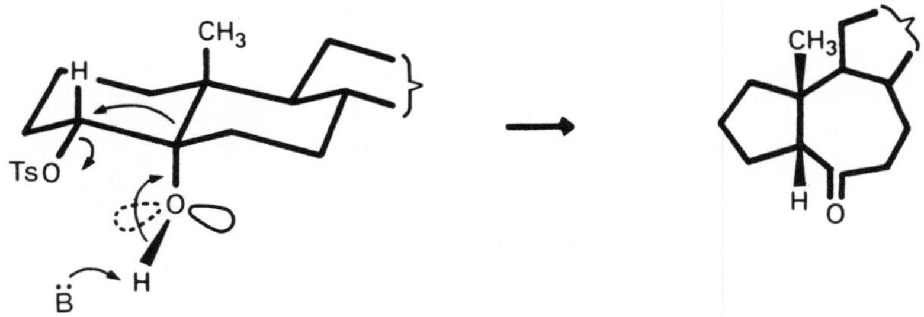

147   **148**

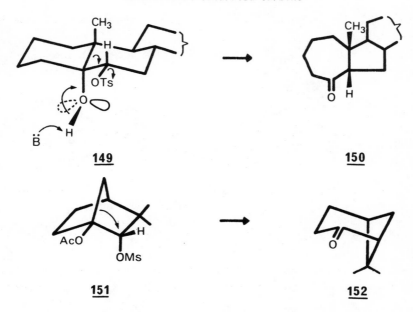

149   **150**

151   **152**

Also, on passage through alumina, the stereocontrolled ring contraction of tosyloxycyclobutanols **153** and **155** occurs in high yield to give **154** and **156** respectively (57).

153   **154**

155   **156**

Interestingly, Yamada and co-workers (58) have shown that the silver ion-promoted rearrangement of the chloronitrile isomers **157** and **159** yields the tricyclic isomeric cyanoketones **158** and **160** respectively.

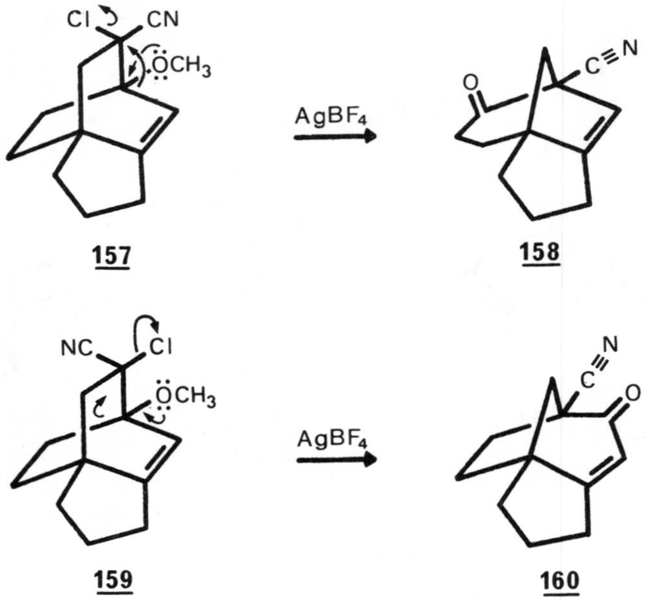

The products obtained from the deamination of vicinal aminoalcohols are determined by the conformation of the ground state, where the migrating bond is always oriented antiperiplanar to the C—N bond (59). Indeed, diazotization with nitrous acid of **161** and **163** gave only the cyclopentane aldehyde **165** via the diazohydroxide intermediates **162** and **164** respectively. Diazotization of compound **166** gave the cyclohexanone **168** (98%) via intermediate **167** which undergoes a stereocontrolled internal hydride migration. Finally, the fourth isomer, i.e. **169**, gave the cis epoxide **171** via intermediate **170**. Favre and Gravel (60, 61) had previously observed that when the amino group can take more than one conformation as in 1-aminomethyl-cyclohexanols, product formation from the diazotization reaction is directly related to the conformation of the methyldiazonium group.

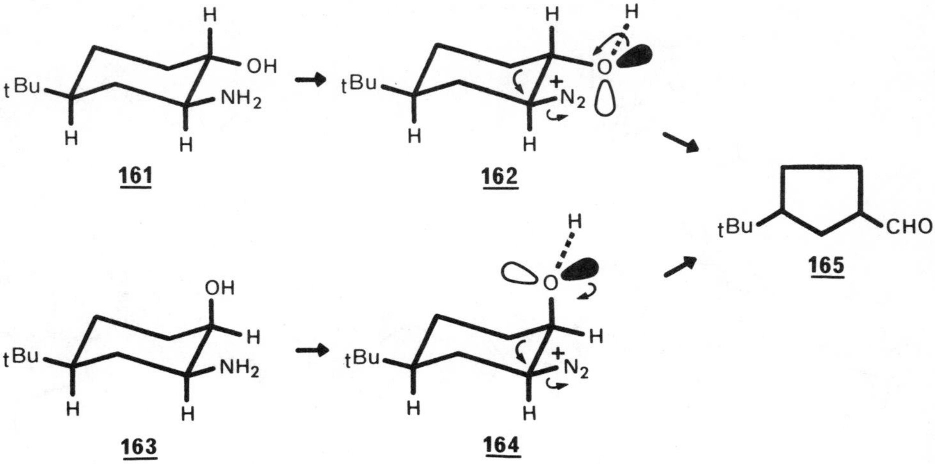

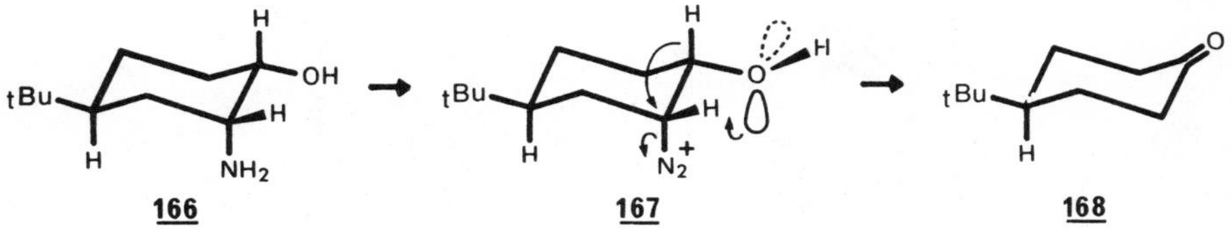

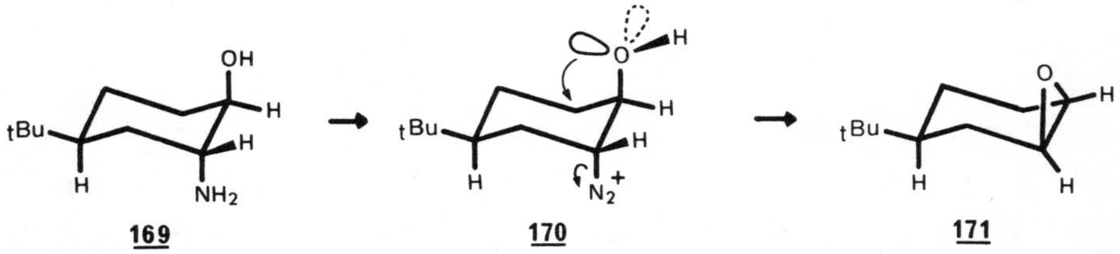

In the Woodward synthesis of prostaglandin (62), intermediate **172** formed in situ from the corresponding amine was smoothly transformed into bicyclic aldehyde **173**. Seebach and co-workers (63) have also observed several stereospecific rearrangements using the same reaction. For example, diazotization of amine **174** gave specifically the cis-cyclopentane **175** which was then epimerized into the more stable trans-cyclopentane **176**.

An epoxide oxygen can serve as the leaving group in these rearrangements. For instance, Marshall and Kerschen (64) have found the BF₃ catalyzed transformation **177** → **178**.

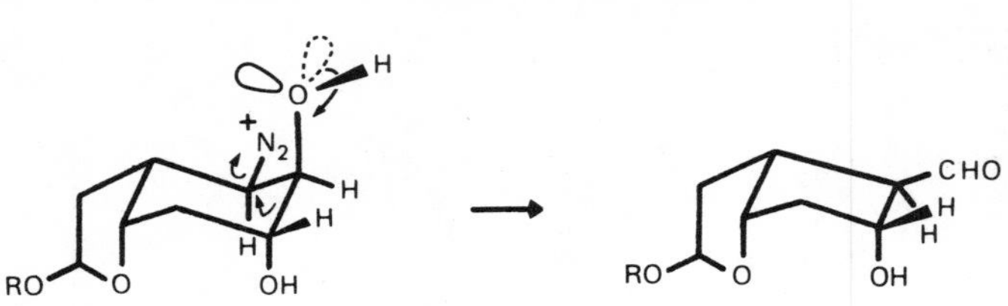

172   173

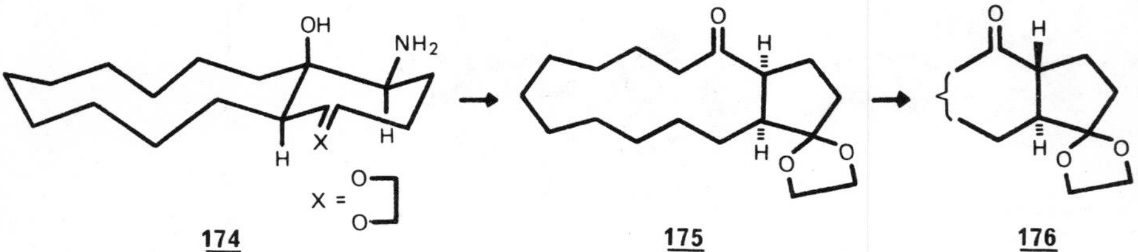

174   X = $\begin{matrix} O- \\ O- \end{matrix}$ 175   176

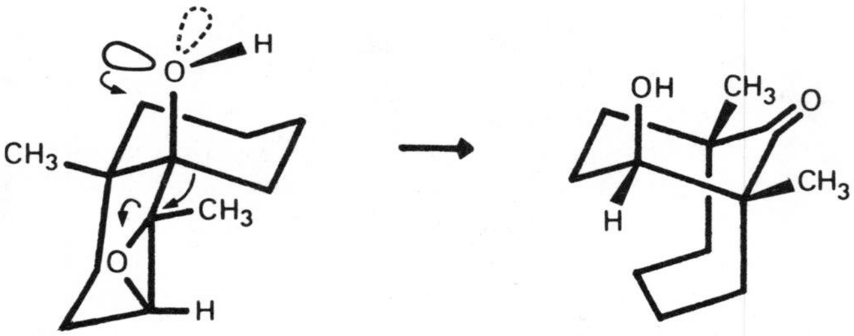

177   178

On the basis of these results, the benzyl-benzylic acid rearrangement 179 →
180 → 181 should occur with stereoelectronic control. The stereochemistry
of the transposition should be as depicted by 182 → 183.

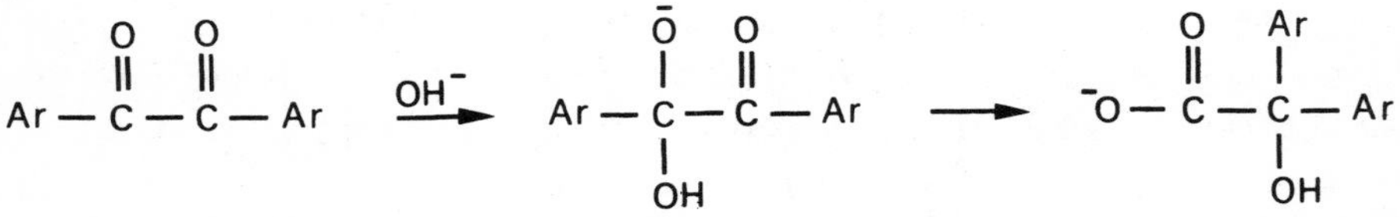

179   180   181

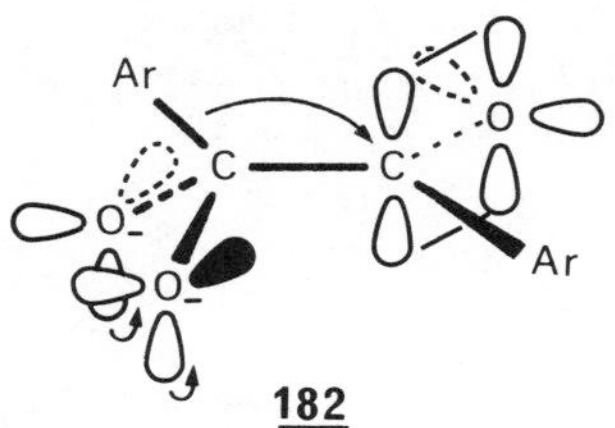

182   **183**

A similar situation should prevail in the quasi-Favorski rearrangement. Indeed, it has been observed (65) that refluxing compounds **184** and **187** in xylene in the presence of sodium hydroxide gave the isomeric carboxylic acid derivatives **186** and **189** respectively. Thus, there is a neat inversion of configuration of the carbon atom which was initially bearing the halogen atom (cf. **185** and **188**).

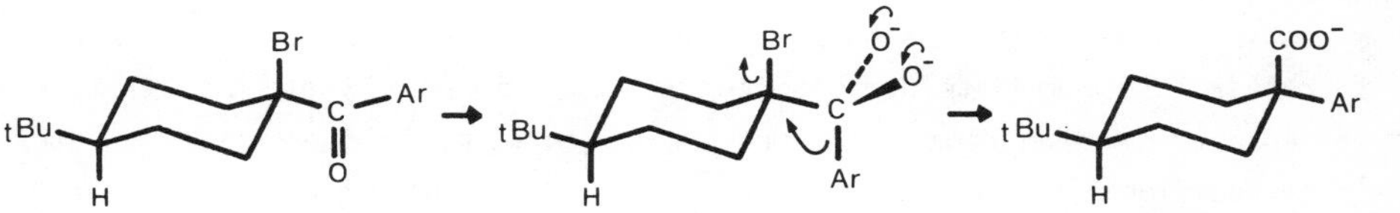

184   **185**   **186**

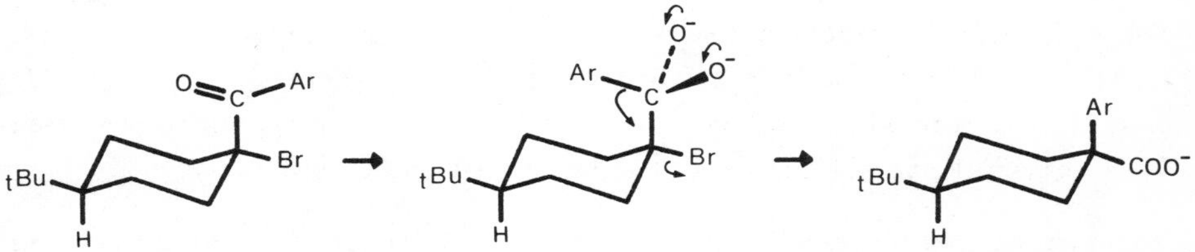

187   **188**   **189**

It has also been reported (66, 67) that trans-2-bromo-3-methylcyclobutanone **190** undergoes a stereocontrolled ring contraction to form trans-2-methyl-cyclopropane derivatives **192** by using either NH_3, aqueous sodium carbonate or water. Inversion of configuration (cf. **191**) is again observed.

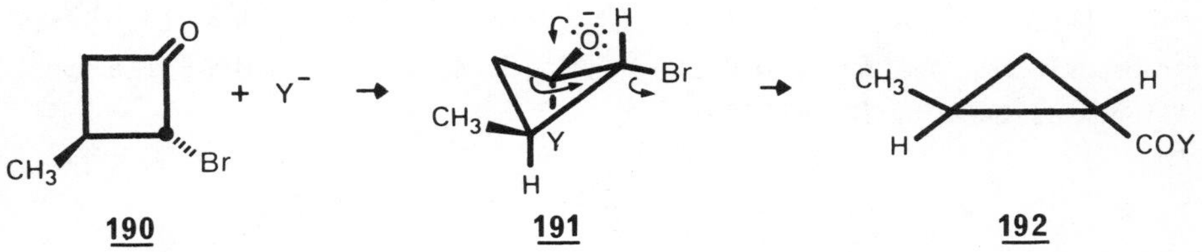

190   **191**   **192**

The double Favorski-like contraction of dibromodiketone **193** (KOH, H_2O; H^+) which yielded p-cubanedicarboxylic acid (**194**) is another example of this class of ring contraction reaction. This reaction was used by Eaton and Cole (68) in their synthesis of cubane.

193   **194**

In these last examples, there are two oxygens each having an electron pair oriented antiperiplanar to the migrating group. This factor should ease the migrating process in a manner similar to that observed in the study of tetrahedral intermediate derived from amides and esters. Note also that the presence of the two negatively charged oxygen atoms in these intermediates is an additional parameter which should facilitate the migration reaction.

Reaction with carbonium ion

The reactions of carbonium ions occur _via_ transition states having precise stereochemistry in which the electron pair of the attacking nucleophile must be colinear with the empty p-orbital of the electron-poor carbon atom. Thus, powerful stereoelectronic effects control these reactions.

For instance, the attack of a nucleophile Y^- from above or below the plane of a carbonium ion having the conformation **195** will, if $R_3 \neq R_4$, give two diastereomers in conformations **196** and **197** respectively. Carbonium ion **195** can also form a double-bond ($\rightarrow$**198**) by the loss of a proton because the C — H bond is properly aligned with the p-orbital of the carbonium ion. For the same reason, it can also undergo a migration of the hydrogen atom with its electron pair to give the rearranged carbonium ion **199**. Similarly, in skeletal rearrangement (47, 48) such as the Wagner-Meerwein or the pinacol transposition, the migrating alkyl group must be that which is properly aligned as shown by **200** and **201**.

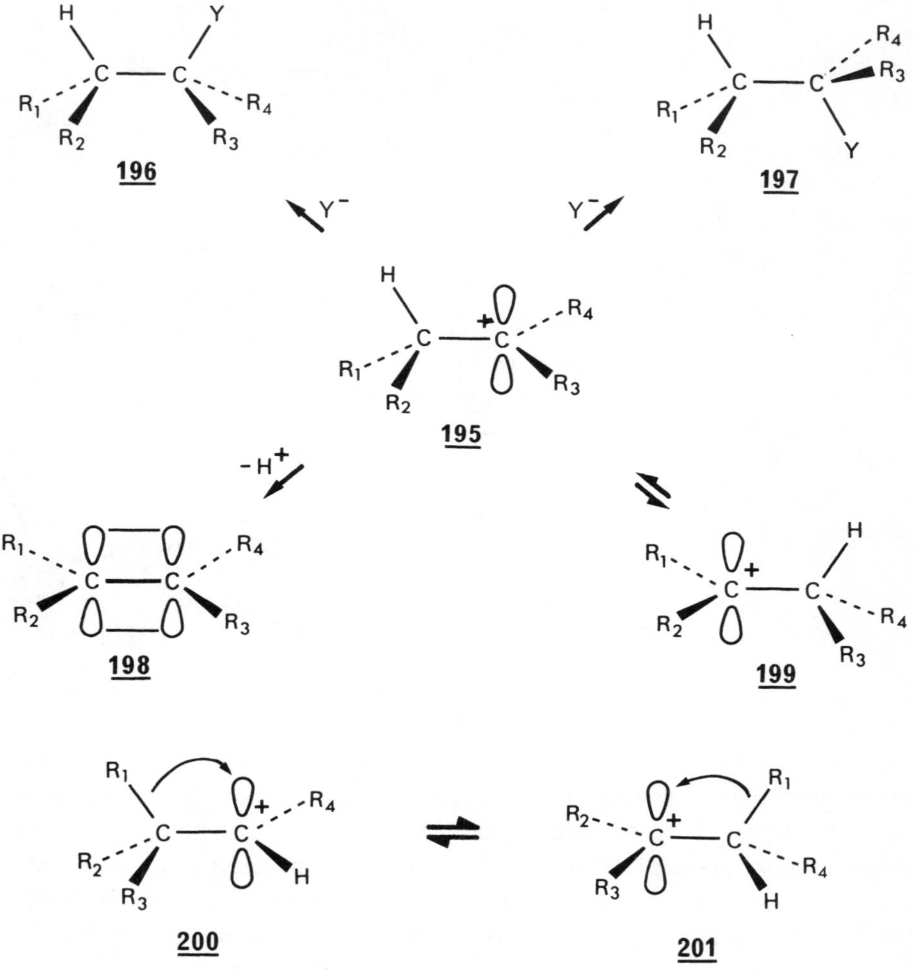

Discrimination can readily be observed between the two possible modes of attack on a carbonium ion (**195** → **196** and **195** → **197**) when the nucleophile is part of the substrate. In such cases, the phenomenon of neighboring group participation is observed (for a review, see ref. 69). For example, solvolysis of the erythro-tosylate isomer **202** in acetic acid gave largely the erythro-acetate isomer **204** via the chiral bridged ion **203**, whereas the threo isomer **205** yielded a racemic mixture of threo products **207A** and **207B** via the achiral intermediate **206** (70).

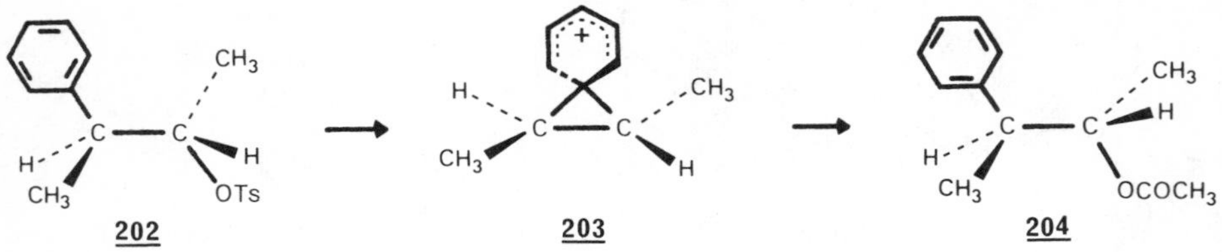

202

203

204

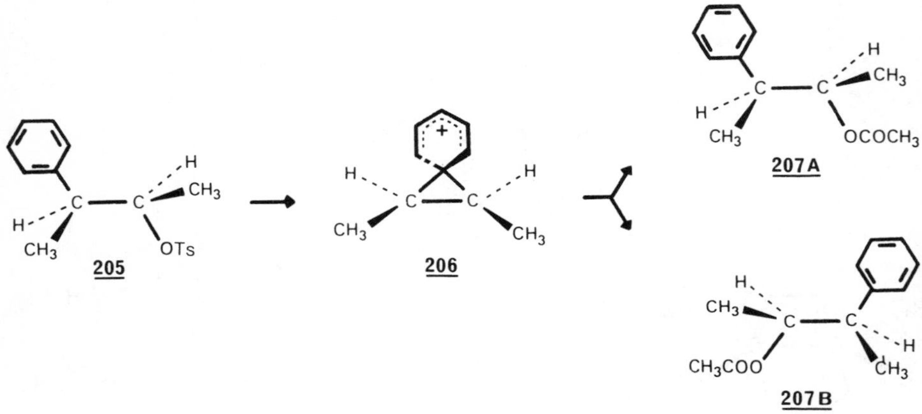

207A

205

206

207B

Solvolysis of cis and trans tosylates 208 and 209 gave the same trans di-acetate 210 (71-73). The cis isomer 208 undergoes the equivalent of a clas-sic SN_2 displacement of the tosylate by acetate ion to give 210. The trans isomer 209 gives first the cyclic acetoxonium intermediate 211 by the par-ticipation of the trans acetate group. Nucleophilic attack of acetate ion on acetoxonium 211 with inversion gives the trans diacetate 210. Interest-ingly, the rate of solvolysis of the trans isomer 209 is approximately 700 times faster than that of the cis isomer 208 indicating the importance of neighboring-group participation (anchimeric assistance).

Thus, by analyzing the different types of nucleophilic substitution reac-tions, the stereoelectronic requirements for reactivity and their resulting effects are clearly illustrated. These effects are also seen in simple and complex rearrangement reactions. For example, the following transposi-tions demonstrate that indeed, the migrating group must be properly aligned as shown by 195 → 199 and 200 → 201. Boron trifluoride catalyzed rearrangement of β-epoxide 212 afforded stereospecifically the trans diketal ketone 213. None of the isomeric more stable cis diketal ketone 214 could be detected. Ireland and Hengartner (74) have interpreted their result as a stereocon-

trolled migration of the α-hydrogen at C_6 in intermediate $\underline{215}$. The stereo-
chemistry of the transition state in $\underline{215} \rightarrow \underline{213}$ corresponds to that in $\underline{195} \rightarrow$
$\underline{199}$ (R_2=O $-\bar{B}F_3$).

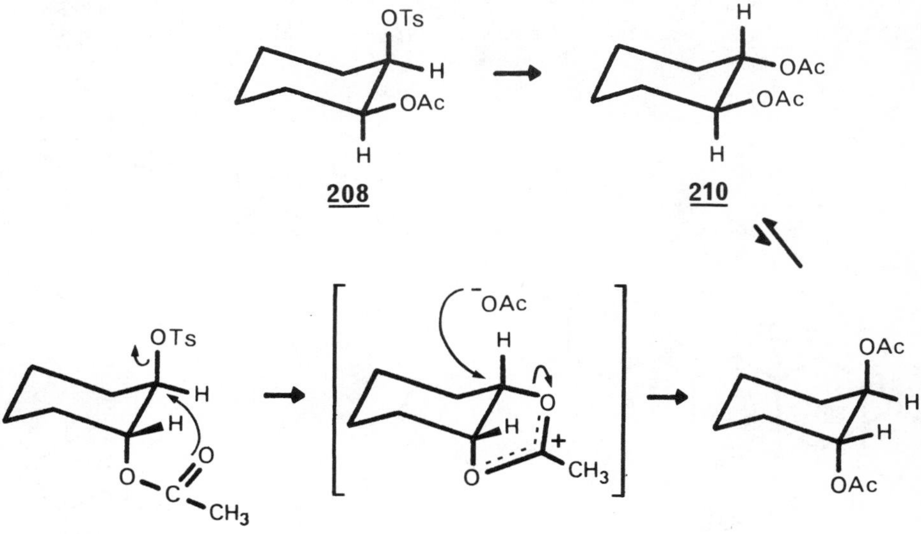

$\underline{\textbf{208}}$ $\underline{\textbf{210}}$

$\underline{\textbf{209}}$ $\underline{\textbf{211}}$

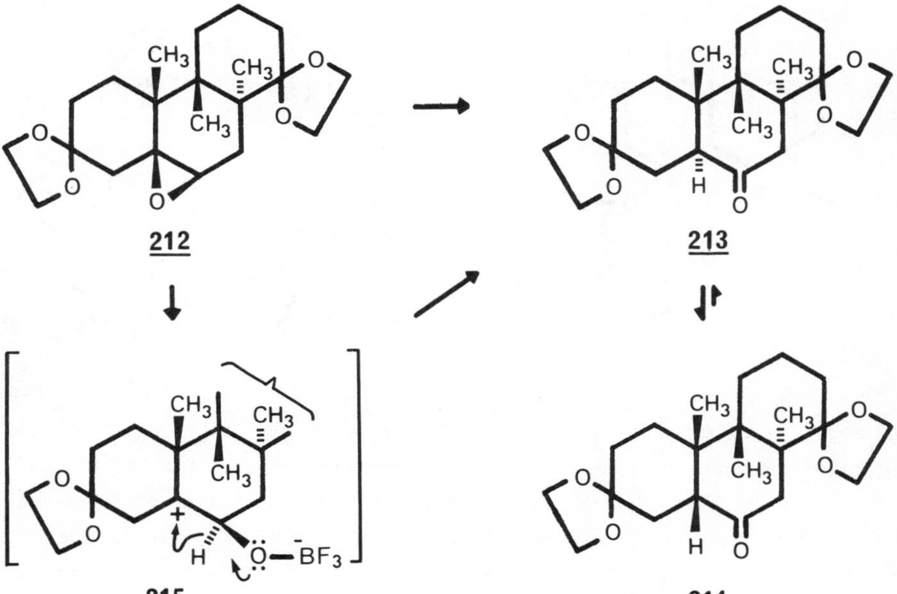

$\underline{\textbf{212}}$ $\underline{\textbf{213}}$

$\underline{\textbf{215}}$ $\underline{\textbf{214}}$

In the case of compound **216**, treatment with boron trifluoride etherate gave a mixture of **217** and **218** (75). The transformation **216 → 217** must have occurred via the intermediate **219** as described above. The formation of **218** is the result of the migration of the methyl group followed by the loss of a proton (**219 → 220 → 218**). These two steps are equivalent to **200 → 201** and **195 → 198**.

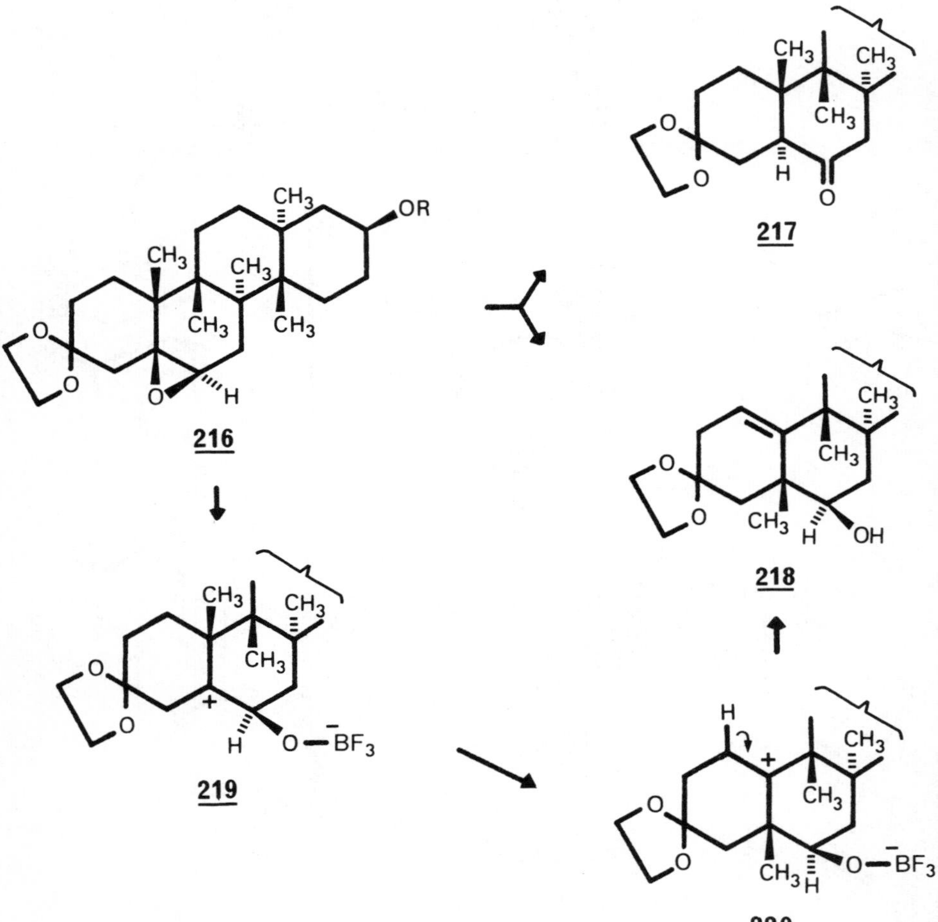

Sometimes, several rearrangements occur consecutively. A spectacular case has been observed (76-79) in the acid-catalyzed transformation of 3-ß-friedelanol (**221**) into 13(18)-oleanene (**222**). In this case, **221** gave presumably the carbonium ion **223** which underwent six stereoelectronically controlled

1,2 shifts to give the carbonium ion **224** which then looses a proton to yield **222**. Also, friedelene, derived from the dehydration of 3-β-friedelanol (**221**) gave compound **222** on acid treatment.

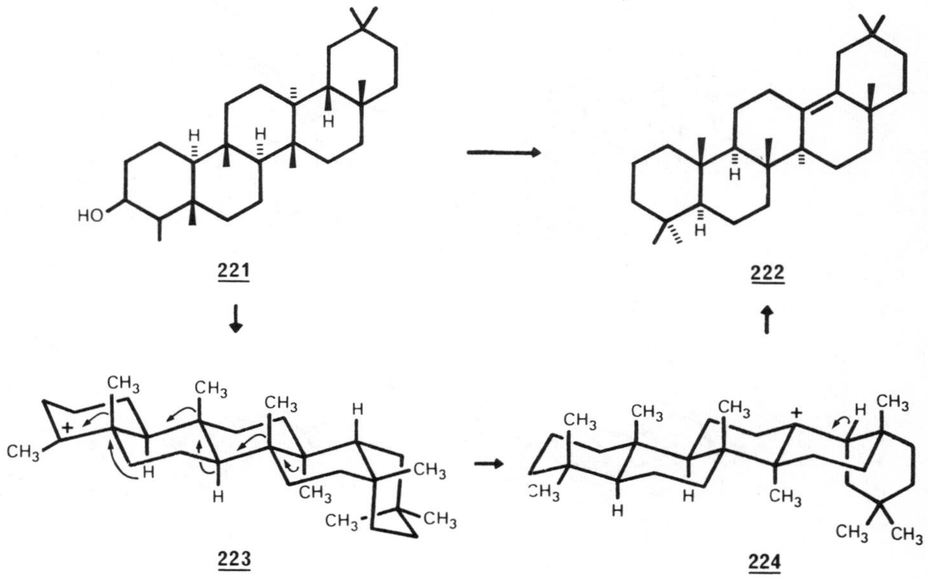

221   **222**

223   **224**

Conclusively, rearrangements of the Wagner-Meerwein type appear to be controlled by powerful stereoelectronic effects.

In the biogenesis of steroids, the enzyme-catalyzed polycyclization of squalene (**225**) produces the tetracyclic substance lanosterol (**226**) which is eventually converted into cholesterol (**227**). Eschenmoser, Stork, and their co-workers (80-82) have proposed that the squalene-lanosterol conversion can be rationalized on the basis of stereoelectronic effects. The stereochemical course of this biological cyclization (83, 84) can be illustrated by considering the transformation of squalene oxide (**228**) (an intermediate in the biosynthesis of cholesterol (83, 84)) into dammaradienol **229**. This transfor-

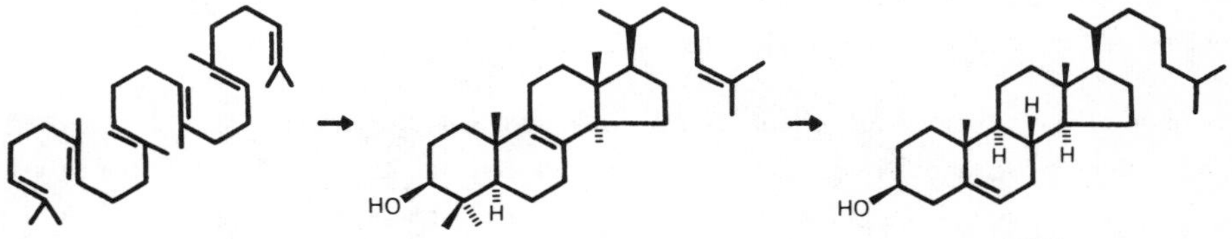

225   **226**   **227**

mation is simpler than the squalene-lanosterol conversion which involves
some rearrangements of carbon atoms.

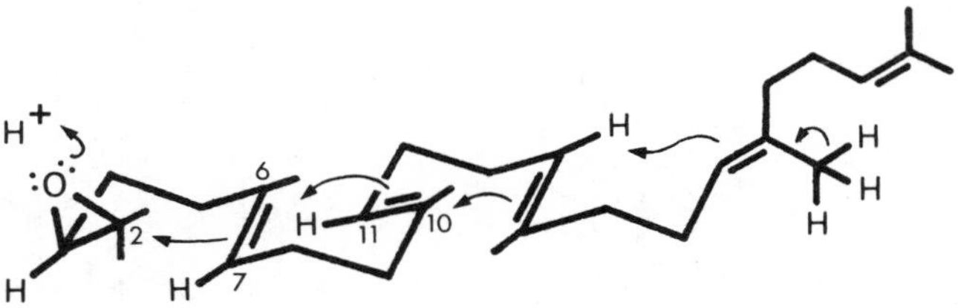

228

↓

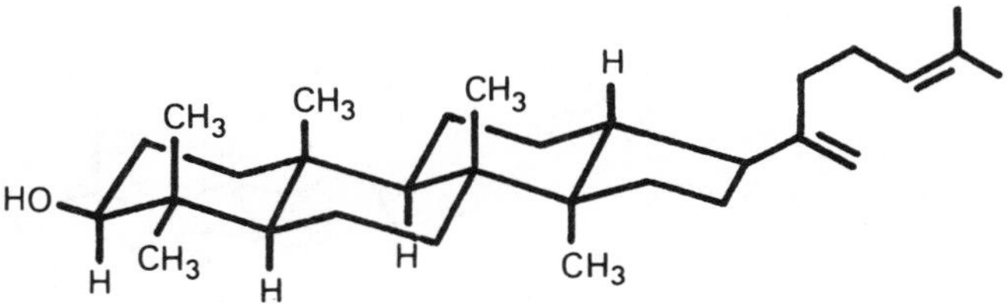

229

The writer has chosen Johnson's summary to describe the basic ideas (85-
87). The process may be regarded as involving trans-antiparallel electrophilic additions to the olefinic bonds in the same stereochemical sense
that bromine adds stereospecifically to alkenes. Thus, protonation of the
oxygen of the squalene-epoxide **228** generates an incipient cationic center
at C-2 which reacts with the π electrons of the 6,7-olefinic bond forming
a sigma bond between C_2 and C_7. The cationic center developing at C-6 ini-
tiates an electrophilic attack on the 10,11-olefinic bond generating the
$C_6 - C_{11}$ sigma bond, etc. Consequently, the all-trans geometry of the olefinic bonds in squalene results in the formation of product **229** having
the four rings trans-fused. This general concept of stereoelectronic control
of polyene cyclization provides a satisfactory rationalization of the stereochemical course of many biological cyclizations.

In a brilliant series of experiments, Johnson and co-workers have discovered
that the treatment of certain polyolefins having trans olefinic bonds in
a 1,5-relationship can produce stereospecific, non enzymic, cationic cyclization products with the all-trans configuration. These transformations

appear to mimic in principle the biogenetic conversion of squalene into polycyclic triterpenoids. This work has been well reviewed (85-87) and only a few representative examples will be described here.

This type of carbon-carbon bond formation occurs through the interaction of double bonds with carbonium ions. It can be viewed as a sort of displacement reaction forming sigma $C - C$ bonds, and it is for this reason that it is described in this Chapter.

Treatment of polyolefinic ketal **230** with stannic chloride in pentane gave a mixture (30% yield) of about equal amounts of the two racemic D-homosteroidal tetracyclic isomers **231** (88). In this cyclization, the first cationic intermediate is not chiral and the two faces of the 5,6-double-bond can react with equal facility with the carbonium ion; as a consequence, the product obtained (**231**) is necessarily racemic. The conversion of the open-chain tetraenic acetal **230** having no chiral centers into a tetracyclic system having seven such centers and producing only two (**231**) out of a possible 64 racemates is a striking tribute to the power of stereoelectronic effects.

230   **231**

Another example is the cyclization of the racemic allylic alcohol **232** at -80°C which furnished the racemic tetracyclic bis-olefin **233** in 70% yield (89, 90). Ozonolysis of **233** gave the bicyclic triketone aldehyde **234** which underwent under acidic conditions a double intramolecular aldol cyclodehydration to produce racemic 16,17-dehydroprogesterone **235**. This represents the first synthesis of a steroid via the now so-called "biomimetic" polyene cyclization method.

Cyclization of the tetraene **236** with trifluoroacetic acid in dichloromethane at temperatures of -50° to -25° gave yields of **237** up to 81% when Ar=α-naphthyl (91). The stereospecific formation of the cis-fused A/B ring junction is a direct consequence of stereoelectronic control. Indeed, in the formation of the cis junction, the newly formed bond (**238 → 239**) is pseudo-axial and is therefore maintained parallel to the π-orbital of the cyclohexenyl double-bond. Such orbital overlap is impossible in forming an A/B trans ring junction as the new formed bond would be pseudoequatorial. The five-membered ring D is produced because the resonance-stabilized benzyl cation **240** is formed in preference to the six-membered ring homobenzylic cation **241**.

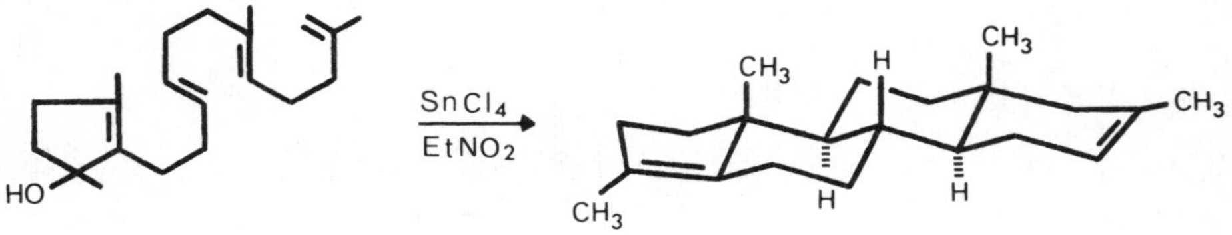

232

$\xrightarrow{\text{SnCl}_4}{\text{EtNO}_2}$

233

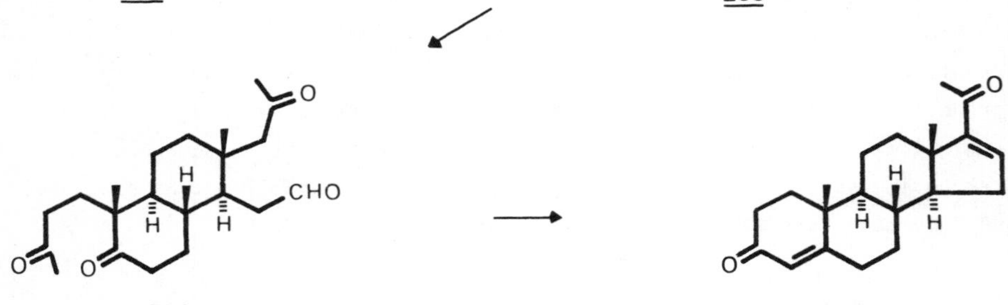

234

235

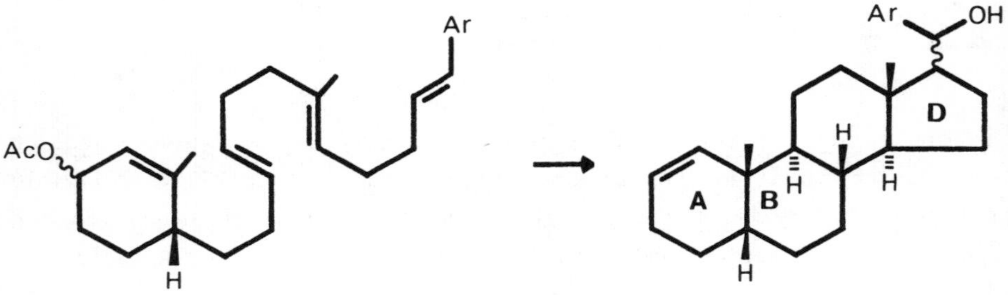

236

237

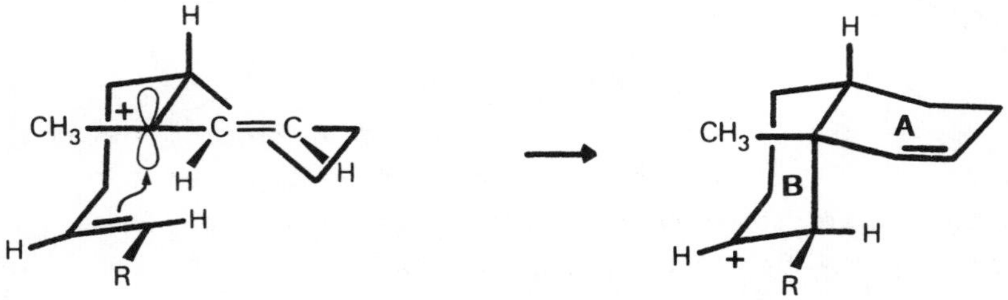

238

239

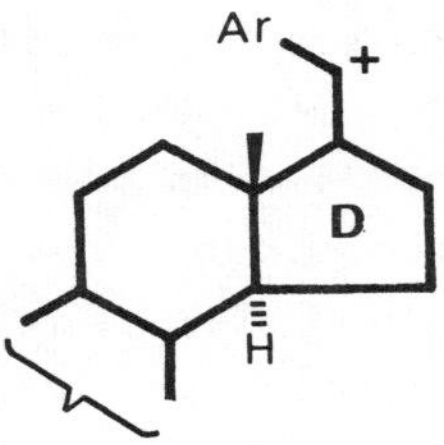

240 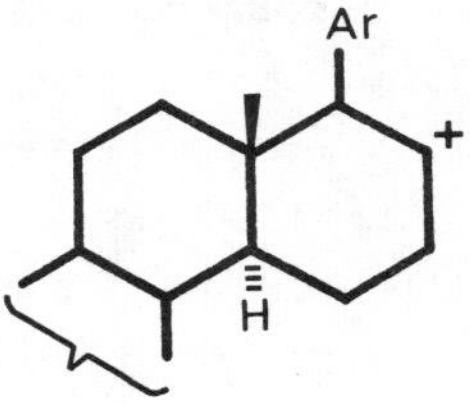   **241**

Johnson and co-workers (92) have recently reported the cyclization of the D-allylic alcohol **242** (optical purity of 91%). When the substrate **242** was treated with trifluoroacetic acid in 1,1-difluoroethane containing ethylene carbonate, a 65% yield of Δ₁-5β-pregnen-20-one (**243**) was obtained with an optical purity of 91%. In a similar fashion, the enantiomer of **242** gave the enantiomer of **243** with an optical purity of 92%. Very little racemization has occurred and the cyclization step is essentially enantiospecific. Again, the A/B ring junction is cis and the process involves essentially total asymmetric synthesis due to the C-6 chiral center in **242**.

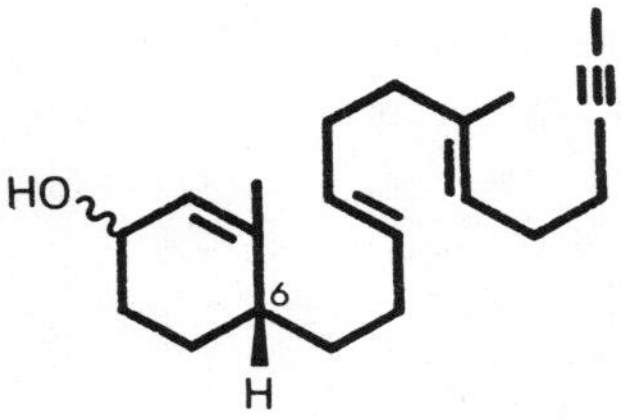

242

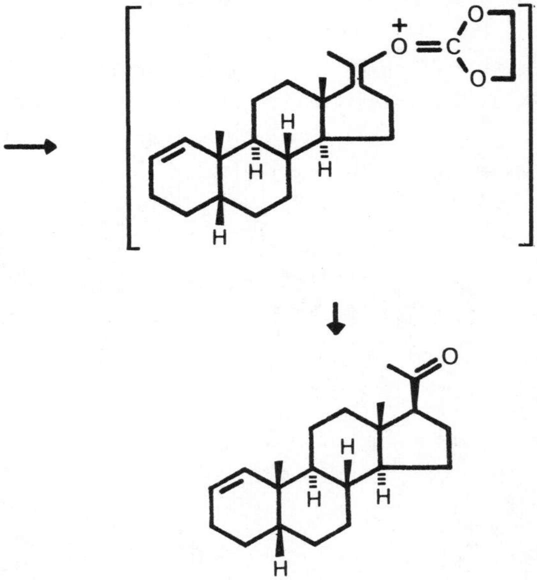

243

In another study (93), cyclization of optically active substrate **244** gave optically active tetracyclic product **245** with the same optical purity. Since, **245** was converted into 11α-hydroxyprogesterone (**246**), this work constitutes a total asymmetric synthesis of that steroid. This remarkable asymmetric control is due to the chiral center at C-10 of **244**: the relative orientation of the hydroxyl group in the transition state of the cyclization process, controlled by stereoelectronic factors, is such that it yields a product (**245**) having an equatorial secondary alcohol.

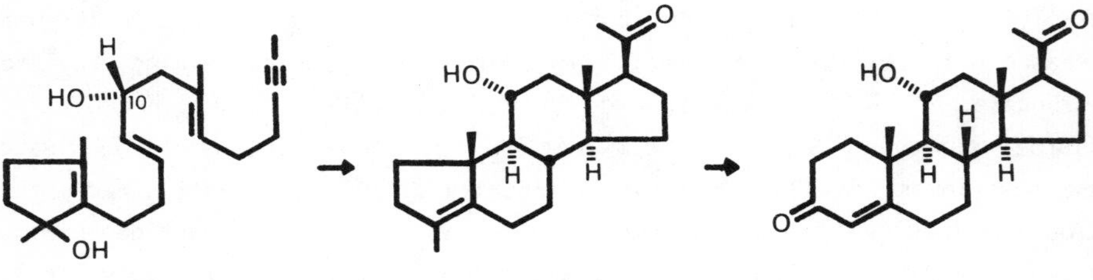

244   **245**   **246**

Van Tamelen has examined extensively the cyclization of monoepoxide-poly-olefins (94). For example, he found (95) that the treatment of racemic epoxide **247** with SnCl₄ in nitromethane provided a major product (35% yield) which was identified as the racemic tetracycle **248**. This represents an over all, close simulation of the squalene → tetracyclic triterpene bioconversion (except for optical activity).

247

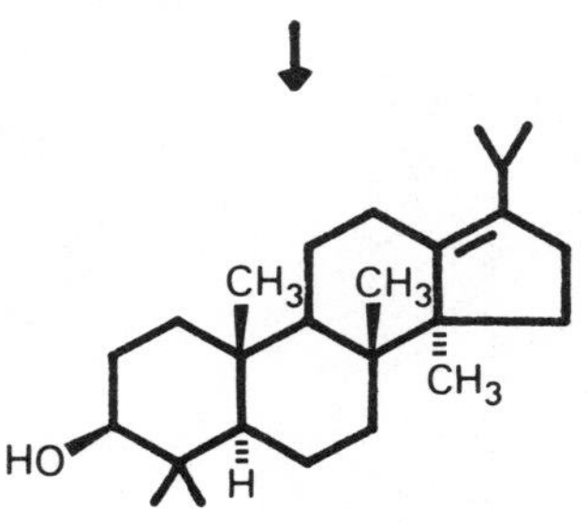

248

In another example, van Tamelen and Loughhead (96) have carried out the cyclization $(BF_3 \cdot Et_2O)$ of ($\pm$)-monocyclic epoxide **249** at low temperature which yielded the tetracyclic alcohol **251** in 25% yield. This can be explained by the formation of cation **250** which is the precise result predicted by the stereoelectronic theory on epoxide opening (producing a <u>cis</u> A/B ring junction) and polyene cyclization (yielding a <u>trans</u> junction for ring B, C, and D). The compound **251** is then formed by the appropriate migrations and proton elimination steps (see arrows in **250**).

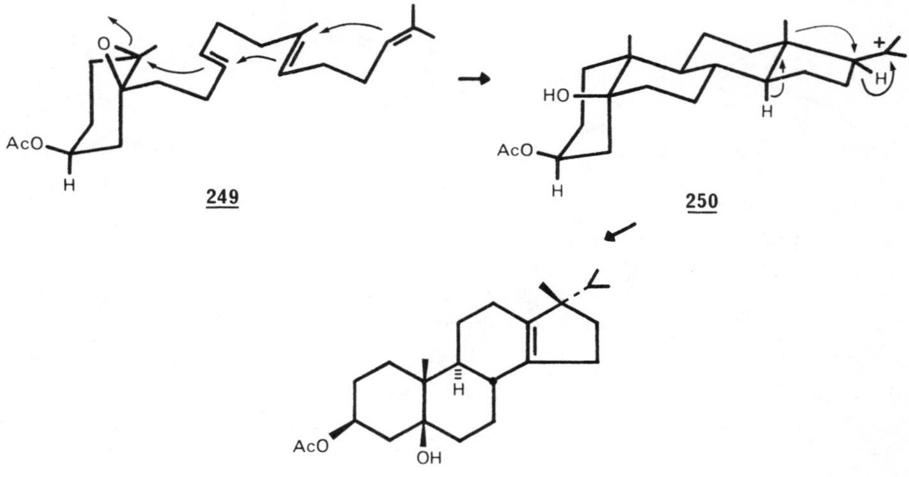

249   **250**

251

An interesting situation arises when a carbonium ion (generated by solvolysis, i.e. **252** → **253**) is next to a bridgehead carbon atom. Very often, one of the C—C bonds of the bridgehead is oriented co-planar with the p-orbital of the cation **253** (or antiperiplanar to the leaving group in the starting product **252**). As a consequence, an electronic delocalization of the electron pair of the C — C bond with the cation (or with the developing cation) is possible. This stereoelectronic effect should therefore strongly influence the formation (and the reactivity) of the resulting carbonium ion.

252   **253**

For example, acetolysis of exo-2-norbornyl brosylate **254** produces exclusively exo-2-norbornyl acetate **255**. The exo-brosylate **254** is more reactive
than the endo-brosylate **256** by a factor of 350 and the acetolysis of optically active exo-brosylate gave completely racemic exo-acetate **255**. Thus,
the carbonium ion produced from exo-**254** is more rapidly (thus more easily)
formed than that from endo-**256**. These results were originally rationalized
in term of a bridged (nonclassical) cation **257** (Winstein approach) (97) or
as the rapidly equilibrating classical carbonium ions **258** and **259** (Brown
approach (98, 99)).

It is presently accepted that there is an electronic delocalization and the
problem is to describe precisely the nature of that delocalization. In one
view, it is described by C—C hyperconjugation involving little geometrical
change (cf. **260**) (100-103) or more recently by 1,3-bridging (cf. **261**) (104-
106). The other view (107) is that in some cases, a partially bridged structure exists in which significant but unequal bonding to a second atom is
taking place (unsymmetrical bridging, cf. **262**) (108, 109) and in others,
strong and equal bonding of the bridging atom to two atoms occurs (symmetrical bridging, cf. **263**) (110, 111).

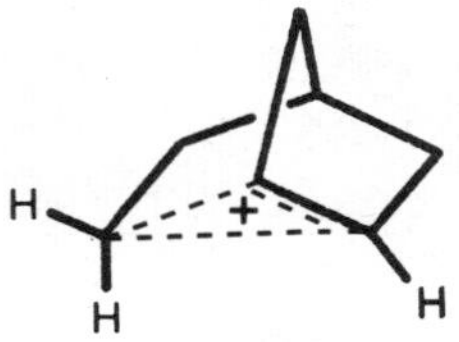

262 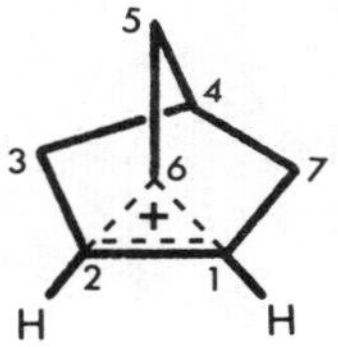   **263**

It is not the intention here to decide between these views but to indicate that either approach involves strong stereoelectronic requirements. As pointed out recently by Grob and co-workers (112), the extent of electronic delocalization must depend on the alignment of the participating sigma bond with the leaving group. The Newman projections **264** and **265** illustrate the bond alignment in **254** and **256** respectively. In projection **264**, the $C_1 - C_6$ bond is perfectly antiperiplanar to the leaving group and it can therefore assist the ejection of that group effectively. In **265**, the $C_1 - C_7$ bond is not antiperiplanar and provides less stabilization in the transition state, so the _endo_ isomer **256** is solvolyzed at a slower rate than the _exo_ isomer **254**.

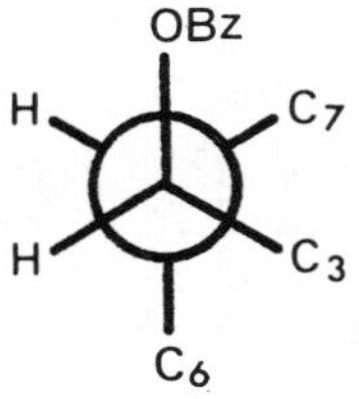

264 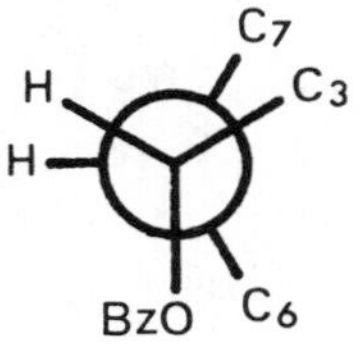   **265**

REFERENCES

(1) Ingold, C.K. "Structure and Mechanism in Organic Chemistry"; 2rd Ed.: Cornell University Press: Ithaca, N.Y., 1954

(2) March, J. "Advanced Organic Chemistry"; 2rd Ed.; McGraw-Hill Inc.: New York, N.Y., 1977; p. 265.

(3) Tenud, L.; Farooq, S.; Seibl, J.; Eschenmoser, A. Helv. Chim Acta 1970, _53_, 2059.

(4) Corey, E.J. J. Am. Chem. Soc. 1954, _76_, 175.

(5) Angyal, S.J. Chem. and Ind. 1954, 1230.

(6) Valls, J.; Toromanoff, E. Bull. Soc. Chim. Fr. 1961, 758.

(7) Barton, D.H.R.; Cookson, R.C. Quart. Rev. 1956, _10_, 44.

204 Reactions at Saturated Carbons

(8) Stork, G.; Cama, L.D.; Coulson, D.R. J. Am. Chem. Soc. 1974, 96, 5268.
(9) Stork, G.; Cohen, J.F. J. Am. Chem. Soc. 1974, 96, 5270.
(10) Lallemand, J.Y.; Onanga, M. Tetrahedron Lett. 1975, 585.
(11) Decesare, J.M.; Corbel, B.; Durst, T.; Blount, J.F. Can. J. Chem. 1981, 59, 1415.
(12) Barnett, W.E.; Sohn, W.H. Tetrahedron Lett. 1972, 1777.
(13) Barnett, W.E.; Sohn, W.H. J. Chem. Soc., Chem. Commun. 1972, 472.
(14) Barnett, W.E.; Needham, L.L. J. Org. Chem. 1975, 40, 2843.
(15) Coulson, C.A.; Moffitt, W.E. Philos. Mag. 1949, 40, 1.
(16) Bernett, W.A. J. Chem. Educ. 1967, 44, 17.
(17) Pauling, L. "Nature of the Chemical Bond"; 3rd Ed.; Cornell University Press: Ithaca, N.Y., 1960; pp. 136-142.
(18) Robinson, E.R.; Gillespie, R.J. J. Chem. Educ. 1980, 57, 329.
(19) Baldwin, J.E. J. Chem. Soc., Chem. Commun. 1976, 738.
(20) Dewar, M.J.S.; Dougherty, R.C. "The PMO Theory of Organic Chemistry"; Plenum Press: New York, N.Y., 1962; p. 256.
(21) Streitwieser, A. Jr. "Solvolytic Displacement Reactions"; McGraw-Hill: New York, N.Y., 1962; pp. 11-29.
(22) King, J.F.; Tsang, G.T.Y. J. Chem. Soc., Chem. Commun. 1979, 1131.
(23) Magid, R.M. Tetrahedron 1980, 36, 1901.
(24) De Wolfe, R.H.; Young, W.G. Chem. Rev. 1956, 56, 753.
(25) De Wolfe, R.H.; Young, W.G. "The Chemistry of Alkenes"; Chap. 10; S. Patai, Ed.; Interscience: New York, 1964.
(26) Young, W.G.; Webb, I.D.; Goering, H.L. J. Am. Chem. Soc. 1951, 73, 1076.
(27) Goering, H.L.; Josephson, R.R. J. Am. Chem. Soc. 1962, 84, 2779.
(28) Goering, H.L.; Vlazny, J.C. J. Am. Chem. Soc. 1979, 101, 1801.
(29) Stork, G.; White, W.N. J. Am. Chem. Soc. 1953, 75, 4119; 1956, 78, 4609.
(30) Stork, G.; Kreft, III, A.F. J. Am. Chem. Soc. 1977, 99, 3850, 8373.
(31) Dobbie, A.A.; Overton, K.H. J. Chem. Soc., Chem. Commun. 1977, 722.
(32) Kirmse, W.; Scheidt, F.; Vater, H.-J. J. Am. Chem. Soc. 1978, 100, 3945.
(33) Magid, R.M.; Fruchey, O.S. J. Am. Chem. Soc. 1977, 99, 8368; 1979, 101, 2107.
(34) Oritani, T.; Overton, K.H. J. Chem. Soc., Chem. Commun. 1978, 454.
(35) Danishefsky, S. Acc. Chem. Res. 1979, 12, 66.
(36) Danishefsky, S.; Dynak, J.; Yamamoto, M. J. Chem. Soc., Chem. Commun. 1973, 81.

(37) Danishefsky, S.; Tsai, M.Y.; Dynak, J. J. Chem. Soc., Chem. Commun. 1975, 7.

(38) Danishefsky, S.; McKee, R.; Singh, R.K. J. Am. Chem. Soc. 1977, 99, 4783.

(39) Corey, E.J.; Fuchs, P.L. J. Am. Chem. Soc. 1972, 94, 4014.

(40) Clark, R.D.; Heathcock, C.H. Tetrahedron Lett. 1975, 529.

(41) Trost, B.M.; Taber, D.F.; Alper, J.B. Tetrahedron Lett. 1976, 3857.

(42) Taber, D.F. J. Am. Chem. Soc. 1977, 99, 3513.

(43) Kondo, K.; Umemoto, T.; Takahatake, Y.; Tunemoto, D. Tetrahedron Lett. 1977, 113.

(44) Isobe, M.; Iio, H.; Kawai, T.; Goto, T. J. Am. Chem. Soc. 1978, 100, 1940.

(45) Danishefsky, S.; Singh, R.K. J. Am. Chem. Soc. 1975, 97, 3239.

(46) Singh, R.K.; Danishefsky, S. J. Org. Chem. 1976, 41, 1668.

(47) Redmore, D.; Gutsche, C.D. "Advances in Alicyclic Chemistry"; Hart, H. and F.J. Karabatsos, Eds; Academic Press: New York, N.Y., 1971.

(48) Gutsche, C.D.; Redmore, D. "Carbocyclic Ring Expansion Reactions"; Academic Press: New York, N.Y., 1968.

(49) MacSweeney, D.F.; Ramage, R. Tetrahedron 1971, 27, 1481.

(50) Posner, G.H.; Loomis, G.L. Tetrahedron Lett. 1978, 4213.

(51) Hedden, P.; MacMillan, J. Tetrahedron Lett. 1971, 4939.

(52) Tokoroyama, T.; Matsuo, K.; Kanazawa, R.; Kotsuki, H.; Kubota, T. Tetrahedron Lett. 1974, 3093.

(53) Kido, F.; Uda, H.; Yoshikoshi, A. Chem. Commun. 1969, 1335.

(54) Bates, R.B.; Büchi, G.; Matsuura, T.; Shaffer, R.R. J. Am. Chem. Soc. 1960, 82, 2327.

(55) Nussim, M.; Mazur, Y. Tetrahedron 1968, 24, 5337.

(56) Paukstelis, J.V.; Macharia, B.W. Tetrahedron 1973, 29, 1955.

(57) Paukstelis, J.V.; Kao, J.-L. Tetrahedron Lett. 1970, 3691.

(58) Yamada, Y.; Kimura, M.; Nagaoka, H.; Ohnishi, K. Tetrahedron Lett. 1977, 2379.

(59) Chérest, M.; Felkin, H.; Sicher, J.; Šipoš, F.; Tichý, M. J. Chem. Soc. 1965, 2513.

(60) Favre, H.; Gravel, D. Can. J. Chem. 1961, 39, 1548.

(61) Favre, H.; Gravel, D. Can. J. Chem. 1963, 41, 1452.

(62) Woodward, R.B.; Gosteli, J.; Ernest, I.; Friary, R.J.; Nestler, G.; Raman, H.; Sitrin, R.; Suter, Ch.; Whitesell, J.K. J. Am. Chem. Soc. 1973, 95, 6853.

(63) Weller, T.; Seebach, D.; Davis, R.E.; Laird, B.B. Helv. Chim. Acta
     1981, 64, 736.

(64) Marshall, J.A.; Kerschen, J.A. Synth. Commun. 1980, 10, 409.

(65) Baudry, D.; Bégué, J.P.; Charpentier-Morize, M. Tetrahedron Lett.
     1970, 2147; Bull. Soc. Chim. Fr. 1971, 1416.

(66) Conia, J.-M.; Salaün, J. Tetrahedron Lett. 1963, 1175.

(67) Conia, J.-M.; Salaün, J. Bull. Soc. Chim. Fr. 1964, 1957.

(68) Eaton, P.E.; Cole, Jr. T.W. J. Am. Chem. Soc. 1964, 86, 962, 3157.

(69) Capon, B.; McManus, S.P. "Neighboring Group Participation"; Vol. 1;
     Plenum Press: New York and London, 1976.

(70) Cram, D.J. J. Am. Chem. Soc. 1949, 71, 3863.

(71) Winstein, S.; Grunwald, E.; Buchles, R.E.; Hanson, C. J. Am. Chem.
     Soc. 1948, 70, 816.

(72) Winstein, S.; Hanson, C.; Grunwald, E. J. Am. Chem. Soc. 1948, 70,
     812.

(73) Winstein, S.; Hess, H.V.; Buckles, R.E. J. Am. Chem. Soc. 1942, 64,
     2796.

(74) Ireland, R.E.; Hengartner, U. J. Am. Chem. Soc. 1972, 94, 3652.

(75) Ireland, R.E.; Giger, R.; Kamata, S. J. Org. Chem. 1977, 42, 1276.

(76) Corey, E.J.; Ursprung, J.J. J. Am. Chem. Soc. 1956, 78, 5041.

(77) Dutler, H.; Jeger, O.; Ruzicka, L. Helv. Chim. Acta 1955, 38, 1268.

(78) Brownlie, G.; Spring, F.S.; Stevenson, R.; Strachan, W.S. J. Chem. Soc.
     1956, 2419.

(79) Coates, R.M. Tetrahedron Lett. 1967, 4143.

(80) Stork, G.; Burgstahler, A.W. J. Am. Chem. Soc. 1955, 77, 5068.

(81) Stadler, P.A.; Eschenmoser, A.; Schinz, H.; Stork, G. Helv. Chim.
     Acta 1957, 40, 2191.

(82) Eschenmoser, A.; Ruzicka, L.; Jeger, O.; Arigoni, D. Helv. Chim. Acta
     1955, 38, 1890.

(83) van Tamelen, E.E.; Willett, J.D.; Clayton, R.B.; Lord, K.E. J. Am.
     Chem. Soc. 1966, 88, 4752.

(84) Corey, E.J.; Russey, W.E.; Ortiz de Montellano, P.R. J. Am. Chem. Soc.
     1966, 88, 4750.

(85) Johnson, W.S. Acc. Chem. Res. 1968, 1, 1; Chimia 1975, 29, 310.

(86) Johnson, W.S. Angew. Chem. Int. Ed. 1976, 15, 9.

(87) Johnson, W.S. Bioorg. Chem. 1976, 5, 51.

(88) Johnson, W.S.; Wiedhaup, K.; Brady, S.F.; Olson, G.L. J. Am. Chem. Soc.
     1968, 90, 5277.

(89) Johnson, W.S.; Semmelhack, M.F.; Sultanbawa, M.U.S.; Dolak, L.A. J. Am. Chem. Soc. 1968, 90, 2994.

(90) Johnson, W.S.; Li, T.-T.; Harbert, C.A.; Bartlett, W.R.; Herrin, T.H.; Staskun, B.; Rich, D.H. J. Am. Chem. Soc. 1970, 92, 4461.

(91) Johnson, W.S.; Dominguez, J.; Garst, M.E. Unpublished observations.

(92) Johnson, W.S.; McCarry, B.E.; Markezich, R.L.; Boots, S.G. J. Am. Chem. Soc. 1980, 102, 352.

(93) Johnson, W.S.; Brinkmeyer, R.S.; Kapoor, V.M.; Yarnell, T.M. J. Am. Chem. Soc. 1977, 99, 8341.

(94) van Tamelen, E.E. Acc. Chem. Res. 1968, 1, 111; 1975, 8, 152.

(95) van Tamelen, E.E.; Milne, G.M.; Suffness, M.I.; Rudler Chauvin, M.C.; Anderson, R.J.; Achini, R.S. J. Am. Chem. Soc. 1970, 92, 7202.

(96) van Tamelen, E.E.; Loughhead, D.G. J. Am. Chem. Soc. 1980, 102, 869.

(97) Winstein, S.; Trifan, D.S. J. Am. Chem. Soc. 1949, 71, 2953; 1952, 74, 1147, 1154.

(98) Brown, H.C. Acc.Chem. Res. 1973, 6, 377 and references quoted therein.

(99) Brown, H.C. (with comments by P.v.R. Schleyer) "The Nonclassical Ion Problem"; Plenum Press: New York, N.Y., 1977.

(100) Grob, C.A.; Schiess, P.W. Angew. Chem. Int. Ed. 1967, 6, 1.

(101) Grob, C.A. Angew. Chem. Int. Ed. 1976, 15, 569.

(102) Jensen, F.R.; Smart, B.E. J. Am. Chem. Soc. 1969, 91, 5688.

(103) Traylor, T.G.; Hanstein, W.; Berwin, H.J.; Clinton, N.A.; Brown, R.S. J. Am. Chem. Soc. 1971, 93, 5715.

(104) Grob, C.A.; Gunther, B.; Hanreich, R.; Waldner, A. Tetrahedron Lett. 1981, 22, 835.

(105) Grob, C.A.; Hanreich, R.; Waldner, A. Tetrahedron Lett. 1981, 22, 3231.

(106) Grob, C.A.; Waldner, A. Tetrahedron Lett. 1981, 22, 3235.

(107) Schleyer, P.v.R.; Lenoir, D.; Mison, P.; Liang, G.; Prakash, G.K.Surya; Olah, G.A. J. Am. Chem. Soc. 1980, 102, 683.

(108) Radom, L.; Pople, J.A.; Schleyer, P.v.R. J. Am. Chem. Soc. 1972, 94, 5935.

(109) Olah, G.A.; Prakash, G.K.Surya; Liang, G. J. Am. Chem. Soc. 1977, 99, 5683.

(110) Olah, G.A.; DeMember, J.R.; Lui, C.Y.; Porter, R.D. J. Am. Chem. Soc. 1971, 93, 1442.

(111) Nordlander, J.E.; Neff, J.R.; Moore, W.B.; Apeloig, Y.; Arad, D.; Godleski, S.A.; Schleyer, P.v.R. Tetrahedron Lett. 1981, 22, 4921.

(112) Fisher, W.; Grob, C.A.; von Sprecher, G.; Waldner, A. <u>Tetrahedron Lett.</u> 1979, 1905.

CHAPTER 6

REACTIONS ON SP$_2$ TYPE
UNSATURATED SYSTEMS

NUCLEOPHILIC ADDITION

Ketones and aldehydes

We have already discussed in Chapter 2 that nucleophilic addition to a carbonyl group is controlled by stereoelectronic effects. Both X-ray data and theoretical calculations indicate a clearly defined path (cf. p. 32) for the attack of a nucleophile on a carbonyl group. Baldwin (1) has also used a vector analysis approach to predict the stereochemistry of the addition products.

The reactions of hydrides and Grignard reagents with simple open-chain aldehydes and ketones 1 (L, M, S, and R being groups containing carbon and hydrogen only) are known to lead predominantly to the diastereoisomers 2 as predicted by Cram's rules (2-4). A modification of the interpretation of these experimental results was published by Chérest, Felkin, and Prudent (5).

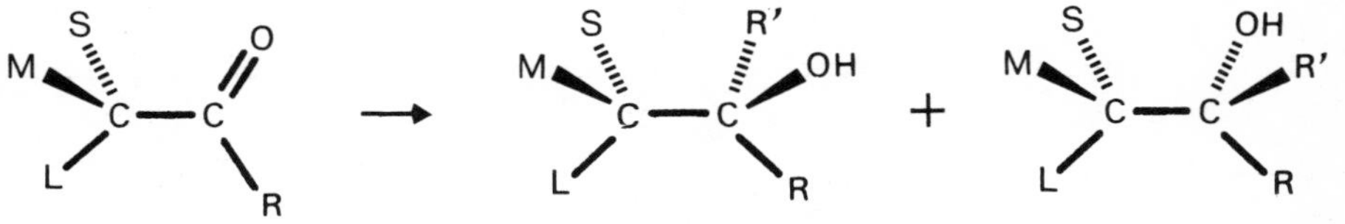

$$ \underline{1} \qquad\qquad \underline{2} \qquad\qquad \underline{3} $$

Both interpretations are based on the fact that the nucleophile must attack the carbonyl group perpendicularly (a consequence of stereoelectronic control), and the preference for the formation of diastereoisomer **2** is due to steric effects.

In the Felkin model (5), the important steric interactions involve R'^- and R rather than the carbonyl oxygen as assumed by Cram (2-4) and also Karabatsos (6). On this basis, the preferred mode of attack is **4** → **5** yielding **6**, the least strained of six possible staggered conformations (three staggered conformations are possible for each of the diastereoisomers **2** and **3**; **6** is equivalent to the least strained conformation of **2**). Recently, Anh and Eisenstein (7) have concluded from their ab initio calculations that the transition **4** → **5** → **6** does indeed correspond to the minimum energy transition state.

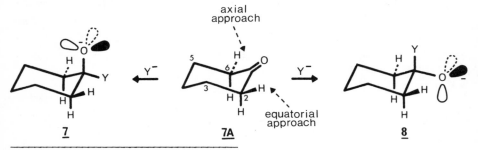

When the carbonyl group is exo to a ring as in cyclohexanone, both the equatorial and the axial approach can lead to a chair intermediate (cf. **7** and **8**) with stereoelectronic control because, in each case, the oxygen atom has an electron pair antiperiplanar to the C—Y bond. So, in this case, one process is favored over the other solely on the basis of steric effects and torsional strain due to the presence of the ring (8).*

*It has however been suggested by Cieplak (9) that the stereochemistry of nucleophilic addition to cyclohexanone is determined by a combination of steric and stereoelectronic effects. According to this interesting model, steric hindrance favors the equatorial approach while electron donation favors the axial approach. The stereoelectronic effect favors the axial approach because the axial C—H bonds next to the carbonyl group (C_2—H_a and C_6—H_a) are better electron donors than the C_2—C_3 and C_5—C_6 α bonds (cf. **7A** → **7** and **7A** → **8**).

The situation is however different when the carbonyl group is part of a ring as in oxonium ion **9**. The attack from the bottom face of **9** leads to a chair intermediate **10** while that from the top face leads to a twist-boat **11**. The attack from the bottom face of **9** is therefore strongly preferred for stereoelectronic and steric reasons.

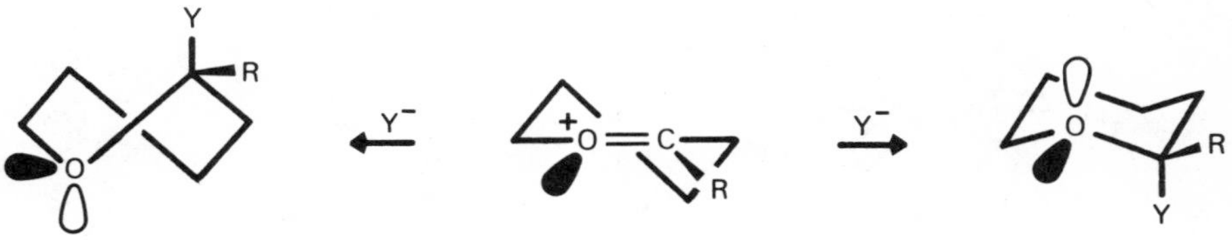

| **11** | **9** | **10** |

The oxygen atom in **9** can be replaced by sulfur, and the same prediction is maintained. We have already seen that this is indeed the case for cyclic oxonium and sulfonium ions (Chapter 2). It is essentially for the same reason that lactones, thionolactones (Chapter 3) as well as lactams and their derivatives (Chapter 4) behave in exactly the same manner. Indeed, an axial attack on **12** (X=O, S, or NR) leads to an intermediate having a chair conformation (**12** → **13**) while an equatorial attack necessarily leads to the less favorable boat conformation (**12** → **14**).

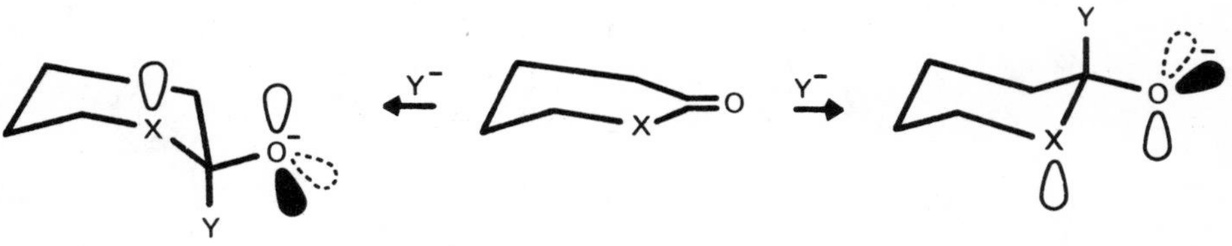

| **14** | **12** | **13** |

Iminium salts

The reaction of nucleophiles with the conformationally rigid piperidinium ion **15**, like that with cyclic oxonium ions, can also be controlled by stereoelectronic effects. On that basis, the addition of a nucleophile on the upper face of **15** must lead to the boat-like intermediate **16** whereas that from the lower face must lead to the chair-like intermediate **17**. The transition state leading to **16** must be less favorable than that leading to **17** and product **17** should therefore be favored.

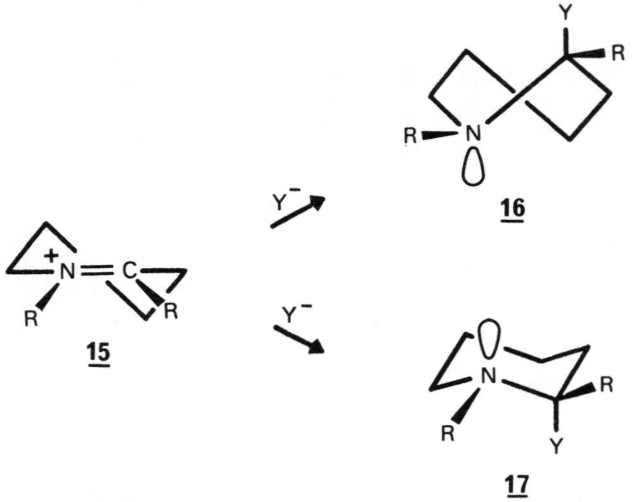

In the total synthesis of reserpine, Woodward and collaborators (10) have reported that the quaternary iminium salt 18 was reduced with aqueous methanolic sodium borohydride to methyl O-acetyl isoreserpate (19). This is the anticipated product whether the stereochemical sense of the reaction is subject to steric or thermodynamic control as pointed out by Woodward. It is also the expected one on the basis of stereoelectronic control.

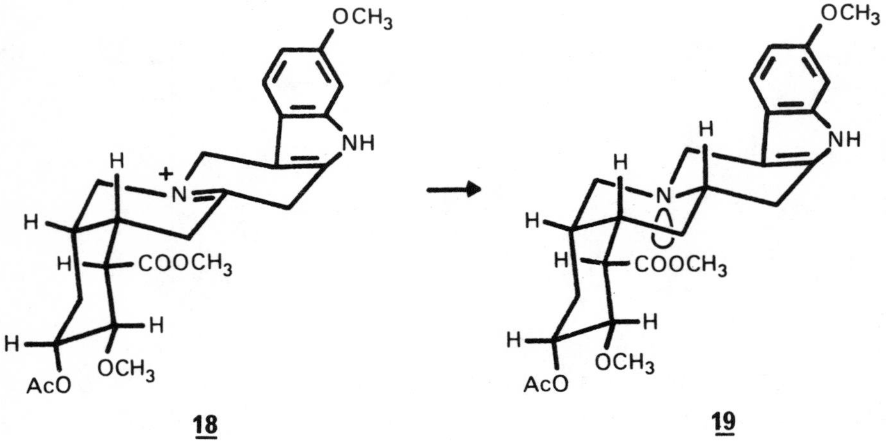

Bohlmann and co-workers (11, 12) have shown that the borohydride reduction of iminium salts of type 20 gave product 21 in preference to its isomer 22. These results indicate that there is stereoelectronic control in these reductions as discussed by Toromanoff (13).

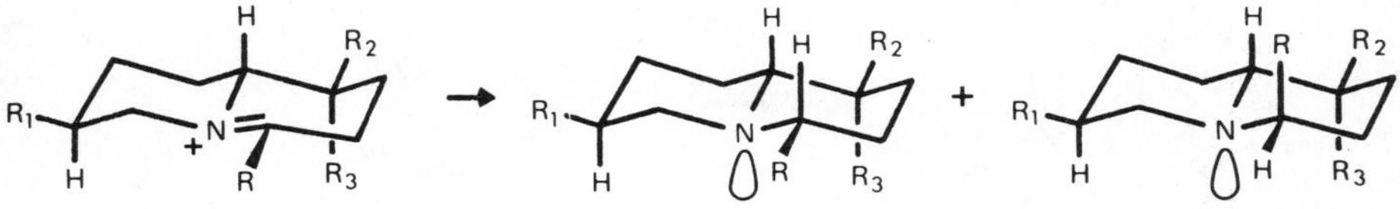

20   **21**   **22**

This is confirmed by Wenkert and collaborators (14) who showed that the borohydride reduction of compound **23** gave **24**. Similarly, Stork and Guthikonda
(15) observed that the reduction of **25** (produced in situ) gave (±)-yohimbine
(**26**).

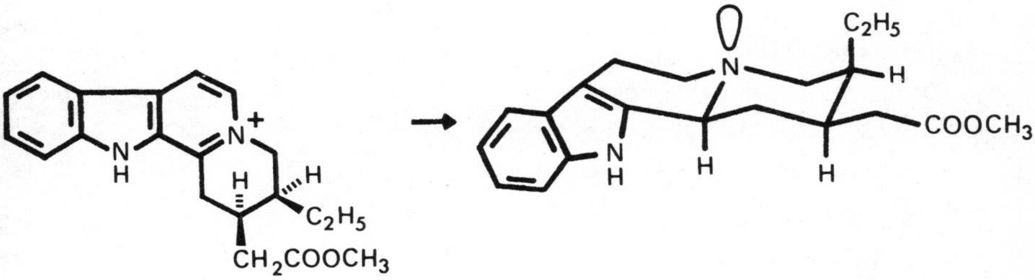

23   **24**

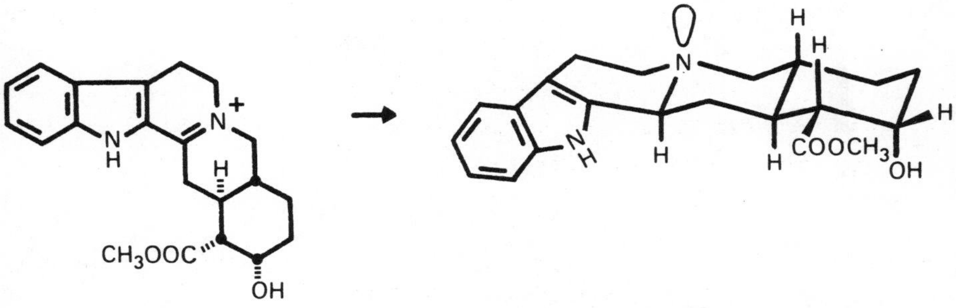

25   **26**

Stevens and Lee (16) have recently completed a stereospecific synthesis
of (±)-monomorine (**28**). In the last step of this synthesis, the piperidinium
ion **27** was reduced with sodium cyanoborohydride to give only (±)-monomorine
(**28**).

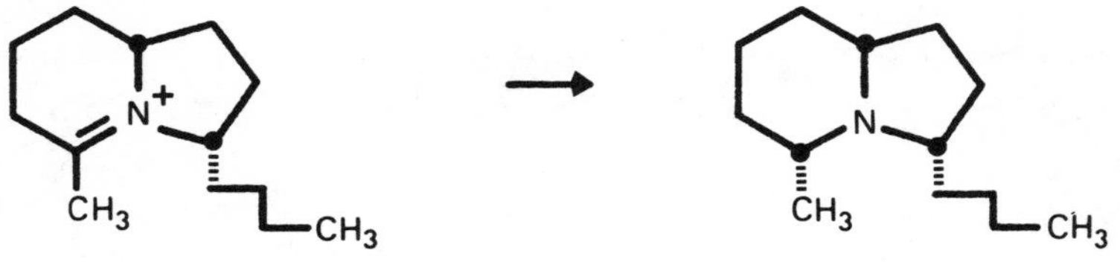

27   28

There are four possible transition states in the reduction of 27 wherein
maximum orbital overlap can be maintained with respect to the attacking
hydride reagent and the developing electron pair on nitrogen. Two of these
(cf. dotted arrow in 29 and 30) require boat-like transition states in order
to satisfy the stereoelectronic requirements and are unfavorable kinetically. Of the two possible chair-like transition states (cf. solid arrow
in 29 and 30) the latter suffers from a strong steric interaction between
the nucleophile and the C-8 pseudo-axial hydrogen. The process 30 → 32 is
thus disfavored by comparison with the process 29 → 31.

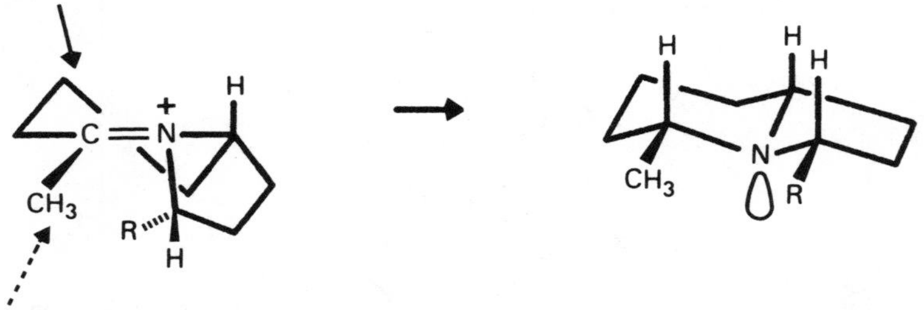

29   31

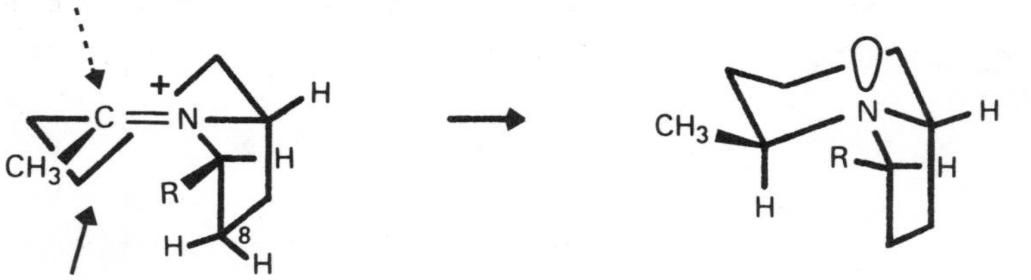

30   32

In another interesting study which led to a stereospecific synthesis of
one of the stereoisomer of gephyrotoxin-223, Stevens and Lee (17) found that
the reaction of piperidinium ion 33 with cyanide led to axial cyanoamine
34 in 96% yield. The cyanoamine 34 also served as a latent form of salt 33.

The reaction of **34** with excess Grignard reagent (CH$_3$MgBr or CH$_3$CH$_2$CH$_2$MgBr) led stereospecifically to the axially oriented isomer **35** (R=CH$_3$ or CH$_3$CH$_2$CH$_2$) in high yield. The high degree of stereoselectivity observed is again in complete agreement with the preceding argument (**29** → **31**) based on the principle of stereoelectronic control.

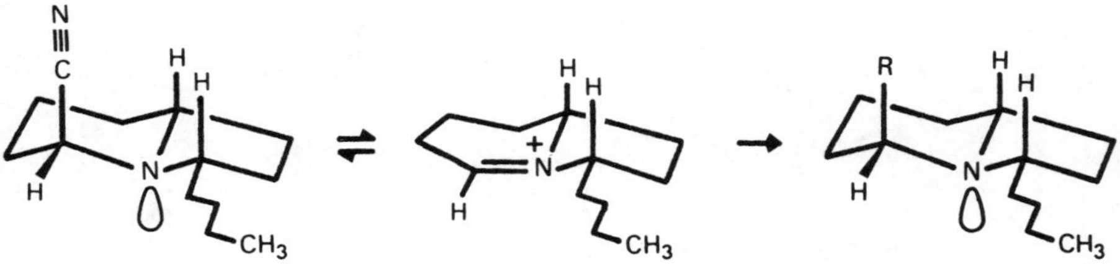

The reaction of **34** with excess Grignard reagent (CH$_3$MgBr or CH$_3$CH$_2$CH$_2$MgBr)

34   **33**   **35**

Reduction of vinylogous carbamate **36** with sodium cyanoborohydride in acidic methanol gave exclusively the equatorial aminoester **38**. Eschenmoser and co-workers (18) have explained this result by invoking a stereoelectronically controlled antiperiplanar addition of hydride ion on the iminium ion **37**.

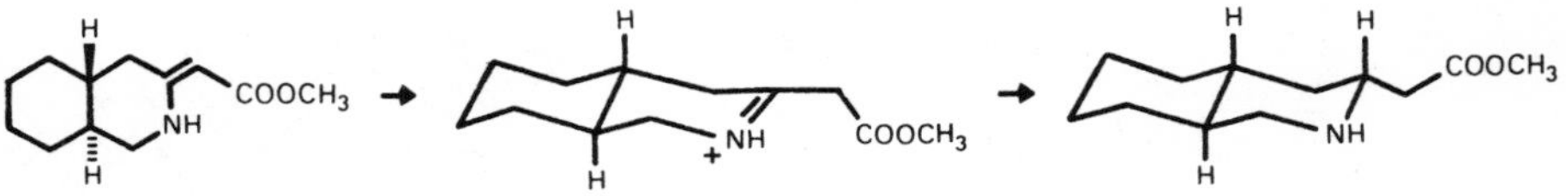

36   **37**   **38**

Petrzilka, Felix, and Eschenmoser (19) have shown that the reaction of cyanide ion on the ion **39** gave mainly the addition product **40**. Similarly, Riediker and Graf (20) observed that the addition of cyanide on the ion **41** gave preferentially **42** which is the result of a stereoelectronically controlled reaction of the most hindered face of the iminium ion. Stereospecific addition of cyanide to enamine **43** via the iminium salt **44** to yield compound **45** has also been reported (21).

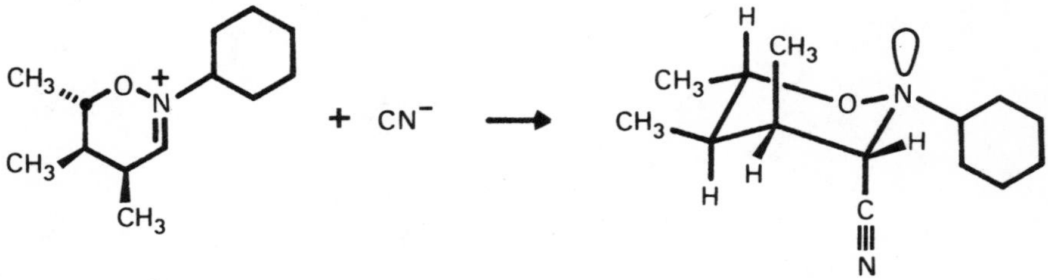

39   + CN$^-$ ⟶   **40**

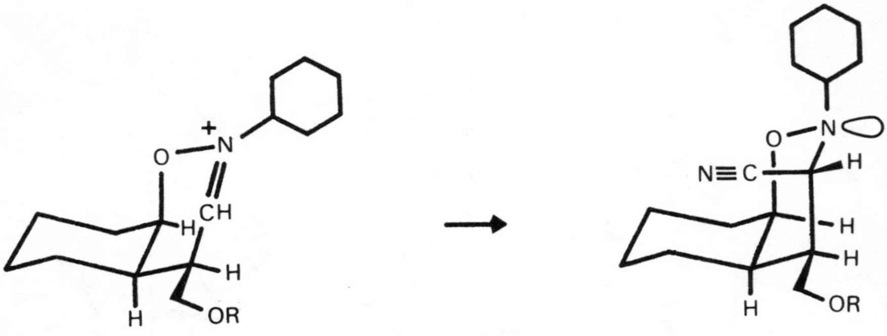

41   **42**

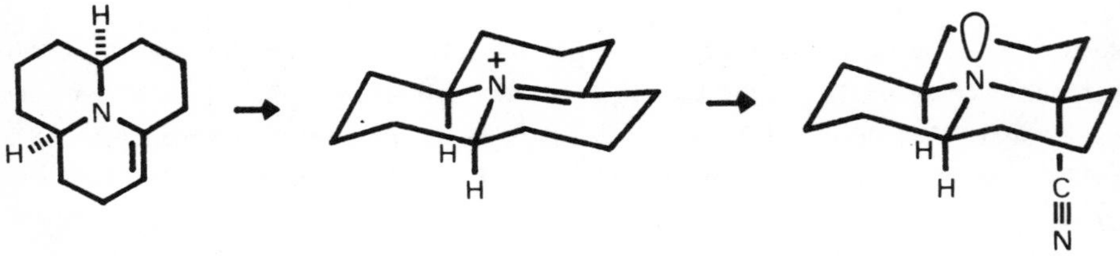

43   **44**   **45**

Stork and Guthikonda (15) have shown that the cyclization of iminium ion **46** (produced in situ) gave (±)-β-yohimbine (**47**) and Wenkert and co-workers (22) have further observed that the acid-catalyzed cyclization of enamines of type **48** yielded exclusively product **50** via the cyclization of iminium ion **49**. In the last two examples, formation of the C—C bond is the result of a trans-addition on the iminium double-bond in agreement with the principle of stereoelectronic control.

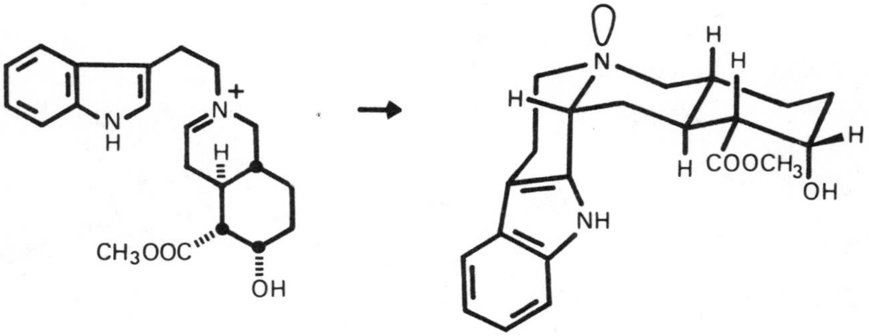

46   **47**

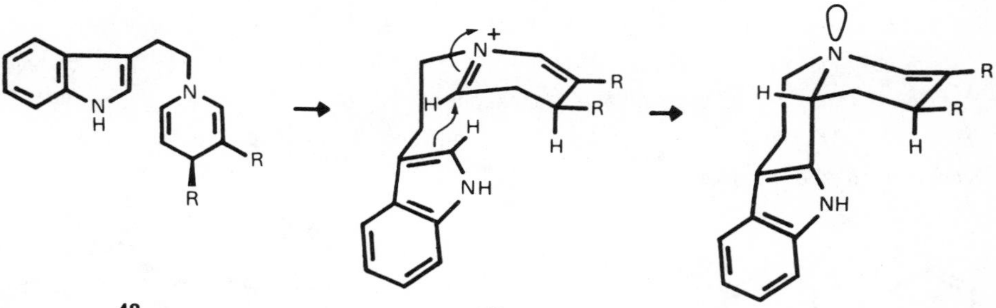

48   **49**   **50**

Dean and Rapoport (23) have reported the stereospecific cyclization of iminium ion **51** to give the cis isomer **52**. Similarly, cyclization of iminium
ion **53** gave **54** exclusively. Cyclization must therefore occur as shown by
55.

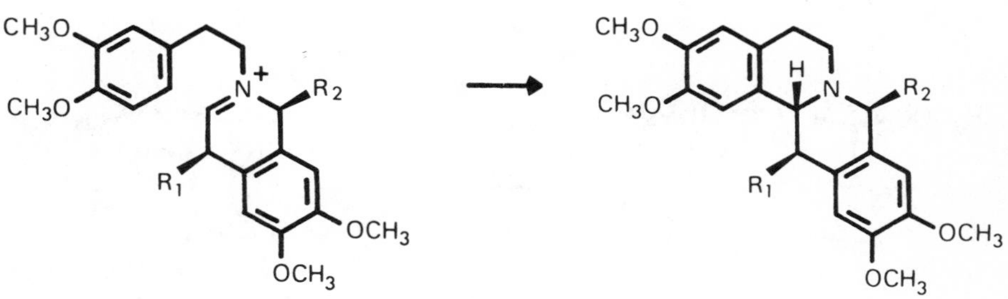

51   R$_1$=CH$_3$, R$_2$=H   **52**
53   R$_1$=H,   R$_2$=CH$_3$ **54**

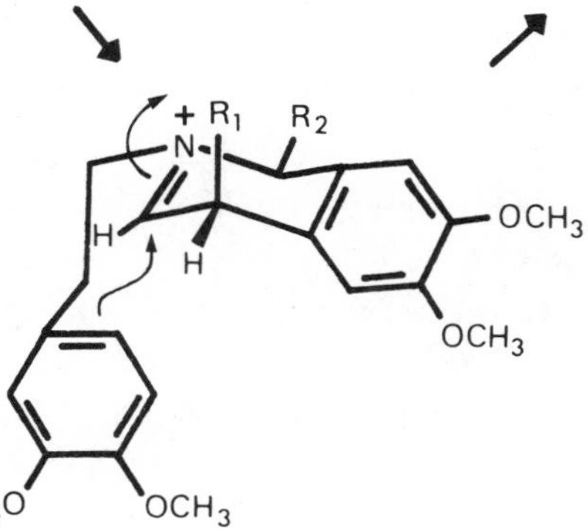

55

Cook and collaborators (24) have shown that the Pictet-Spengler condensation of iminium salts of structure **56** (R=alkyl or aryl group) gave only the _trans_ isomer **57** and in high yield. This result can be explained if the stereochemistry of the transition state corresponds to **58** → **59** where the principle of stereoelectronic control is respected and where the steric effects are reduced to the minimum.

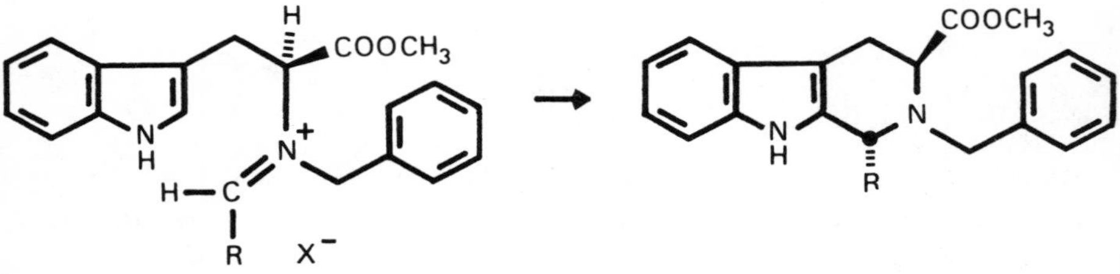

56   **57**

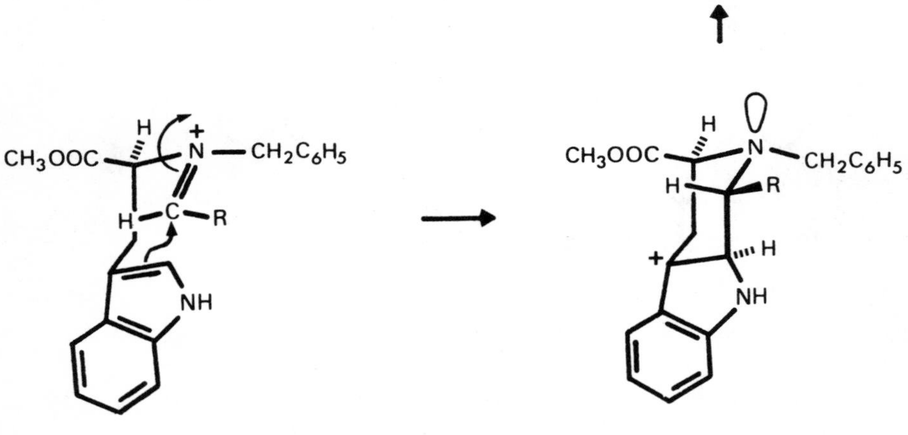

58   **59**

Overman and Fukuya (25) have also observed the unexpected preference of organolithium and Grignard reagents which add to iminium ion **60** from the more sterically congested α face yielding **61**. The authors explained their results by a strong stereoelectronic control during the addition of the alkyl group (cf. **62** → **61A** → **61**). The other conformation **63** was eliminated because of a strong A^{1,2} steric interaction (26) between the N-alkyl group and ring A. Similarly, lithium aluminium hydride reduction of **64** gave **65** as the major isomer (27).

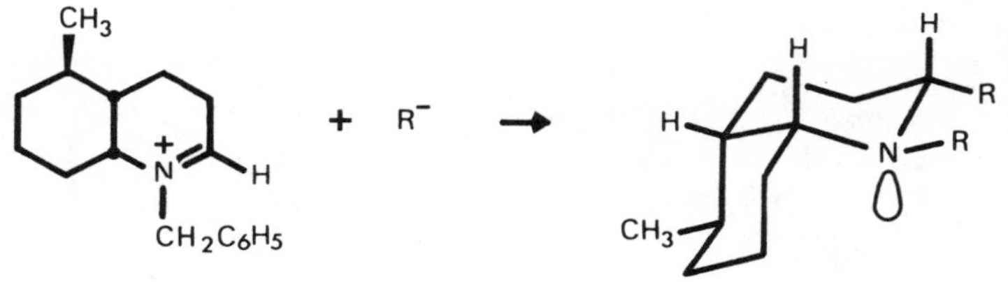

Stevens and Lee (28) have reported an elegant synthesis of coccinelline (**69**). Treatment of **66** at pH=1 gave intermediate **67** which was then treated with dimethyl acetone dicarboxylate at pH=5.5 to give a single tricyclic isomer, the ketodiester **68**, in 75% yield. Compound **68** was then converted into coccinelline (**69**). This result shows that the Robinson-Schopf reaction (29, 30) can take place with a remarkable control of stereochemistry.

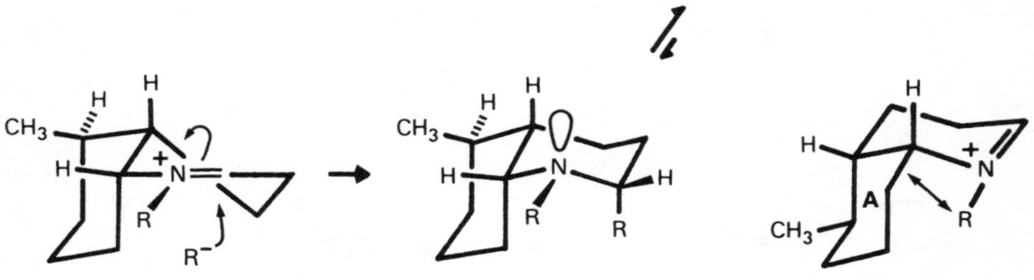

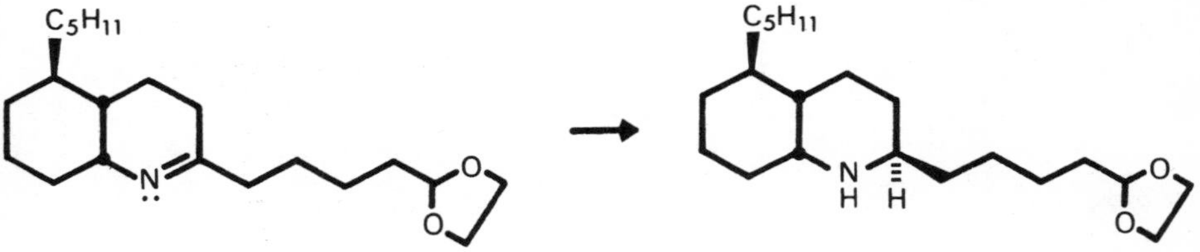

In principle, there are four possible transition states for the attack of a nucleophile on a cyclic iminium ion such as **67**. Two of these, **71** and **72**, are boat-like transition states and are kinetically disfavored. Of the two possible chair-like transition states, **70** and **73**, the latter suffers from an unfavorable 1,3-diaxial interaction between the R group and the incoming nucleophile. Transition state **70** is therefore favored.

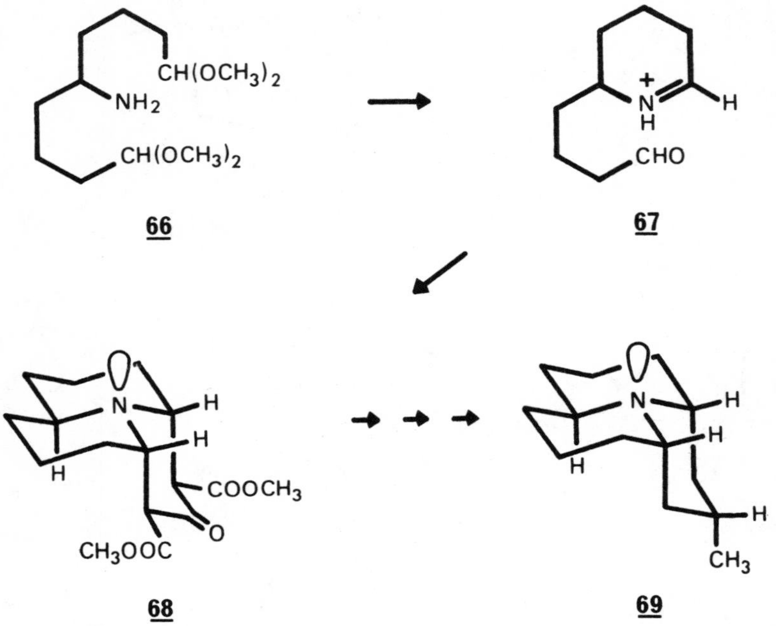
CH(OCH3)2
NH2
CH(OCH3)2
66
N
H
H
CHO
67
N
H
H
H
COOCH3
O
CH3OOC
68
N
H
H
H
H
CH3
69

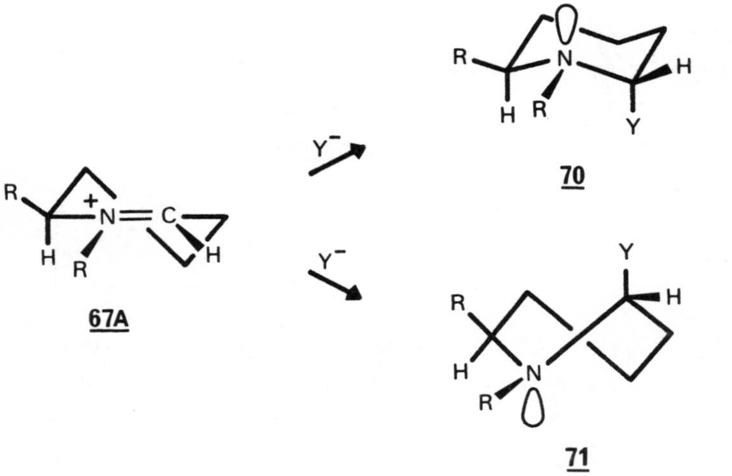
R
N=C
H
R
H
67A
Y−
Y−
R
N
H
H
R
Y
70
Y
H
R
H
N
R
71

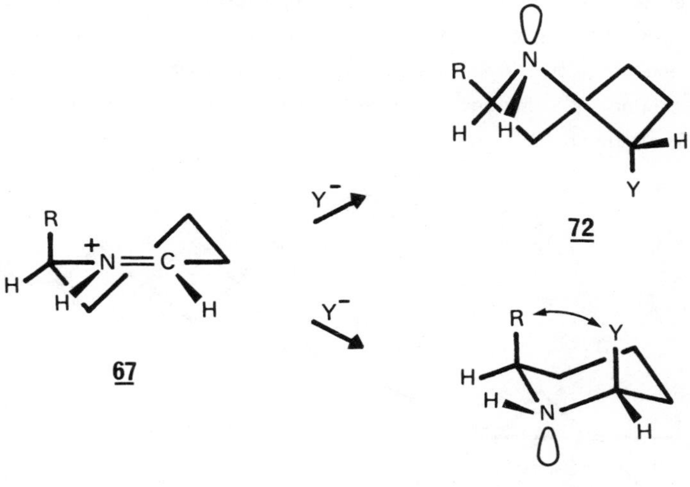

72

67

73

Consequently, the condensation of **67** with dimethyl acetone dicarboxylate must yield the _trans_ intermediate **74** which after conversion into **75** can be transformed, again with stereoelectronic control, into the _cis-trans_ tricyclic ketodiester **68**.

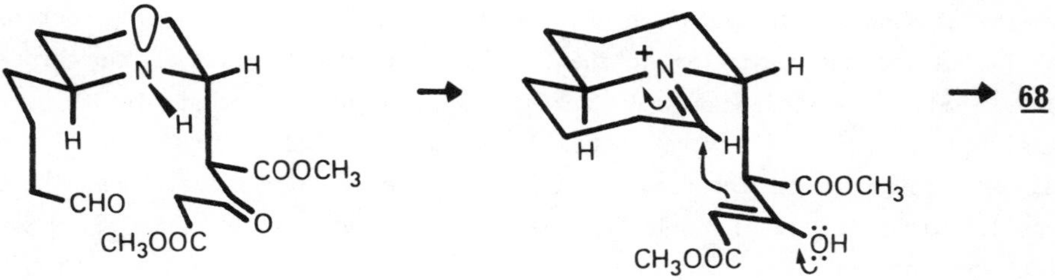

COOCH$_3$

CHO

CH$_3$OOC

74

+N

COOCH$_3$

CH$_3$OOC

ÖH

75

→ **68**

α,β-Unsaturated ketones

Stereoelectronic effects should also play an important role in the nucleophilic 1,4-additions of anions to conjugated systems. These effects should therefore influence the Michael reaction as well as the hydrocyanation of α,β-unsaturated ketones. Studies on these reactions provided evidence that the kinetically controlled addition of a nucleophile to a cyclohexenone derivative is indeed subject to stereoelectronic effects.

Considering a conformationally rigid cyclohexenone such as **76**, an attack by a nucleophile with stereoelectronic control on the top face yields the boat-like enolate ion **77** whereas that on the bottom face gives the chair-like enolate ion **78**. The second process should therefore be favored as suggested by Toromanoff (31).

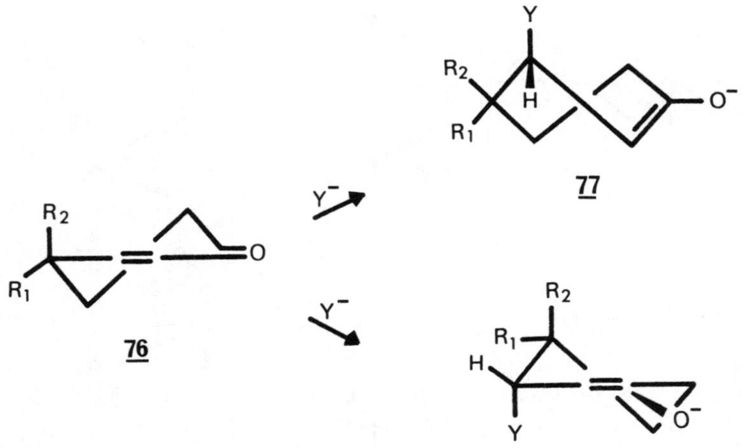

76

77

78

In the conjugate hydrocyanation of (-)-carvone (**79**), Djerassi and co-workers (32) observed the formation of the axial cyanoepimer **80** as the major product. Similarly, Alexander and Jackson (33) found that substrate **81** gave exclusively the axial cyano compounds **82** and **83**. The two epimers **82** and **83** were found to be interconvertible under the reaction conditions. These results indicate clearly that the chair-like enolate process **76** → **78** is the preferred pathway.

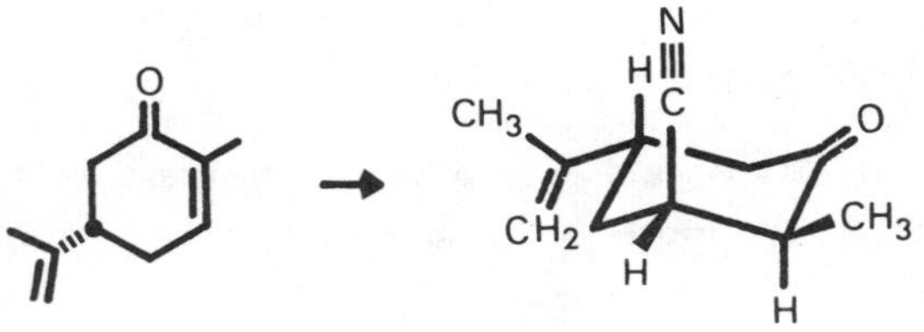

79   **80**

81   **82**   **83**

The bicyclic enone **84** in which ring A can adopt the two conformations **85** and **86** can now be examined. A stereoelectronically controlled attack of cyanide ion on the α face of conformation **85** gives the chair-like interme- diate **87** while that on the β face of conformation **86** yields the chair-like intermediate **88**. An approach from the top face of **85** or the bottom face of **86** is not considered because it leads to intermediates having a boat- like conformation. Intermediate **87** with its trans junction is more stable on the basis of steric effects than intermediate **88** which has a cis ring junction; consequently, the formation of the α-cyano adduct **89** should pre- vail over the β-cyano adduct **90.**

This prediction is supported by several experiments which showed that under kinetically controlled conditions, enones of type **84** give predominantly the α-cyano adduct **89**. For a detailed review, the reader is referred to the recent article of Nagata and Yoshioka (34).

Interestingly, the trans enones **91** and **92** (R=H or CH$_3$) have their ring A essentially conformationally rigid. Consequently, the addition of cyanide ion from the α face, occurring via a chair-like transition state (**93** → **94**) should be preferred over that on the β face which can take place only via a boat-like transition state (**93** → **95**). Agami, Fadlallah and Levisalles (35) have recently observed that in strictly kinetically controlled conditions, only the axial cyano isomer resulting from an attack on the α face was ob- served with the trans enones **91** and **92**.

They have further observed that under thermodynamically controlled condi- tions, the bicyclic α- and β-cyano isomers **96** and **97** can be interconverted to give an equilibrium mixture, and, as predicted on the basis of the stere- oelectronic effects, the axial cyano isomer **96** reaches the equilibrium at a much faster rate.

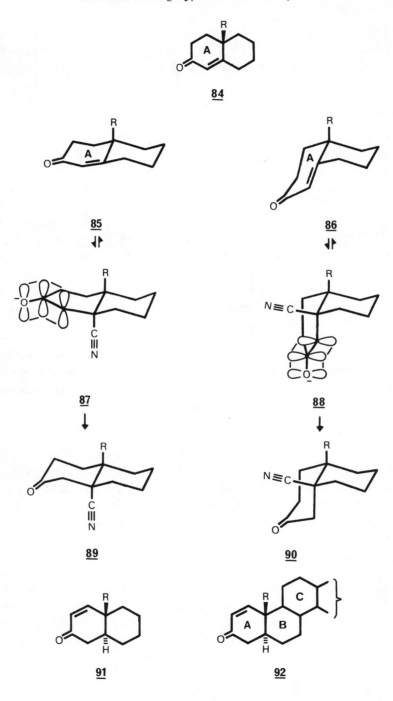

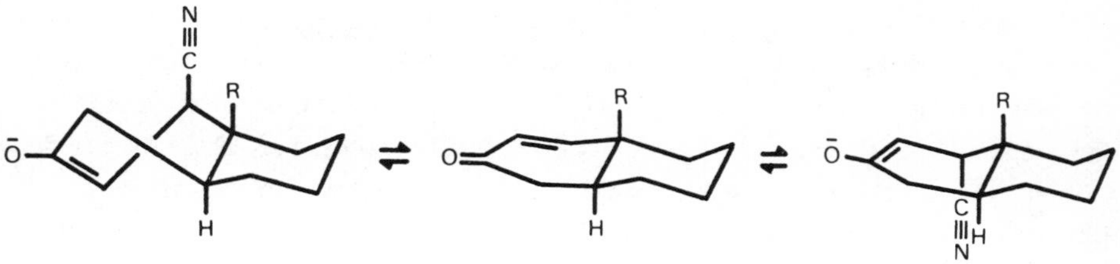

95   93   94

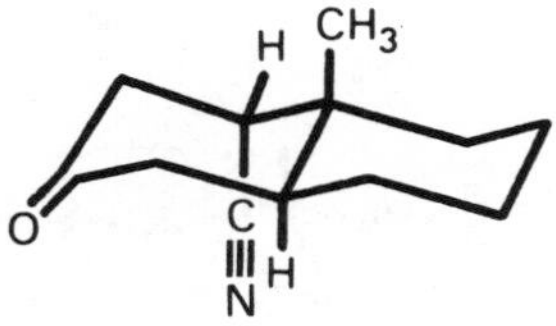

96 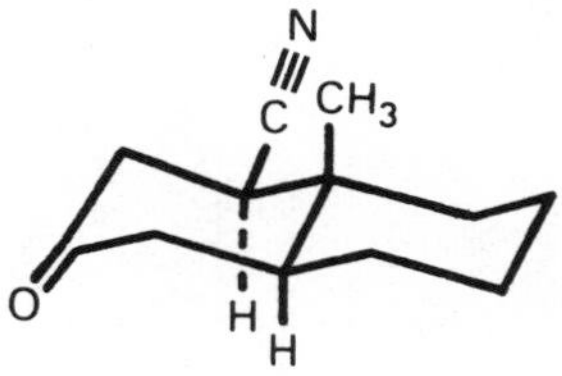   97

Contrary to this bicyclic series, the equilibration of the α- and the β-
cyano isomers derived from the steroid enones 92 (R=H or CH$_3$) does not occur. It has also been impossible to produce the β-cyano isomer from the
hydrocyanation reaction. This result was explained by the fact that in the
steroid series, there is a strong steric interaction between the cyano group
and the C-11 methylene group of ring C in the boat-like conformation 95
(cf. 98). Consequently, addition to the β-face never occurs in these compounds. This was confirmed by the fact that 1α-cyanocholestanone (derived
from 92, R=CH$_3$) readily incorporates labeled cyanide (^{13}C≡N) under the experimental conditions of the hydrocyanation reaction.

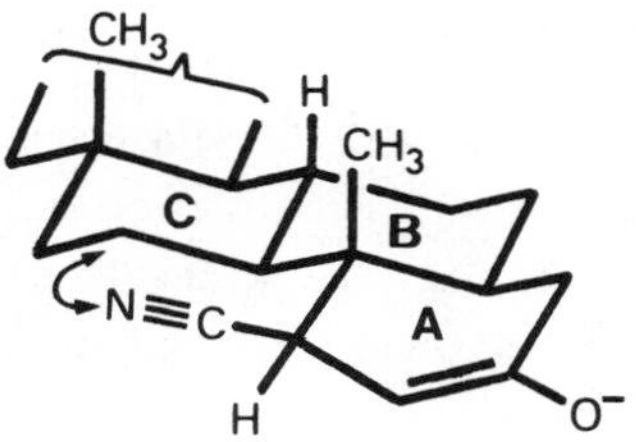

98

Very few studies have been carried out on the stereochemistry of the Michael reaction. However, Abramovitch and Struble (36) have found that compound **101** was the main product when diethyl sodiomalonate (free of ethoxide ion and ethanol) was added to 4-t-butyl-1-cyano cyclohexene (**99**) in boiling toluene. This result can be rationalized by axial attack on **99** to give first **100** having a chair-like conformation which is then transformed into **101** by internal trapping (see arrow). However, when the addition of diethyl malonate anion was carried out in ethanol under thermodynamically controlled conditions, product **103** with an equatorial malonate group was obtained, presumably via the twist-boat intermediate **102**.

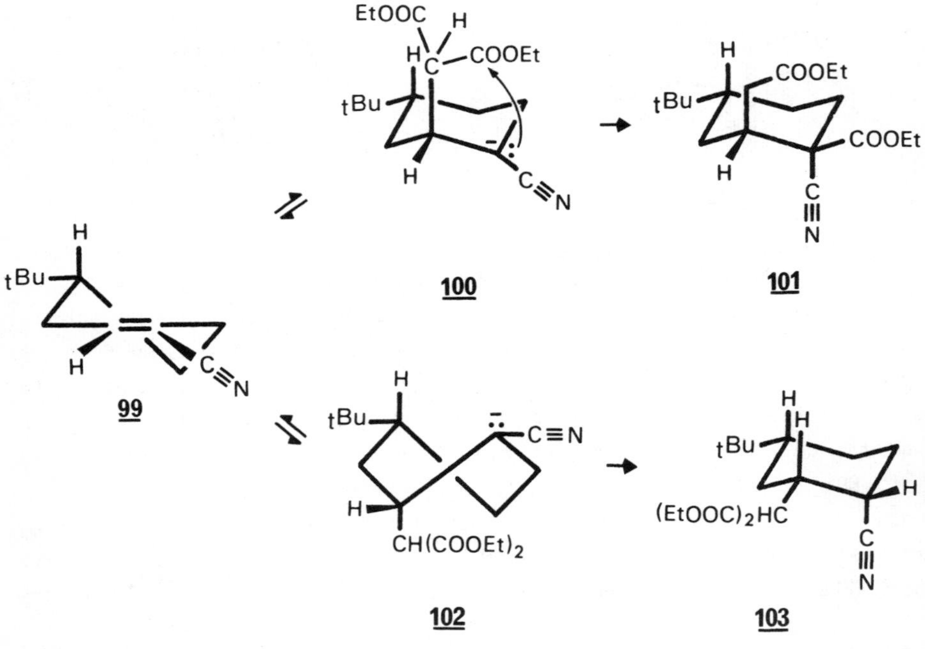

Irie and co-workers (37) have recently observed that the double Michael reaction of dimethyl acetone dicarboxylate on dienone **104** gave the cis decalin product **106**. This result indicates that intermediate **105** underwent a stereoelectronically controlled internal Michael addition to give **106**. Without stereoelectronic control in the Michael reaction, there is no apparent reason to prevent the formation of the trans isomer **107**. However, if this factor is taken into consideration, examination of molecular models indicates that it seems impossible to obtain isomer **107**.

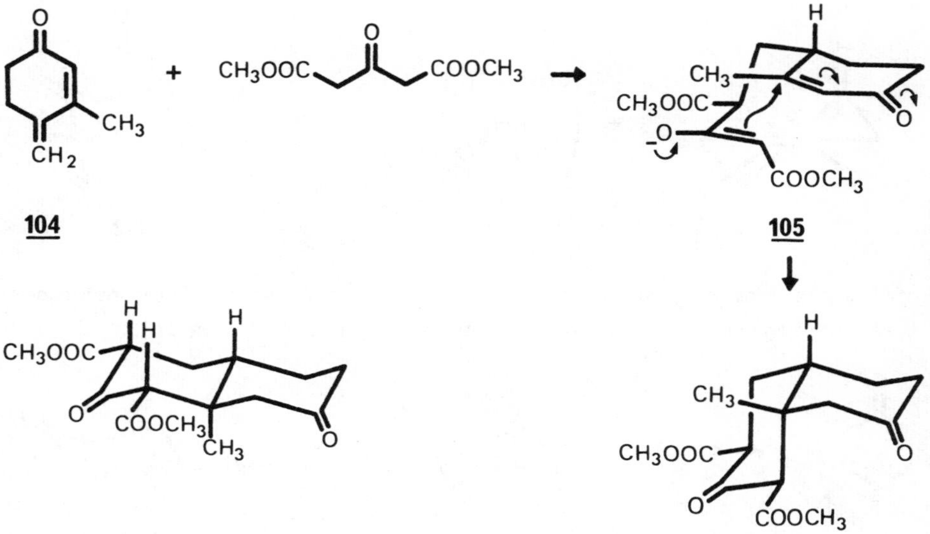

104

105

107

106

House and Fischer (38) have found that lithium dimethyl cuprate reacts with enone 108 and yields a mixture of trans and cis 3,5-dimethyl-cyclohexanones 109 and 110 in a 98:2 ratio. Similar results were observed by Allinger and Riew (39) using methylmagnesium iodide in the presence of copper(I) chloride. In another case, Heathcock and co-workers (40) observed the exclusive formation of the trans isomer 112 from enone 111; no cis isomer was detected. Thus, the preferred mode of approach by cuprate reagent is also 76 → 78 which leads to a chair-like enolate ion.

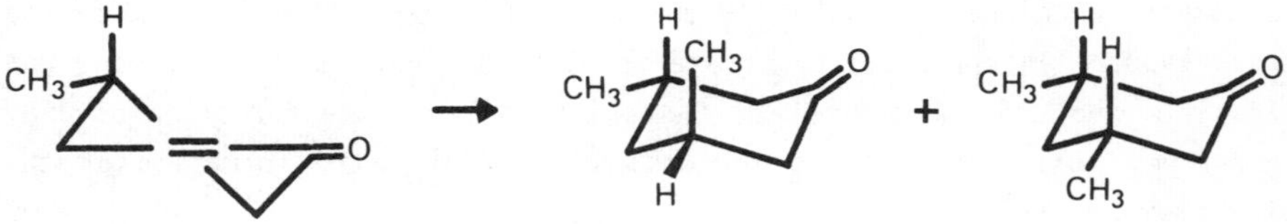

108   **109**   **110**

Luong-Thi and Rivière (41) have found that the conjugate addition of organo-copper reagents to 4-methyl-2-cyclohexenone (113) gave a mixture of trans and cis cyclohexanones 114 and 115 where the trans isomer largely predominates (≈9:1). Stereoelectronic effects predict that an attack of the most stable conformation of 113 yielding a chair-like intermediate enolate ion should give the cis isomer 115. Predominant formation of the trans isomer is therefore a priori unexpected. It is however possible that the formation

111   **112**

of the <u>cis</u> isomer is suppressed because of a strong steric interaction between the organocopper reagent and the methyl group at C-4 in <u>113</u>.

113   **114**   **115**

Enone <u>113</u> can adopt two different conformations <u>116</u> and <u>117</u>. Attack on the
top face of the most stable conformation <u>116</u> gives the chair-like enolate
ion <u>118</u> while an attack from below the plane of the molecule yields the
boat-like enolate ion <u>119</u>. On the other hand, an attack on the bottom face
of the less stable conformation <u>117</u> gives the chair-like intermediate <u>120</u>
while that on the top face gives the boat-like intermediate <u>121</u>. The formation of the boat-like <u>121</u> where the two groups (R and Y) are <u>cis</u> can be
readily eliminated. The chair-like <u>118</u> which leads to the <u>cis</u> isomer has
to compete with the boat-like <u>119</u> and the chair-like <u>120</u> which lead to the
<u>trans</u> isomer. The possibility of steric hindrance between the incoming nu-
cleophile and the alkyl group at C-4 exists only in the formation of <u>118</u>.
Therefore, this extra steric factor would disfavor the formation of the
<u>cis</u> isomer.

Compelling evidence that this is indeed the correct explanation comes from
a study of Rivière and Tostain (42). They have studied the copper-catalyzed
methyl Grignard conjugate addition of 4-substituted cyclohexenones and found
that when the size of the 4-alkyl group is increased from methyl to isopropyl, the <u>trans:cis</u> ratio varies from 72:28 to 89:11. Also using <u>113</u> as
a substrate, they found that on increasing the size of the alkyl group of
the reagent (RMgX, R=CH$_3$, C$_2$H$_5$ and (CH$_3$)$_2$CH), the relative percentage of
the <u>trans</u> isomer goes from 72 to 88%.

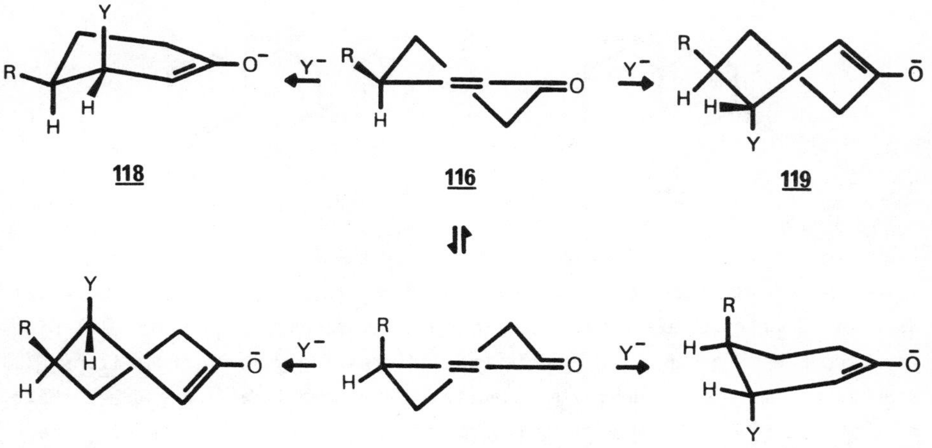

118   **116**   **119**

⇅

121   **117**   **120**

Interestingly, the hydrocyanation of 4-t-butylcyclohexenone gave, as the kinetic product, not the cis (**124**) but the trans cyanoketone **123**. We have seen that there is good evidence that stereoelectronic effects play an important role in the hydrocyanation of conjugated ketones. Consequently, this result can be explained by the above steric argument on the basis of which the formation of the cis isomer **124** is disfavored.

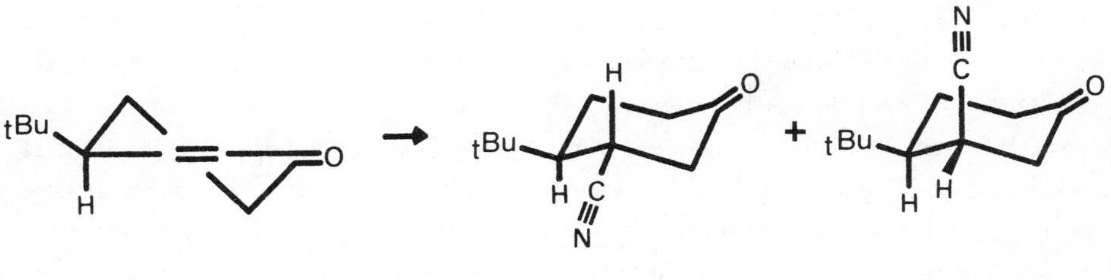

122   **123**   **124**

With bicyclic enones such as **125**, conjugate addition of organocopper reagents gives mainly the corresponding cis decalin product **126**, as described in the review by Posner (43). We have already seen that in this type of bicyclic enone, ring A can take two different conformations (cf. **85** and **86** in p. 224) which can give either a trans (from **85**) or a cis (from **86**) product with stereoelectronic control. With simple nucleophiles, the trans isomer was always favored. The opposite result obtained with organocopper reagents indicates that there are other factors which favor the transition **86** → **88** (CN=R) over the transition **85** → **87** (CN=R).

125   **126**

Piers and co-workers (44) have reported that the cuprous chloride catalyzed
1,4-addition of isopropenylmagnesium bromide to the bicyclic enones **127**
(R=H or CH$_3$) gave exclusively the bicyclic ketones **128** (R=H or CH$_3$) with
the isopropenyl group axially oriented. Interestingly, with their trans A/B
ring junction, these enones are essentially conformationally rigid and their
reactivity is in accord with the predictions made on the basis of stereoelectronic effects.

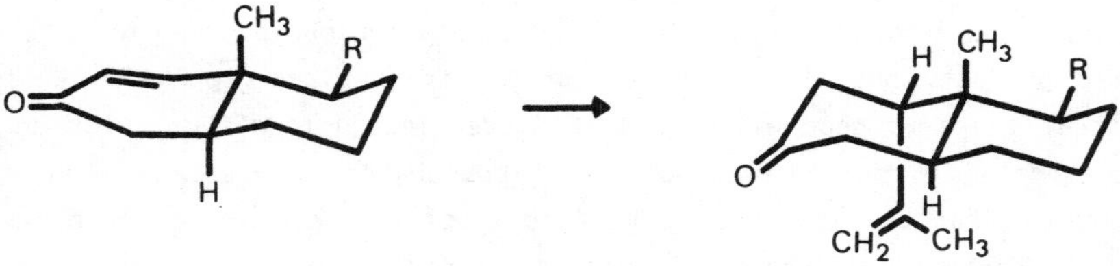

127   **128**

However, the α,α'-dimethyloctalone **129** and the tricyclic enone **132** behaved
differently. Marshall and Andersen (45) obtained from **129** a mixture of isomers **130** and **131** (54% axial and 46% equatorial, when R=isopropyl and 82%
axial and 17% equatorial when R=methyl) while Spencer and collaborators
(46) isolated a ≈1:1 mixture of **133** and **134** from **132**. By comparison with
enones **127** which have an angular methyl group, the replacement of that group
for a hydrogen atom in **129** and **132** must have eased the approach of the reagent on the top face to give products **131** and **134** with the equatorial alkyl
group via a boat-like transition state.

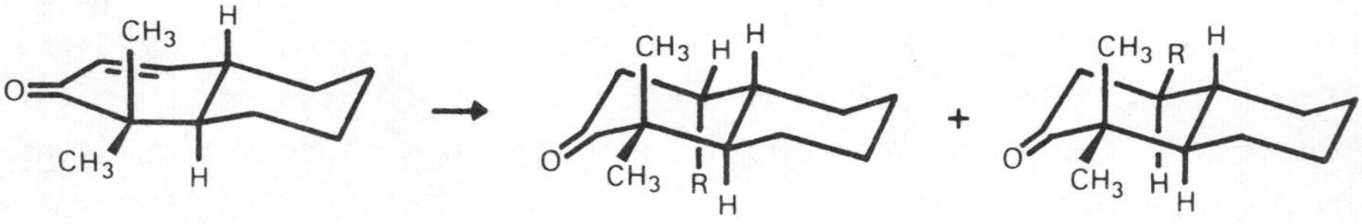

129   **130**   **131**

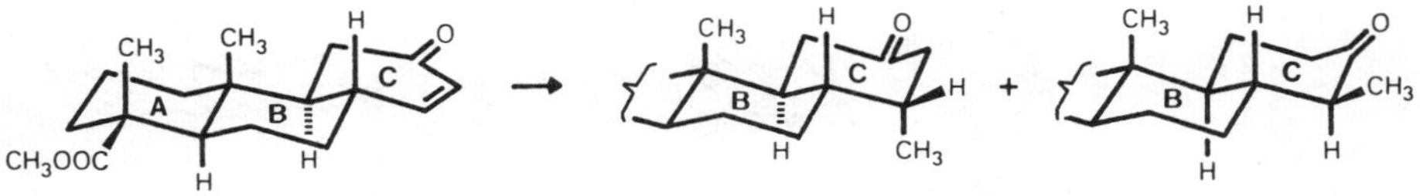

132   133   134

A report of House, Respess and Whitesides (47) showed that the reaction of lithium dimethyl cuprate with the unsaturated ketone **135** gave exclusively the ketone **136** having the methyl <u>trans</u> to the <u>t</u>-butyl group. In this case where the double-bond is exocyclic, stereoelectronic effects allow equal attack from either face. Thus, the exclusive formation of **136** must be due to steric reasons only, the equatorial approach being favored.

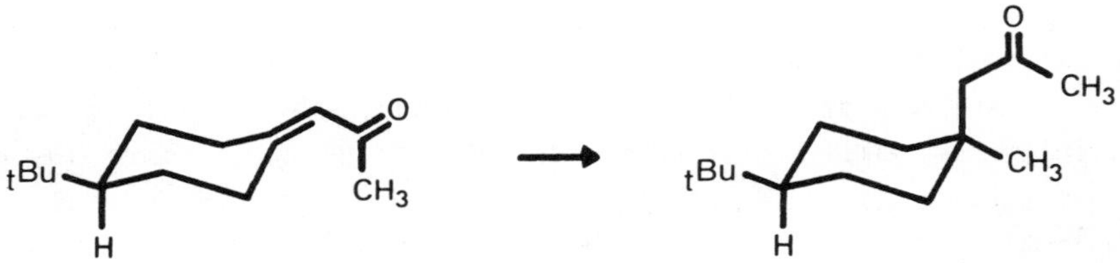

135   **136**

In conclusion, in the case of 1,4-conjugated additions to α,β-unsaturated ketones, some substrates, in order to avoid steric interaction, react through a boat conformation to give an equatorially substituted product, but when there is no steric interaction, the axial attack through a chair conformation is energetically favored. Both processes are however stereoelectronically controlled.

The 1,6-addition to $\alpha,\beta,\gamma,\delta$-dienones is also subject to stereoelectronic effects. Addition on the bottom face of dienone **137** leads to a chair-like intermediate **138** while that on the top face leads to a boat-like intermediate **140** in order to maintain maximum orbital overlap. Also, in **140** the R group encounters an eclipsed 1,2-R/H interaction and more importantly, a 1,4-CH$_3$/R steric interaction which resembles the bowsprit flagpole arrangement of a twist-boat form of cyclohexane. This analysis of Marshall and Roebke (48) predicts that the <u>trans</u> product **139** should prevail over the <u>cis</u> product **141**.

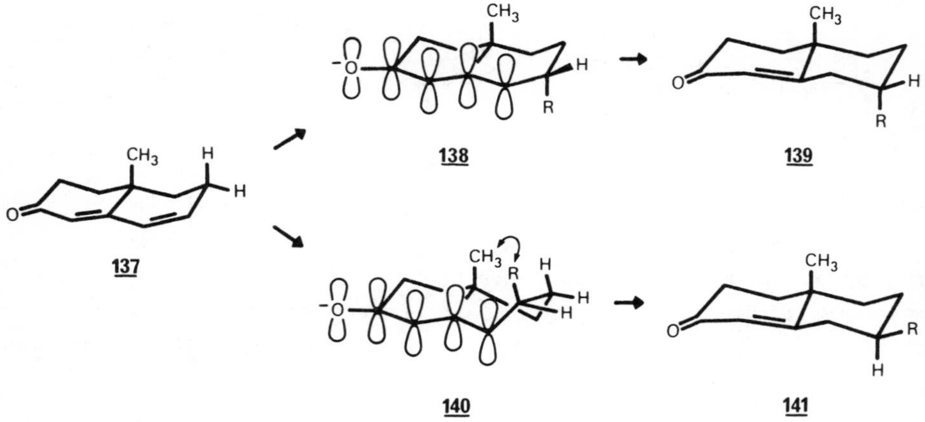

Support for this view comes from their work on the stereochemistry of 1,6-addition (catalyzed by cupric acetate) of methyl-, ethyl-, isopropyl-, and t-butylmagnesium halides on dienone **137**: 93% of the axial isomer **139** was obtained with CH$_3$MgI, 98% with C$_2$H$_5$MgBr and 100% with (CH$_3$)$_2$CHMgBr and (CH$_3$)$_3$CMgCl. Campbell and Babcock (49) have also studied the cuprous catalyzed reaction of methylmagnesium iodide with various steroidal 4,6-dien-3-ones. They found axially substituted C-7 methyl derivatives as the predominant 1,6-adducts.

When the bicyclic dienone **142** was treated with dimethyl malonate and potassium t-butoxide in t-butanol at 25°C for 10 days, the axial isomer **143** was obtained in 51% yield. None of the equatorial isomer **144** was isolated. When the reaction was carried out at reflux, a mixture of **143** and **144** was isolated in which the more stable equatorial isomer predominated (50).

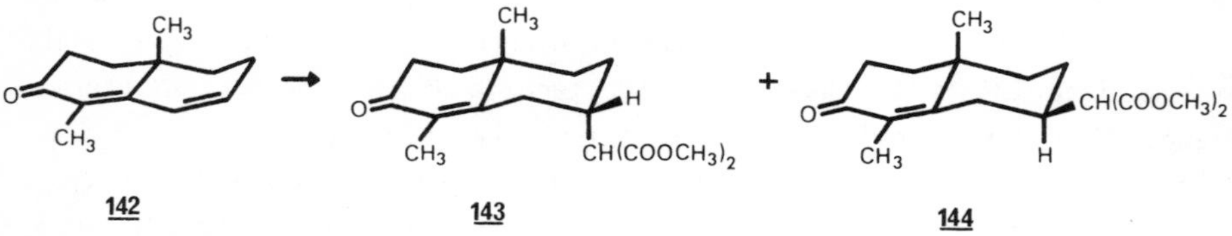

We have already seen evidence from the retrohydrocyanation reaction (p. 222) that the ejection of a leaving group β to a carbonyl which yields an α,β-unsaturated system should also be stereoelectronically controlled, i.e. the C — Y bond of the leaving group should be parallel to the π system of the enolate ion (**145** → **146**).

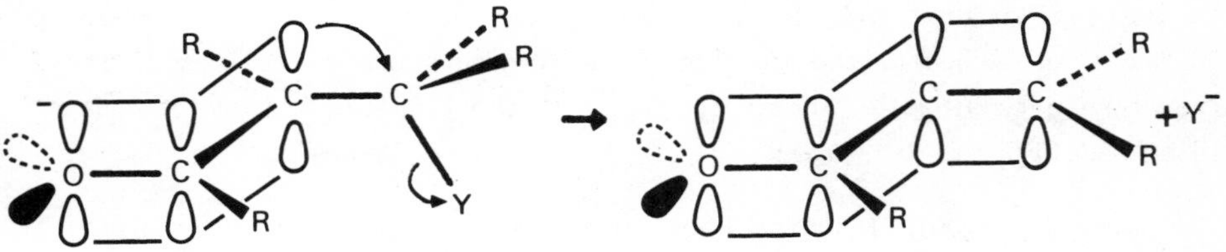

145   **146**

Further evidence that this is indeed the case comes from the dehydration under basic conditions of β-hydroxy ketones (51-54). For example, the dehydration of **147** can yield the <u>cis</u> or the <u>trans</u> conjugated ketones **149** and **151** <u>via</u> the enolate anions **148** and **150** respectively. The formation of the <u>trans</u> product **151** is favored because there is less steric interaction between the planar enolate anion system and the phenyl group at the β-carbon

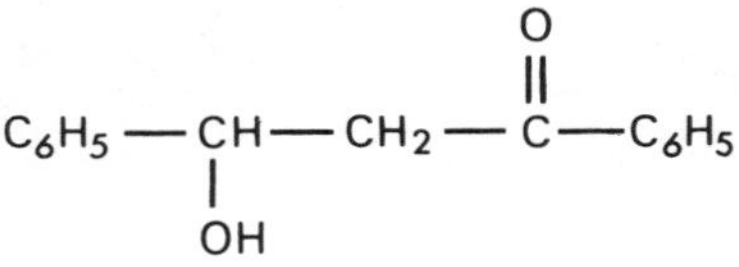

147

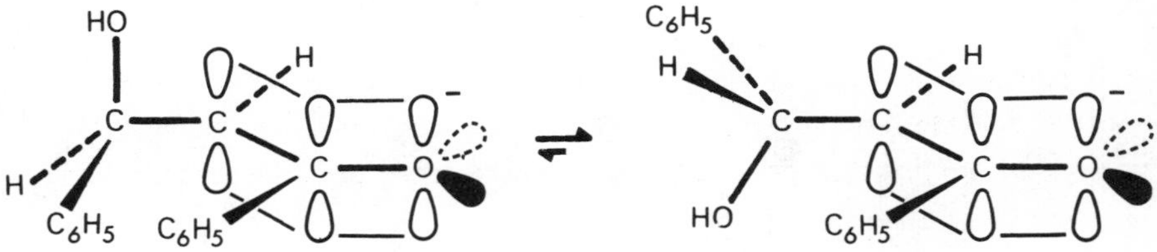

148   **150**

149   **151**

in the conformation **150**. In other words, in the transition state leading
to the <u>trans</u> product, the two large phenyl groups are not eclipsed. It is
therefore a combination of steric and stereoelectronic effects which control
this reaction.

The rules of Baldwin (55) for ring closure in trigonal systems (see p. 171
for an introduction) are the following: 3- to 7-<u>Exo-Trig</u> processes (**152-
156**) are all favored processes. 3- to 5-<u>Endo-Trig</u> (**157-159**) are disfavored
but 6- and 7-<u>Endo-Trig</u> (**160-161**) are favored. The literature is replete
with examples of 3- to 7-<u>Exo-Trig</u>: for instance, lactonization of ω-hydroxy-
acids and esters are of this type, the formation of lactams from ω-aminoacids
and also the Dieckmann cyclization of diesters.

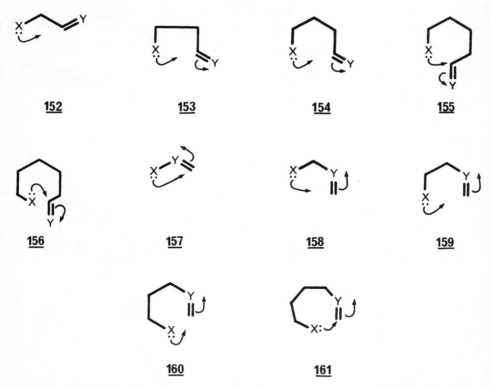

152   **153**   **154**   **155**

156   **157**   **158**   **159**

160   **161**

Baldwin and co-workers (56, 57) have reported that all attempts to cyclize
hydroxy-enones **162** and **164** (R=H or OCH$_3$) under basic conditions failed to
give the corresponding furanones **163** and **165** (R=H or OCH$_3$). The susceptibil-
ity of these systems to conjugate addition of alkoxide nucleophile was verified by conducting the attempted cyclization with sodium methoxide in deute-

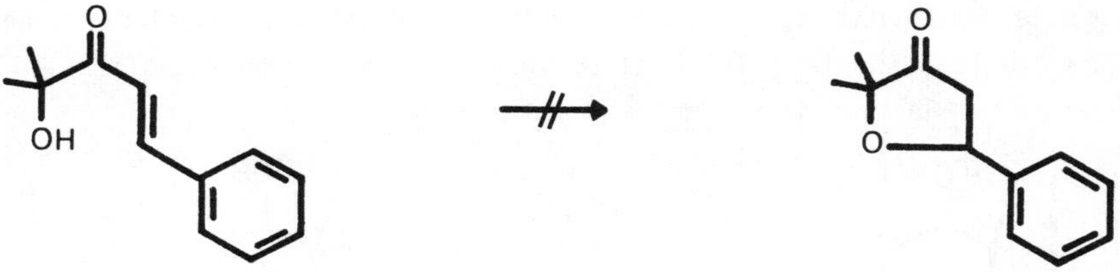

162   **163**

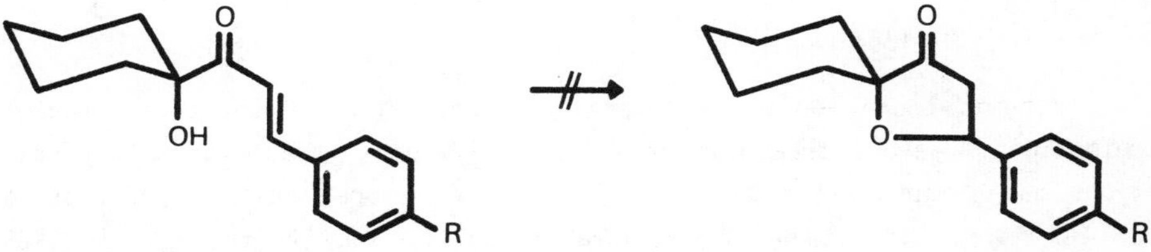

164   **165**

rated methanol. Under these reaction conditions, the α-deuterated analogs
(cf. **166**) of **162** and **164** (R=OCH$_3$) were isolated. The incorporation of a
deuterium atom α to the carbonyl group was rationalized as a consequence
of reversible addition of the methoxide anion giving an adduct such as **167**
which underwent deuterium exchange and subsequent elimination of methanol.

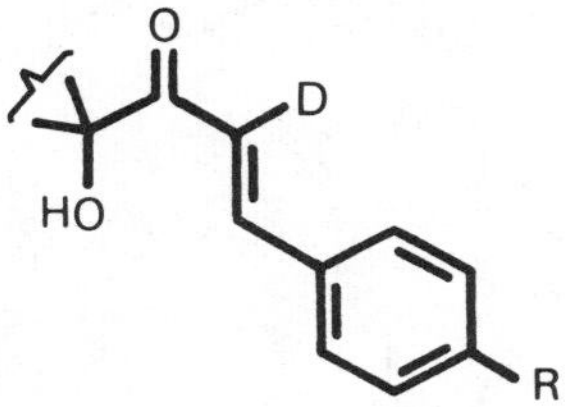

166 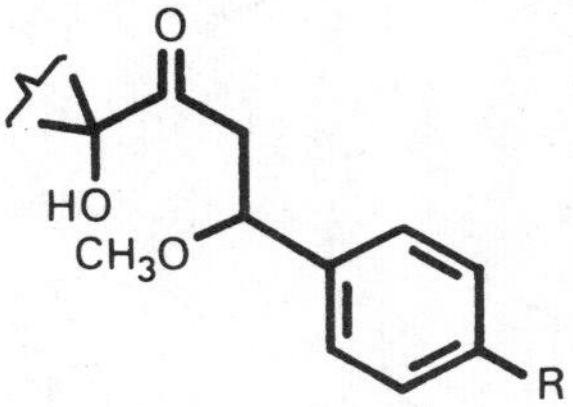   **167**

On the other hand, they have shown that furanones **163** and **165** (R=H or OCH$_3$)
rapidly and efficiently exchange both their α-hydrogen atoms under the same
basic conditions in which **162** and **164** are not converted into **163** and **165**.
This proves that the lack of ring closure or ring opening is a result of a
kinetic rather than a thermodynamic barrier. Thus, the 5-Endo-Trigonal process (**159**) is indeed a geometrically disfavored reaction.

The acid-catalyzed cyclization of hydroxy-enones **162** and **164** (R=H or OCH$_3$)
gives the corresponding furanones **163** and **165**. The success of these reac-

tions is attributed to the reduction in the rotational barrier around the enone double-bond, (cf. **168A** and **168B**), allowing thereby a cyclization (the closure of **168B** is a 5-Exo-Trig allowed process).

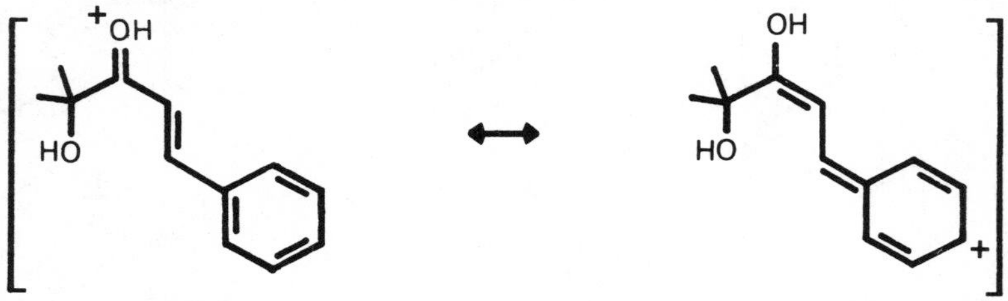

168A	**168B**

The phenolic enone **169** was recovered unchanged from treatment with sodium methoxide; however, treatment of furanone **170** under the same basic conditions gave smoothly the enone isomer **169**. Therefore, the lack of closure of **169** is the result of an unfavorable equilibrium. The process **170 → 169** can be looked at as a 5-Exo-Trigonal process due to the resonance structure **171B**.

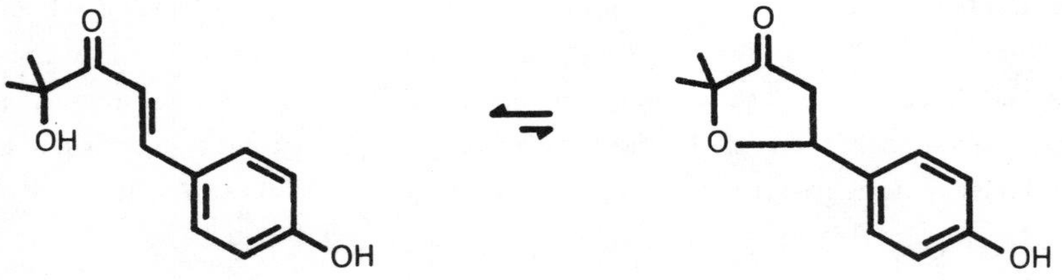

169	**170**

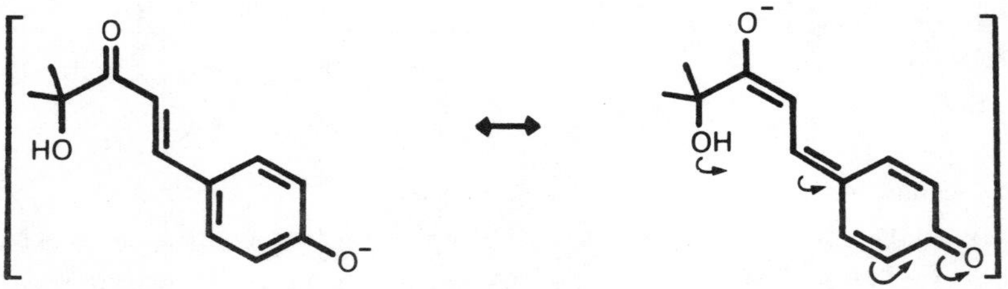

171A	**171B**

In another experiment (56), the disfavored 5-Endo-Trig process (**159**) was placed in competition with the favored 5-Exo-Trig process (**154**). The hydroxy conjugated ester **172** upon treatment with a variety of bases closed efficiently and cleanly to the lactone **173** (5-Exo-Trig process) with no trace of tetrahydrofuran **175** (5-Endo-Trig process). Also, **173** smoothly added meth-oxide ion to yield the ether **174**, showing that the double-bond is very sus-ceptible to Michael type addition. Again, the ester **175** exchanged its α-hydrogen atom under the conditions of conversion of **172** to **174** with no re-version to **172**.

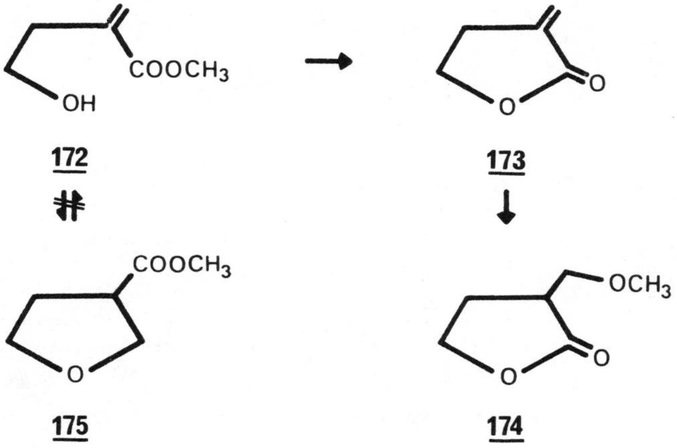

A nitrogen analog of **172** was also studied (56). The amino-diester **176**, upon release from its stable hydrochloride salt, rapidly closed at 25°C to the lactam **177** (100%) via the favored 5-Exo-Trig pathway. The disfavored 5-Endo-Trig process yielding the cyclic amino-diester **178** was not observed. On the other hand, it is known that primary amines undergo a 1,4-addition to α-substituted acrylic esters (**179** → **180**) more rapidly than they are acylated to the α-substituted acrylamides (**179** → **181**).

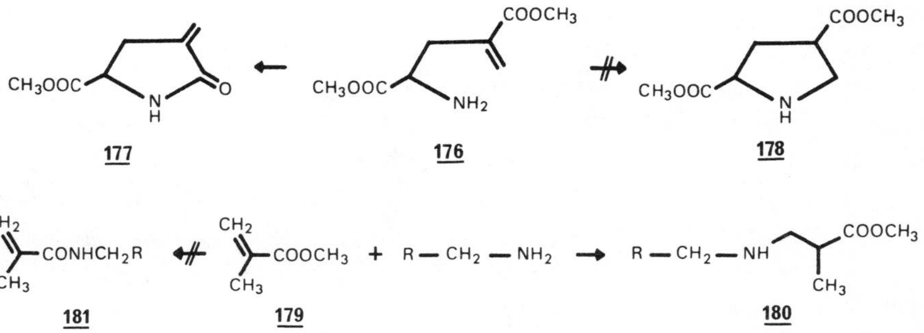

The reactions of cinnamic acid derivatives with hydrazine are also in accord (56) with the above findings. The hydrazide **182** cannot, even at 200°C, be converted into the pyrazolone **185** (5-Endo-Trig process). However, the ester reacts with hydrazine at 65°C to give cleanly **185**, by way of the 1,4-adduct **184**, followed by the favored 5-<u>Exo-Trig</u> closure.

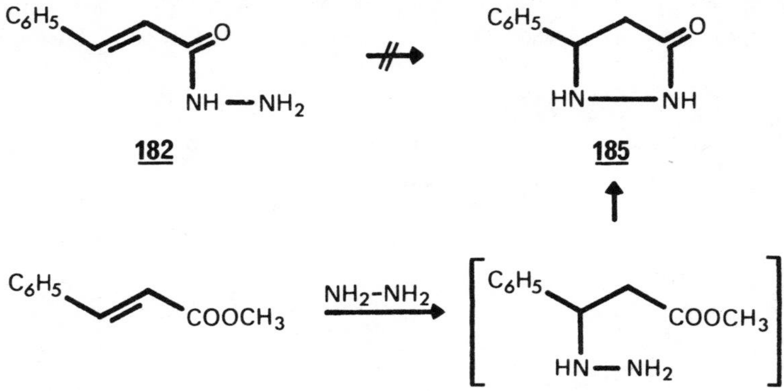

182   **185**

183   **184**

In contrast, to the difficulty of ring closure by the 5-<u>Endo-Trig</u> pathways, the 6-<u>Endo-Trig</u> reaction occurs readily: on treatment with methanolic sodium methoxide the α,β-unsaturated ketones **186** smoothly closed to the 4-chromanones (**187**) (56).

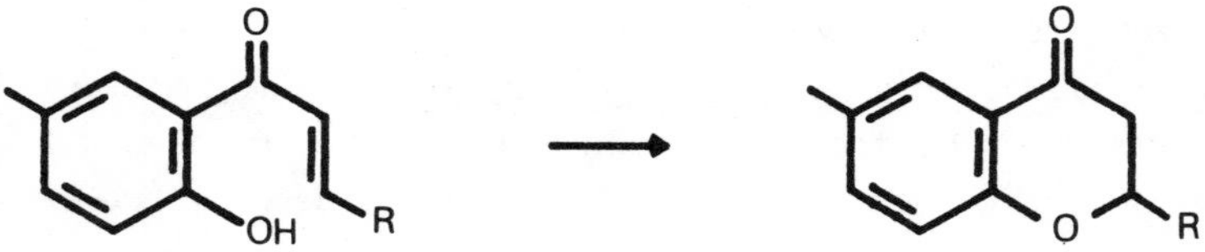

186   **187**

The 6-<u>Exo-Trigonal</u> (**155**) and the 6-<u>Endo-Trigonal</u> (**160**) processes are favored for the formation of six-membered rings. However, when a choice of these two modes of ring closure exists, the <u>Exo</u> pathway is the faster ring closing process as shown by Baldwin and Reiss (58). Treatment of <u>trans</u>-heptenoate ester **188** with a variety of bases (NaH, CH$_3$ONa, <u>t</u>-BuOK) led rapidly to the cyclic tetrahydropyran **189** in quantitative yield <u>via</u> the 6-<u>Exo-Trig</u> pathway. In contrast, the ester **190** cyclized relatively slowly under the same conditions to the α-methylene lactone **191** and gave none of the product of the 6-<u>Endo-Trig</u> mode, <u>i.e.</u> the tetrahydropyran **192**. Also, **192** was found readily to exchange the proton α to the ester function under the cyclization conditions of **190** to **191**. The relatively slower formation of **191** was attributed to the fact that the starting product exists in the preferred S-<u>Trans</u> conformation **193** (rather than the S-Cis **190**) in which the 6-Exo-Trig closure is sterically improbable. This is in contrast with the extremely rapid closure of the ester **194** to the lactone **195** (a 6-Exo-Trig process) even under neutral conditions.

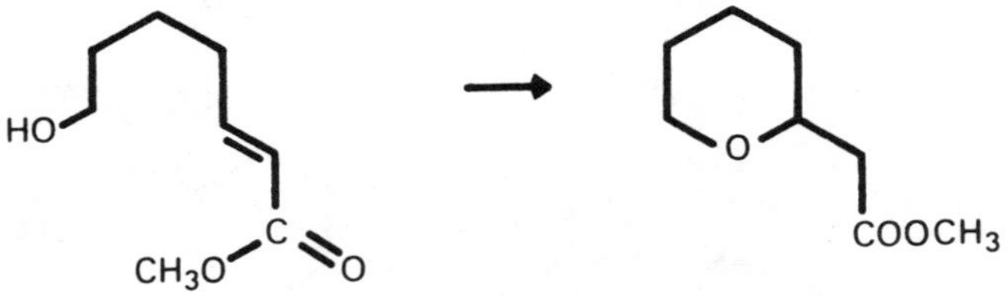

188   **189**

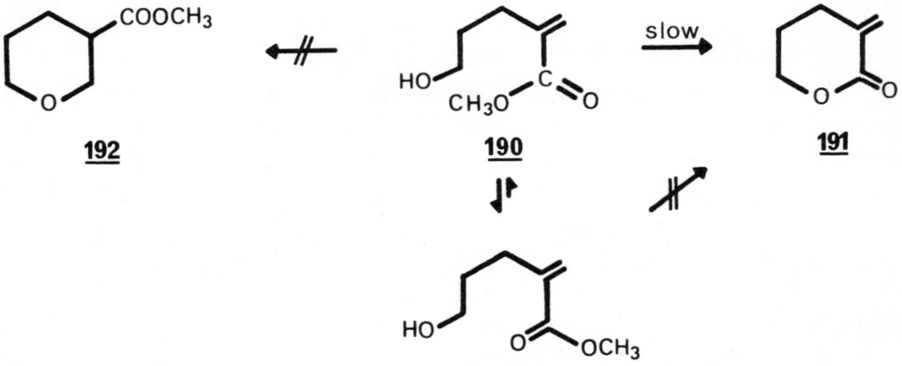

192   **190**   slow   **191**

193

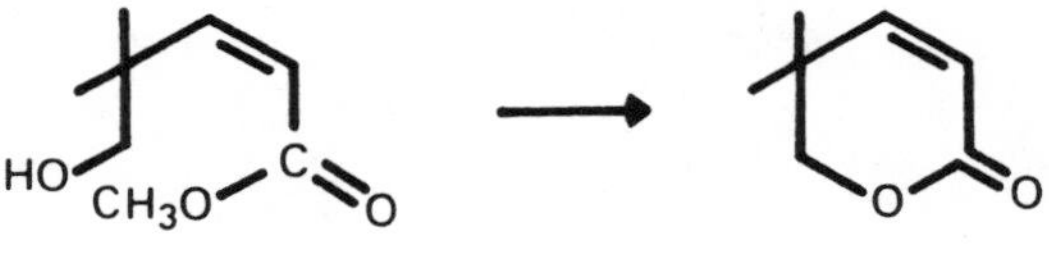

194   **195**

Baldwin and Kruse (57) have found that the enolate salts **196A-B** (M=Li or K) give the enol-ether **197** and no cyclopentanone **198**. In contrast to this result, the enolates **199A-B** (M=Li or K) under the same conditions gave only the cyclohexanone **201**. In this case, the enol-ether **200** was not formed.

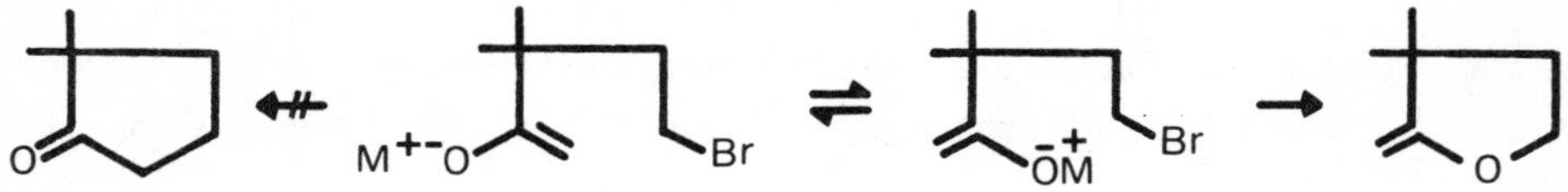

198   **196A**   **196B**   **197**

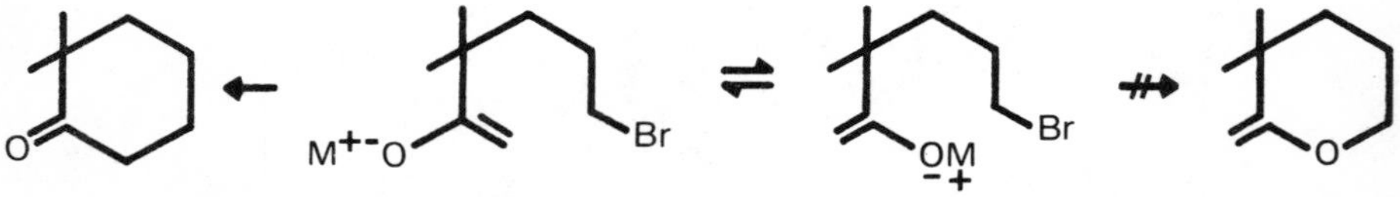

201   **199A**   **199B**   **200**

Baldwin concluded that the remarkable difference between these two cyclizations results from stereoelectronic control of the alkylation of the ambi-
dent nucleophile, i.e. the enolate ion. For such an ion, carbon alkylation
requires approach of the electrophile perpendicular to the plane of the
enolate, whereas oxygen alkylation requires approach in the plane of the
enolate. Consequently, in the five-membered ring case, the C-alkylation
process **196A** → **198** (which can be considered as a 5-Endo-trigonal process)
is sterically difficult, but not the O-alkylation process **196B** → **197** (a 5-
Exo-tetrahedral process).

In a very recent paper, Seebach and Golinski (59) have proposed a series of
rules which explain the preferred formation of the threo-configuration **204**
in the condensation of the Michael acceptor **202** with the Michael donor **203**.
These are:

(a) staggering of all bonds around the newly formed bond is necessary,
(b) the C = C   ond of the donor must be in a gauche (synclinal) arrangement
     between the C = A and the C — H bond of the acceptor,
(c) the H-atom, the smaller substituent on the donor component, must be in
     an anti (antiperiplanar) position with respect to the C = A bond.

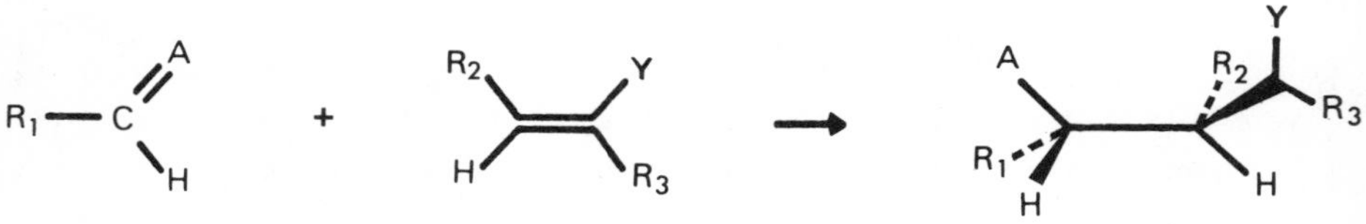

202   **203**   **204**

Therefore, the preferred approach of the two components is illustrated by **205** (or by the Newman projection **206**). When the donor has the geometry **207**, the actual donor and acceptor groups are again situated close to each other (cf. **208**).

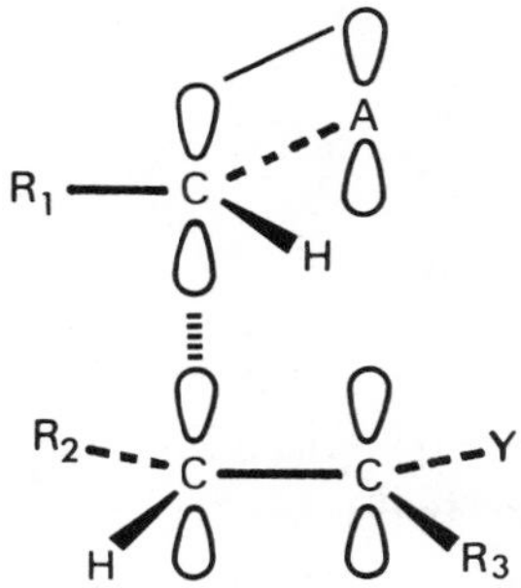

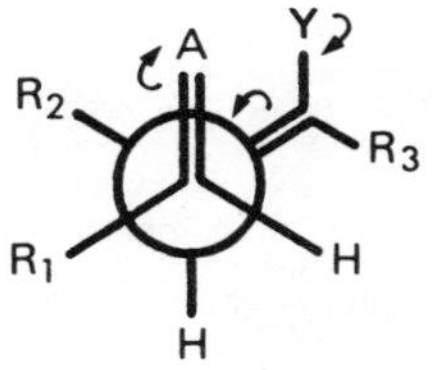

205	**206**

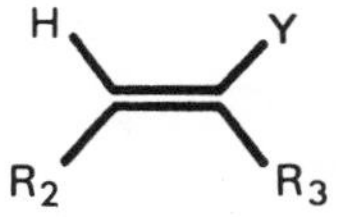

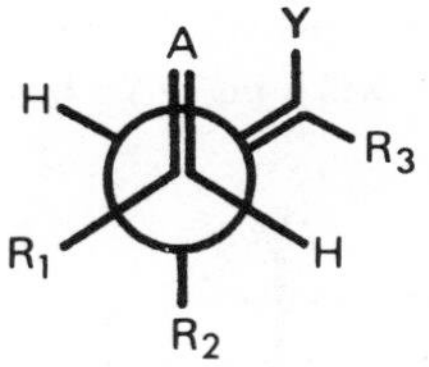

207	**208**

This topological rule readily explained the reaction product **211** (>90% stereoselectivity) of open-chain nitroolefins **209** with open-chain enamines **210**. Seebach and Golinski have further pointed out that several condensation reactions can also be rationalized by using this approach: (a) cyclopropane formation from olefin and carbene, (b) Wittig reaction with aldehydes yielding cis olefins, (c) trans-dialkyl oxirane from alkylidene triphenylarsane and aldehydes, (d) ketenes and cyclopentadiene 2+2-addition, (e) (E)-silylnitronate and aldehydes, (f) syn and anti-Li and B-enolates of ketones, esters, amides and aldehydes, (g) Z-allylboranes and aldehydes, (h) E-alkylborane or E-allylchromium derivatives and aldehydes, (i) enamine from cyclohexanone and cinnamic aldehyde, (j) E-enamines and E-nitroolefins and finally, (k) enamines from cycloalkanones and styryl sulfone.

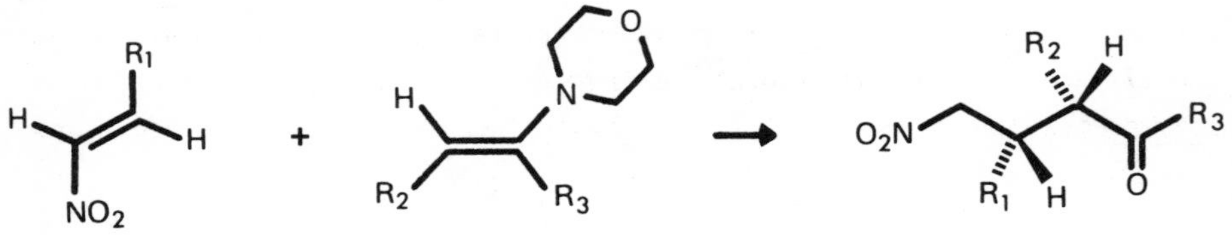

209   **210**   **211**

Reduction of α-β-unsaturated carbonyl compounds

The mechanism of the chemical reduction of enones with metal (Li, Na, etc.) in liquid ammonia can be described by the following equation in which the substrate **212** receives two electrons from the metal to give the dianion intermediate **213**. This intermediate is then successively transformed into the enolate salt **214** and the ketone **215** with an appropriate proton donor source. It can readily be seen that the stereochemical outcome of this reaction depends on the stereochemistry of the protonation step **213** → **214**. An excellent review on this topic has been recently written by Caine (60). This subject will be only briefly discussed here.

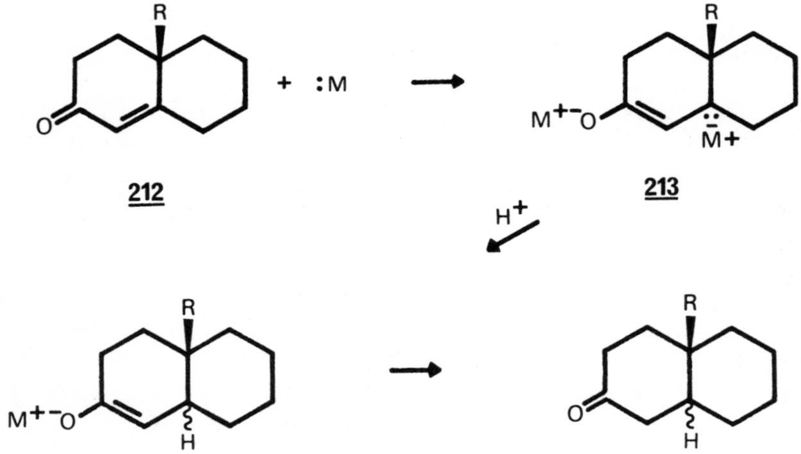

212   **213**

214   **215**

In the reduction of octalone of the type **212**, the resulting enolate dianion **213** can adopt three different half-chair conformations **216**, **217**, and **218**. Of these, only conformations **216** and **217** have the carbanion electron pair parallel to the π orbital of the enolate system allowing an electronic delocalization. Thus, product would come from the protonation of **216** from the bottom face yielding the trans bicyclic ketone **219** or from the protonation of **217** from the top face yielding the cis product **220**. In simple cases, the conformer **216** is more stable than **217** (less steric interaction); the formation of the trans product **219** usually predominates.

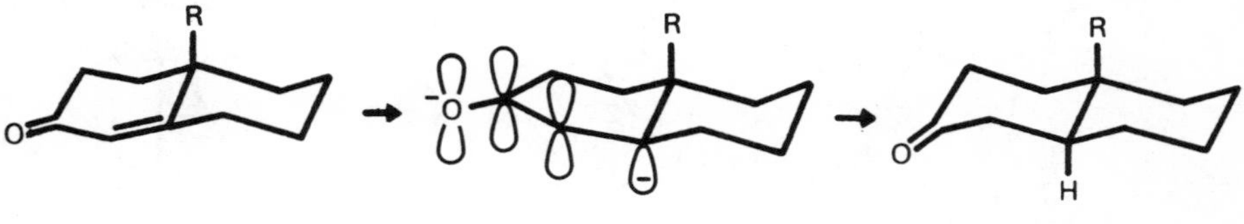

212   **216**   **219**

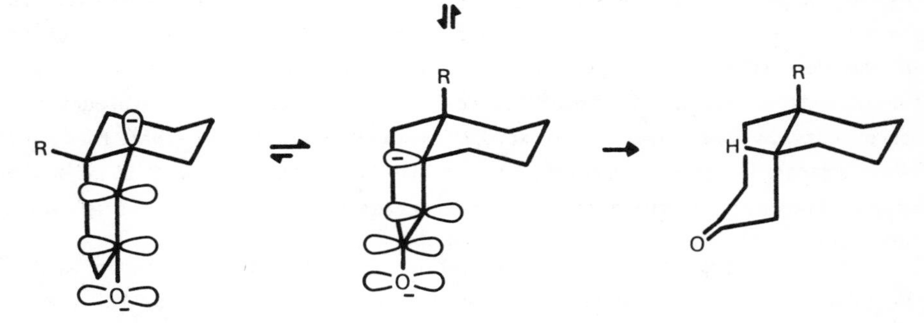

218   **217**   **220**

The above stereoelectronic arguments were proposed by Stork and Darling (61) to explain why the more stable isomer is not necessarily always obtained (62). For example, reduction of the octalone **221** with lithium-ammonia-ethanol followed by oxidation afforded the trans-2-decalone **222** even though the isomeric cis-2-decalone **223** is about 2 kcal/mol more stable than **222**. Conformation **226** of the enolate dianion is the most favored sterically but it is electronically disfavored. Conformations **224** and **225** are both electronically favored but **225** is less favored sterically than **224**. Therefore,

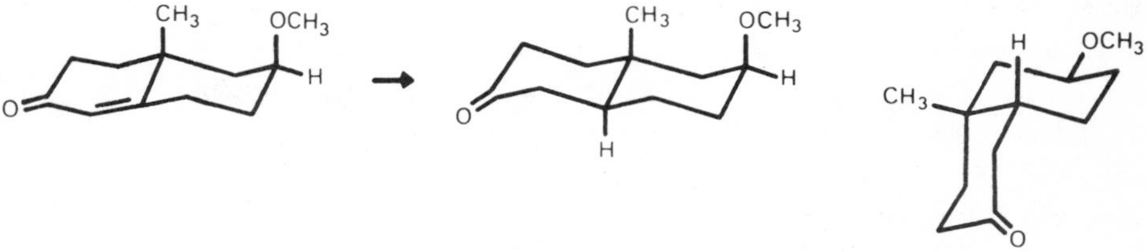

221   **222**   **223**

the enolate dianion intermediate should exist in conformation **224** which is then protonated to give the trans-product **222**.

224   **225**   **226**

The reduction of **212** (R=H) gives a mixture of trans and cis decalones **219** and **220** (R=H) in a 99:1 ratio (63). An analysis of non-bonded interactions in the corresponding enolates **216** and **217** (R=H) indicates that the former should be favored only by about 1.0 kcal/mol, which should correspond to an approximately 80/20 trans/cis ratio. This result indicates that there is a significantly greater preference for the trans species **216** than would be predicted by analysis of non-bonded interactions.

Other factors (charge repulsion, solvation factors, etc.) could influence the position of the equilibrium in favor of enolate dianion **216**. It is also possible that there is a kinetic preference for the formation of dianion **216** and that this species would undergo protonation more rapidly than equilibration. This rule of "axial protonation" of **216** has been found to be widely applicable in many cases. However, in systems in which a significant amount of strain must be introduced in order for protonation to occur axially on **216**, protonation of conformer **217** (and even conformer **218**) becomes important (60).

The hydrogenolysis reaction of α-halo (Y=halogen), α-amino (Y=NR$_2$), α-acyloxy (Y=OCOR) and α-hydroxy (Y=OH) ketones **227** with metals (Li, Ca, Zn, etc.) (64, 65) to yield the corresponding ketone **228** must also be strongly influenced by stereoelectronic effects (66). House (67) has pointed out that it is probably necessary that the α-substituent Y must be able to occupy a conformation in which the π system of the carbonyl group is parallel to the C — Y bond (**229**). In such a conformation, the carbonyl group can accept two electrons to give **230** which can easily eject the Y$^-$ group, generating in this way the intermediate enolate ion **231**.

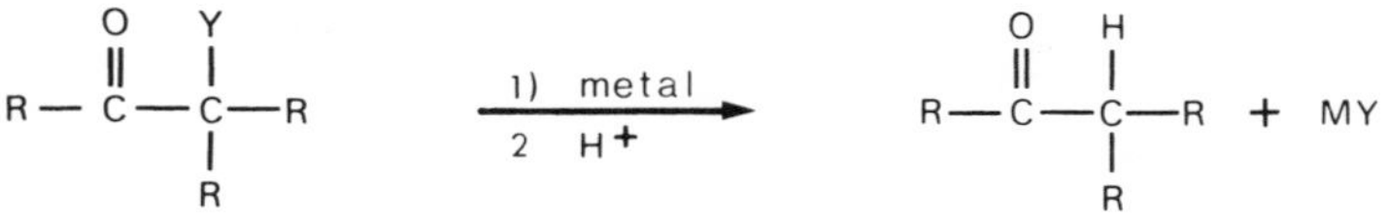

227

228

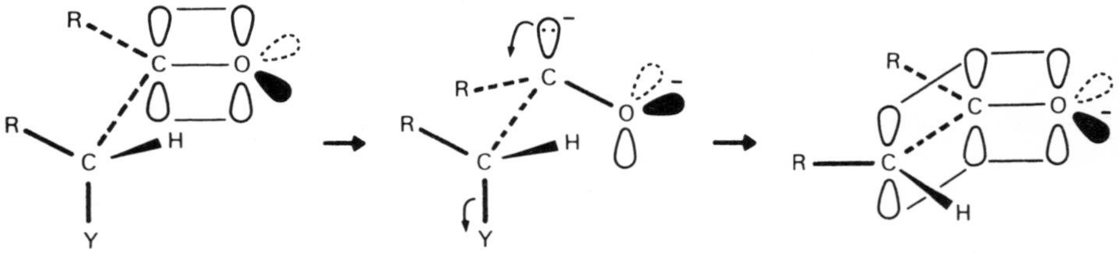

229   230   231

It follows that a conformationally rigid cyclohexanone with a Y group axially oriented should react in its ground state conformation 232 while that
with the Y group equatorially oriented (233) should react via its boat conformation (234). As a consequence, cyclohexanones with axial alpha substituents must be reduced more readily than analogous compounds with equatorial
substituents, especially when the two compounds are essentially conformationally rigid (68).

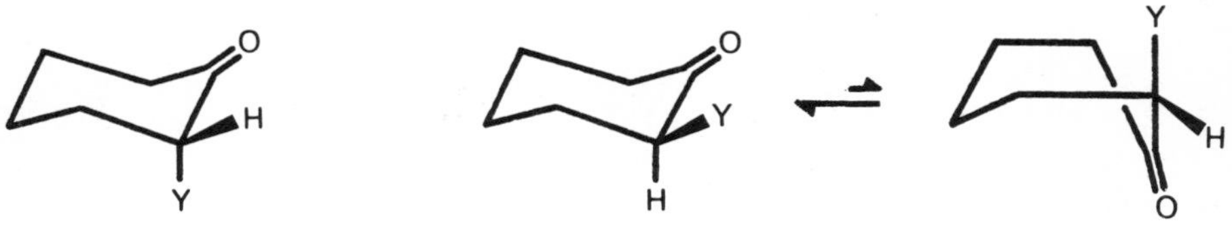

232   233   234

The hydrogenolysis of allylic ether and acetate (235 → 236, R=alkyl or COCH$_3$)
should also take place more easily when the compound can adopt a conformation in which the OR group can become parallel to the π orbital of the double bond (69). The same stereochemical requirement must also be necessary
in the hydrogenolysis of a substituent in a benzylic position (70).

235   236

Reduction of cyclopropyl ketone

The reductive opening of a cyclopropane ring of a conjugated cyclopropyl ketone with lithium in liquid ammonia can be viewed as an overall two elec- trons reduction which yields the equivalent of a carbanion and an enolate ion (cf. __237__). Successive protonations of __237__ then gives the reduced ketone.

$$CH_2 \diagdown \begin{array}{c} CH_2 \\ \end{array} \; CH \diagup \begin{array}{c} O \\ \| \\ C \end{array} R \xrightarrow{2e} CH_2 \diagdown \begin{array}{c} \overset{..}{\overset{..}{C}H_2} \\ \end{array} CH_2 = \begin{array}{c} O^- \\ | \\ C \end{array} R \xrightarrow{2H^+} CH_2 \diagdown \begin{array}{c} CH_3 \\ \end{array} CH_2 \begin{array}{c} O \\ \| \\ C \end{array} R$$

__237__

Norin (71) and Dauben and collaborators (72-74) have shown that the bond of the cyclopropane ring which is reductively cleaved corresponds to the bond which better overlaps with the π system of the adjacent carbonyl group. Thus, when the cyclopropyl ketone exists preferentially in the cisoïd and transoïd conformations described by the Newman projections __238__ and __239__, the $C_1 - C_2$ bond is always cleaved in preference to the $C_1 - C_3$ bond.

__238__   __239__

A few representative examples will be reported here. The reduction of lumi- cholesterone (__240__) with lithium in liquid ammonia gives the spiroketone __241__ in high yield (70). Thus, the $C_1 - C_{10}$ bond which better overlaps with the π system of the carbonyl groups is cleaved. Norin (71) as well as Dauben and Deviny (72) have reported the reduction of (±)-carone (__242__) into __243__. In this case, the $C_1 - C_7$ bond was cleaved for the same stereoelectronic reason. The isomeric 3α,5- and 3β,5-cyclocholestan-6-ones __244__ and __246__ gave respectively cholestan-6-one (__245__) and 3α-methyl-5,β-A-norcholestan-6-one (__247__). Examination of models clearly showed that in the 3α-5 isomer __244__, the $C_3 - C_5$ bond, and in the 3β-5 isomer __246__, the $C_4 - C_5$ bond were the better overlapping bonds. The isomeric 4,5-methano steroids __248__ and __250__ were also studied and the same arguments for the direction of cleavage as mentioned for carone can be applied here. Both materials upon reaction with lithium in liquid ammonia give in high yield a single product __249__ and __251__ respec- tively.

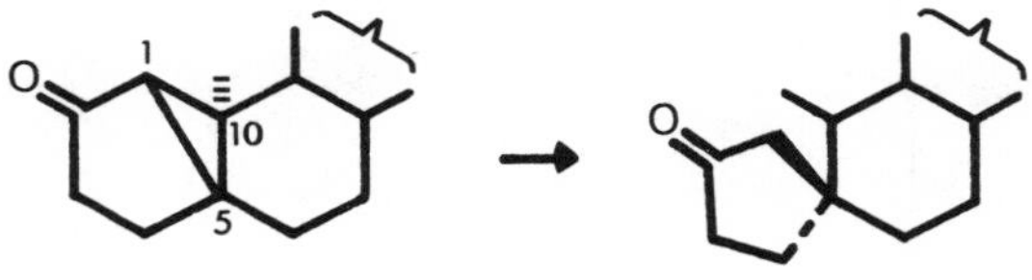

240   241

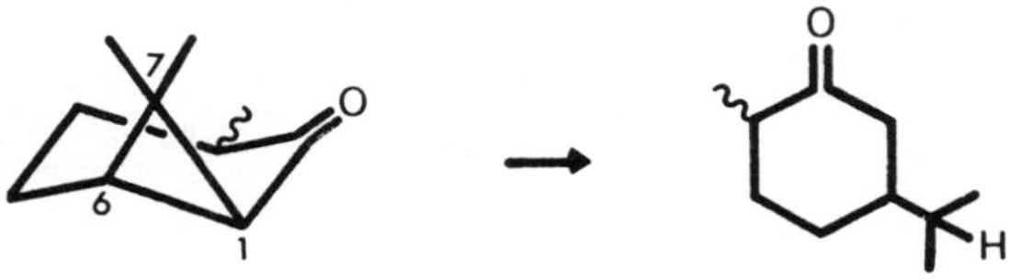

242   243

244   245

246   247

248   249

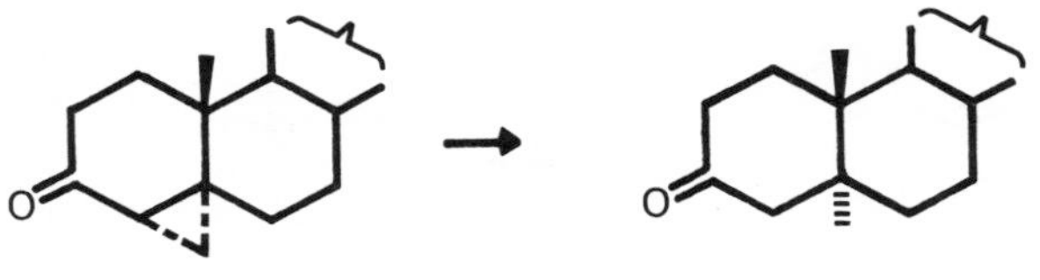

250   251

Dauben and Wolf (73) have also studied the chemical reduction of a series
of acyclic cyclopropyl ketones (252-254) with lithium in liquid ammonia.

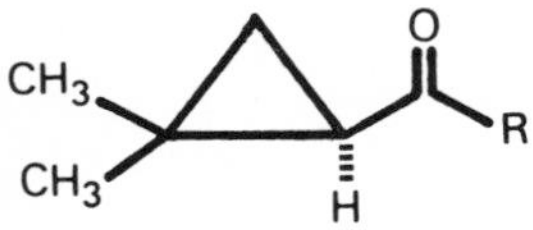

252 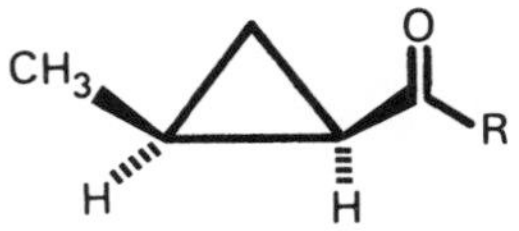   253 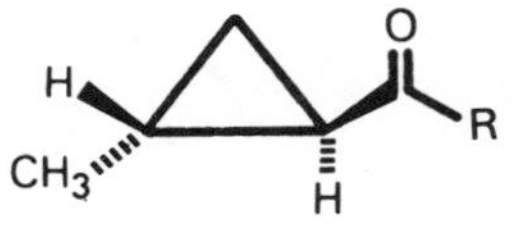   254

These compounds can give two different types of products (path a: 255 → 256 →
257 and path b: 255 → 258 → 259).

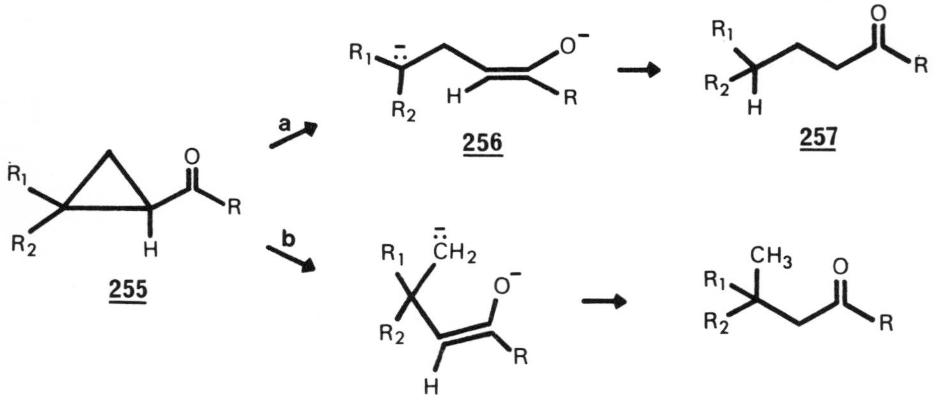

256   257

255

258   259

The preferential formation of either 257 or 259 can be rationalized via a
specific cleavage of the cyclopropyl C—C bond which has maximum overlap with
the carbonyl group in the conformation which has the least steric interactions.

The cisoïd conformation (cf. 238) is more reactive than transoïd conformation (cf. 239) (vide infra), so we will consider only the cisoïd conformations 260 and 261 for compounds having the general structure 255. Reductive
cleavage of conformations 260 and 261 must give respectively compound 257
(via path a) and compound 259 (via path b).

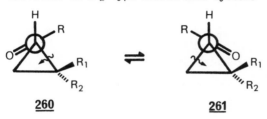

The dimethylketone **252** must exist preferentially in conformation **260** (R$_1$=R$_2$= CH$_3$) which has much less steric interaction than conformation **261** (R$_1$=R$_2$= CH$_3$). The reduction of **252** must therefore give mainly compound **257** via path **a**. Experimentally, an 8:2 ratio of **257** and **259** (R$_1$=R$_2$=CH$_3$) was indeed obtained. The cis ketone **263** must also exist preferentially in conformation **260** (R$_1$=CH$_3$; R$_2$=H), so, product **257** via path **a** should be formed predominantly. This was confirmed experimentally as a 9:1 mixture of **257** and **259** (R$_1$= CH$_3$, R$_2$=H) was formed. The trans ketone **254** must exist as an approximately equal mixture of conformations **260** and **261** (R$_1$=H, R$_2$=CH$_3$) so, an equal proportion of products from path **a** and path **b** should be expected. Experimentally, it was found that path **b** is preferred in a 9:1 ratio. The preferred formation of **259** (R$_1$=CH$_3$, R=H) could be explained because path **b** yields intermediate **258** (R$_1$=H, R$_2$=CH$_3$), a primary carbanion which is more stable than the secondary carbanion **256** (R$_1$=H, R$_2$=CH$_3$).

In another investigation, Dauben and Wolf (74) have shown that the cisoïd conformation **238** is reduced in preference to the transoïd conformation **239**. This was proven by trapping the enolate ion intermediate with acetic anhydride. Reduction of the cisoïd conformation **238** yields an enolate dianion which is protonated by ammonia to give the trans enolate ion **262**. This ion is then trapped by acetic anhydride to yield the trans enolacetate **263**. Similarly, the transoïd conformation **239** must give the cis enolacetate **265** via the cis enolate ion **264**. The compounds chosen for this study were **266**, **267**, and **268**. Accordingly, the experimental results gave a trans:cis ratio for the corresponding enolacetates of 82:18, 88:12, and 70:30 respectively.

It is also possible to reductively cleave a C—C bond (which is not part of a cyclopropane ring) when this bond is parallel to the π system of two adjacent carbonyl groups. The pentacyclic diketone **269** is a perfect example for this stereochemical arrangement and Wenkert and Yoder (75) found that **269** is readily reduced to the tetracyclic diketone **270** with zinc in acetic acid. Paquette and co-workers have reported the transformations **271** → **272**

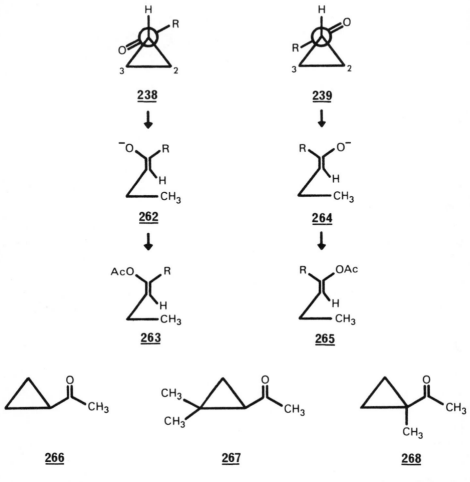

(Na, NH$_3$) (76) and the two consecutive ring cleavages $\underline{273} \rightarrow \underline{274} \rightarrow \underline{275}$ using zinc (77, see 78, 79 for other similar transformations).

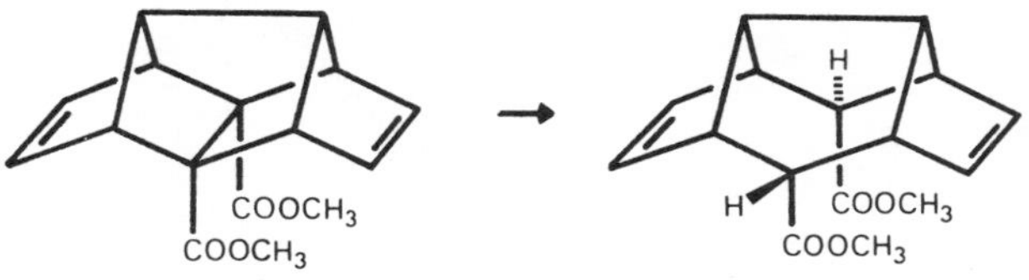

271   **272**

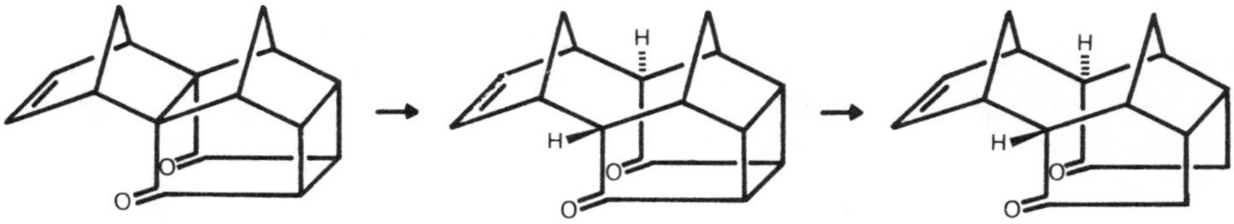

273   **274**   **275**

A similar C — C bond cleavage **276** → **278** (using zinc) has been observed by Eaton and co-workers (80). However, what is more interesting is that compound **278** can be produced directly from the bis-enone **277** upon treatment with zinc. This unusual reaction was successful because the π systems of the two double-bonds are perfectly oriented to induce the C — C bond formation.

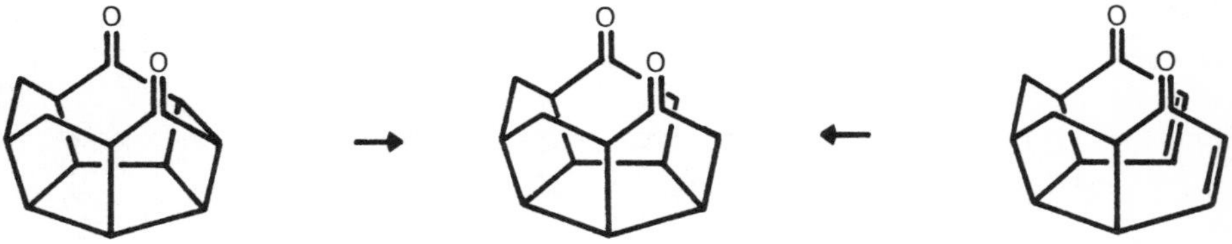

276   **278**   **277**

Reductive cleavages of C — C bonds can also take place when one and even the two carbonyl groups are replaced by other easily reduced functional groups. The only requirement appears to be the proper syn- or antiperiplanar orientation of the reacting functional groups with the C — C bond to be cleaved. The following transformations **279** → **280** and **281** → **282** reported by Baker and Davis (81) and Paquette and Wyvratt (82) respectively constitute representative cases.

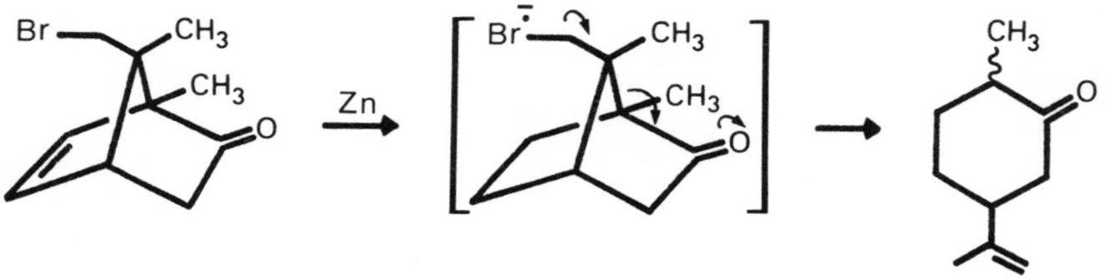

Br
CH3
CH3
O
Zn
Br
CH3
CH3
O
CH3
O

279 280

CH2OMs
CH2OMs
NaI/HMPT
CH2
CH2

281 282

DOUBLE-BOND FORMATION

E2 and E1cB

We have previously discussed (Chapter 5) that the formation of a double-bond from a carbonium ion requires that the bond to the hydrogen on the adjacent carbon should be parallel to the p-orbital of the positively charged carbon atom. This topic will not be further discussed in this Chapter which is concerned with double-bond formation under basic conditions.

In principle, stereoelectronic effects should play an important role in the formation of double-bonds in base-promoted eliminations of HX.

H
C
C
X
base
C=C

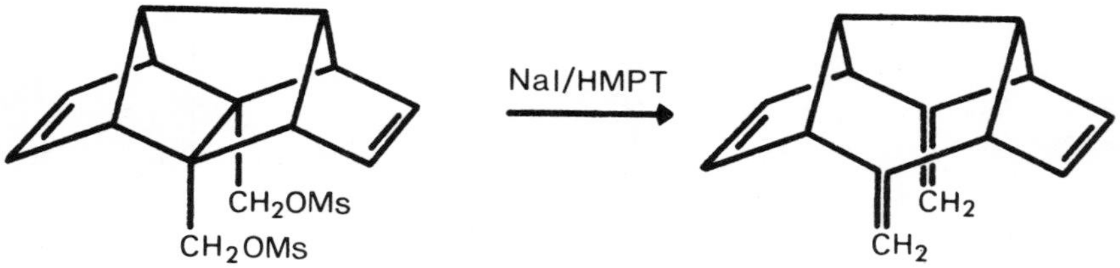

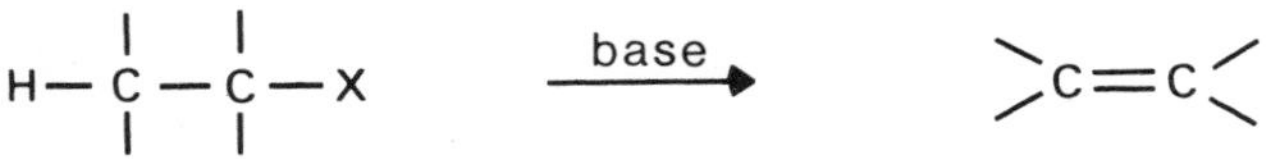

This reaction can take place by either a step-wise or a concerted mechanism. In the non-concerted mechanism (E1cB), the C—H bond is ruptured prior to the scission of the C—X bond. Thus, strong stereoelectronic effects should be observed depending on the relative orientation of the electron pair and the C—X bond in 283; indeed, when the electron pair is oriented antiperiplanar to the C—X bond, it should ease the formation of the double-bond.

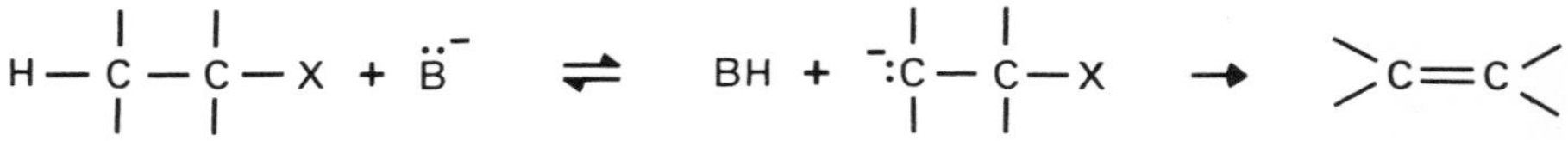

$$H - \overset{|}{\underset{|}{C}} - \overset{|}{\underset{|}{C}} - X + \overset{..}{B}{}^{-} \quad \rightleftharpoons \quad BH + \overset{-}{:}\overset{|}{\underset{|}{C}} - \overset{|}{\underset{|}{C}} - X \quad \rightarrow \quad \overset{}{\underset{}{>}}C{=}C\overset{}{\underset{}{<}}$$

283

A great deal of experimental results have been rationalized on that basis; for instance, compound **284** gives only the olefin **286** via an anti process while the isomer **285** gave a mixture of olefin **286** (syn mode) and **287** (anti mode).

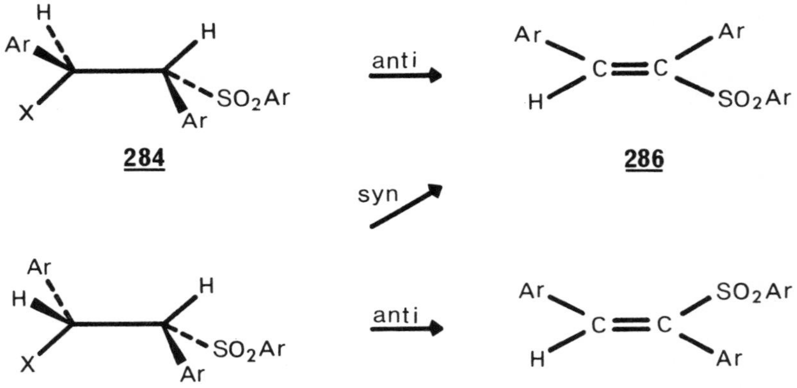

284 anti→ **286**

syn ↗

285 anti→ **287**

This result has been clearly explained by Bartsch and Závada as quoted from their recent review on the olefin-forming E2 and E1cB eliminations (83).

"Proton abstraction from the erythro substrate **284** leads to a mixture of rapidly equilibrating pyramidal carbanions **288-290**. However, the elimination proceeds only via carbanion **288** because its electron pair is ideally situated for expulsion of the leaving group and at the same time, Ar-ArSO$_2$ steric interactions are minimized in carbanion **288**. Therefore a clean anti elimination takes place."

"In elimination from the threo compound **285**, the situation is different because the stereoelectronic and conformational factors are not cooperative. Since expulsion of the leaving group from the stereoelectronically preferred carbanion **291** is rendered difficult by severe Ar-ArSO$_2$ steric interactions, the anti elimination is slowed down and syn elimination via carbanions **292** and **293** become competitive."

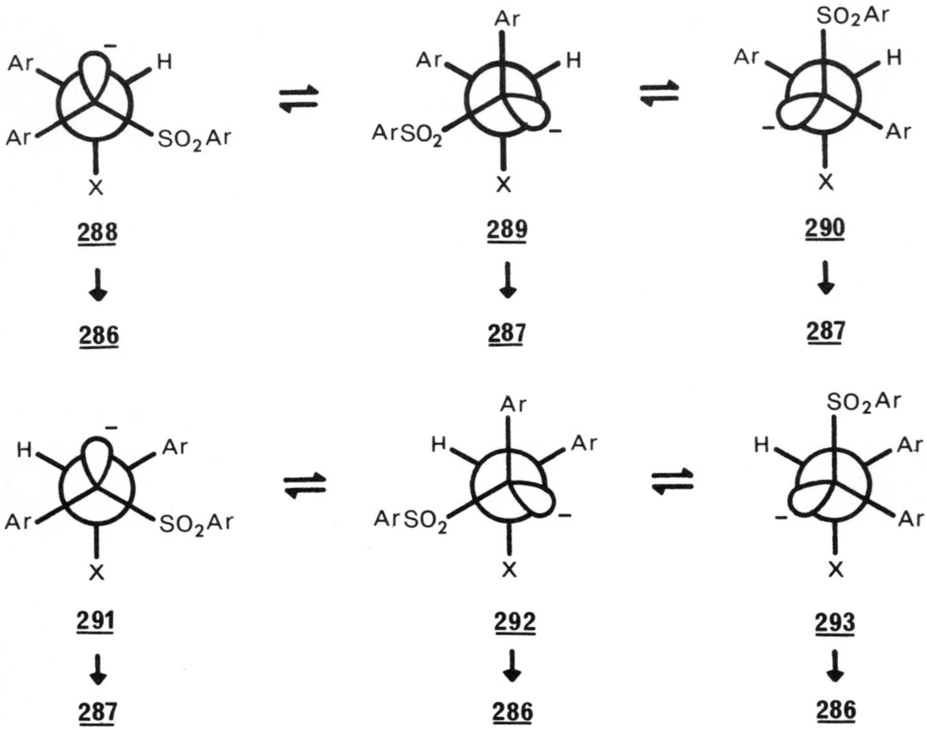

The same authors have further pointed out that the syn elimination from **292** and **293** might require a configurational inversion of the carbanion, so that the electron pair becomes antiperiplanar to the leaving group.

In the concerted mechanism (E2), both the C—H and the C—X bonds are cleaved simultaneously via a single transition state (**294**). The concerted 1,2-elimination depends however upon the dihedral angle θ between the C—H and the C—X bonds, which are going to be broken in the activated complex. Experiments on rigid systems have shown that the activation energy for elimination has two minima corresponding to the antiperiplanar (θ=180°C, cf. **295**) and synperiplanar (θ=0°C, cf. **296**) arrangements of the departing groups. This is a strong indication that stereoelectronic effects play an important role in this reaction.

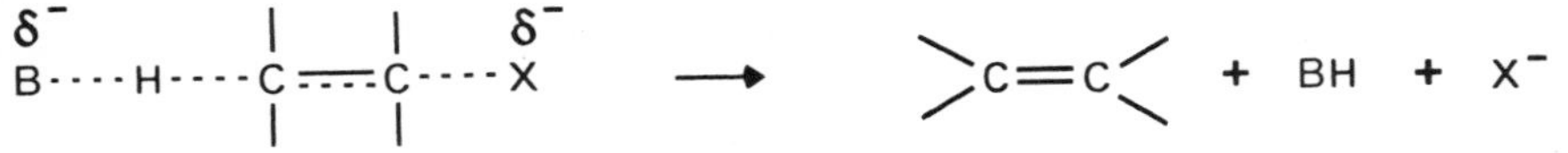

294

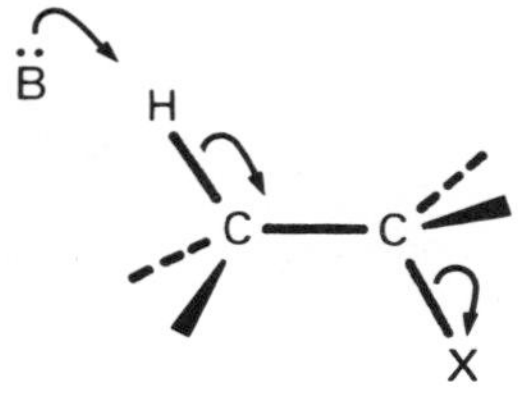

295 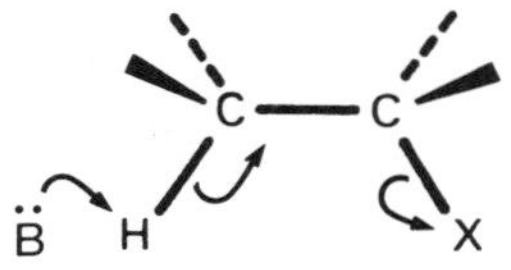   **296**

In conformationally mobile systems, both syn and anti eliminations are theoretically possible. The anti elimination should be favored electronically
over the syn elimination because the electron pair of the C−H bond is antiperiplanar to the leaving group. It has also been suggested (84, 85) that
the syn elimination might require a configurational inversion at the C−H
bond, so the electron pair of that bond becomes antiperiplanar to the C−X
bond (**297** → **298**).

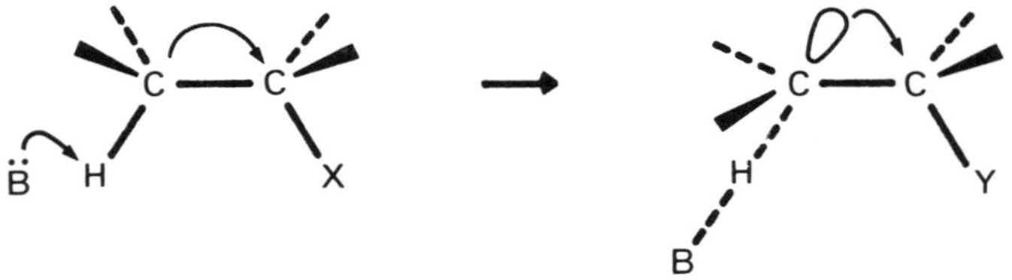

297   **298**

It would appear safe to conclude that where stereoelectronic effects alone
are operating, the anti elimination process is favored over the syn. There
are however several other parameters which are also important, such as the
effects of the nucleophile, the solvent, the alkyl structure of the substrate and the nature of the leaving groups. Any of these variables is capa-
ble of completely reversing the stereochemical course of a concerted elimination reaction (83).

The stereochemical implication for the ejection of a leaving group in the
β-position of a carbonyl group which yields an α,β-unsaturated system has
been treated elsewhere (see p. 233). It may be pointed out here that double
bond formation through the opening of a cyclopropane ring should also take
place following the same stereoelectronic principle. One example of such a
reaction is the transformation of β,γ-cyclopropyl-δ-hydroxyketone **299** which
is smoothly converted into the dienone **300** (86) under acid conditions.

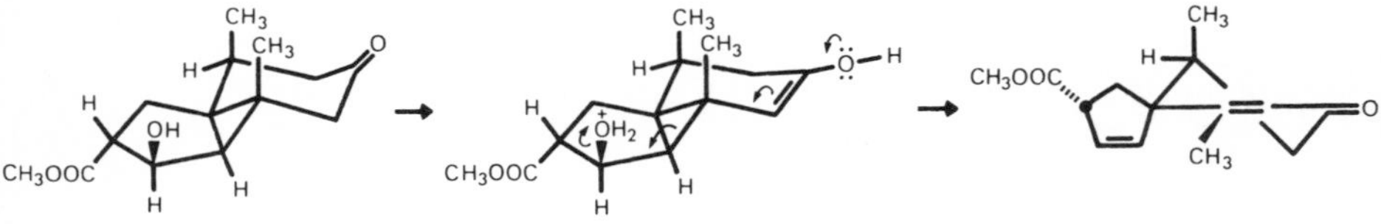

299   **300**

Julia and co-workers (87, 88) have developed a novel synthesis of homoallyl-
ic bromides by rearrangement of cyclopropyl carbinols on treatment with
hydrobromic acid. For instance, the secondary cyclopropyl carbinols **301**
have been converted into trans-bromo olefins **302** with 90-95% stereoselectivity.

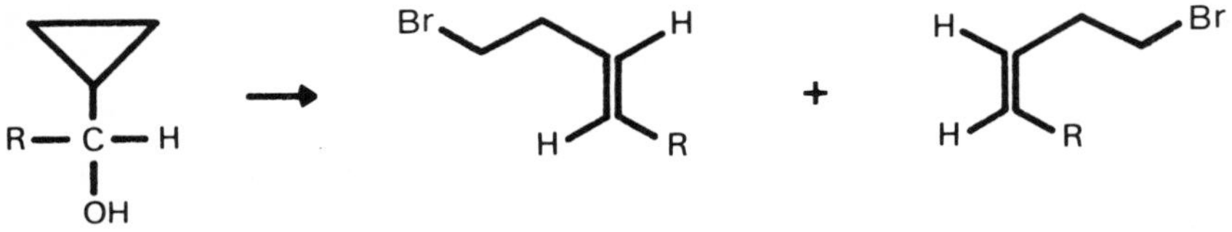

301   **302**   **303**

304   **305**

The Newman projections **304** and **305** illustrate the required transition state
geometries for a concerted process leading to the trans and cis-bromo olefins
302 and **303** respectively. Models show that severe non-bonded steric interactions between the C_1 and C_3 methylene groups and the R group strongly
destabilize arrangement **305** relative to **304** so that the predominant formation of the trans olefin **302** (from **304**) is readily explained.

In a modification of Julia's procedure, Johnson and co-workers (89) have
shown that the treatment of cyclopropyl bromide (**301**, OH=Br) with anhydrous
zinc bromide in ether gave smoothly the trans-homoallylic bromide **302**. Only
trace amounts of the cis isomer **303** were formed. This result can be ratio-

nalized by using a combination of the steric and stereoelectronic arguments as discussed above.

1,4-Elimination

Grob (90) has carried out an extensive study on the fragmentation of γ-amino alcohol derivatives. He has shown that a one-step synchroneous fragmentation takes place only when the compound can adopt conformation **306** where the leaving group is antiperiplanar to the C_2-C_3 bond and the nitrogen electron pair is oriented antiperiplanar to the C_2-C_3 bond. When these γ-amino alcohol derivatives have a stereochemical arrangement other than **306**, substitution of the leaving group and elimination reactions take place, but fragmentation is not observed.

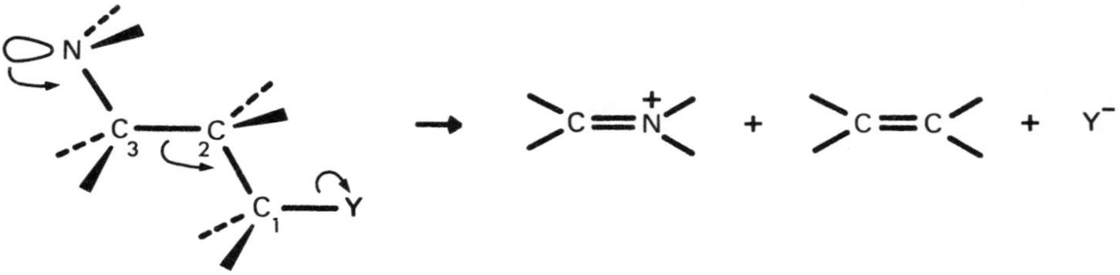

306

For instance, the equatorial chloroisomer **307** gave smoothly the cyclic iminium ion **308**. On the other hand, the axial isomer **309** yielded substitution and elimination products but no **308**. Also, **307** reacts 13,500 times faster than the carbon analog of **307** (NCH_3 replaced by CH_2).

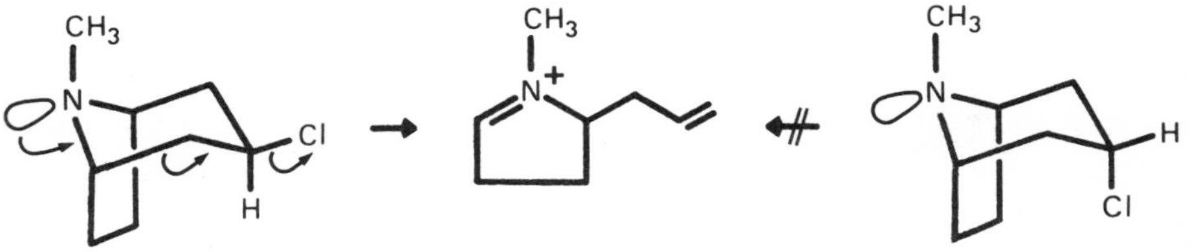

307   **308**   **309**

Similarly, compound **310** gave the iminium ion **311** in quantitative yield, while the isomer **312** reacted at a slower rate to give a complex mixture of products. Also, **310** is 35 times more reactive than the carbon analog of **310** (($CH_3)_2N$ replaced by ($CH_3)_2CH$).

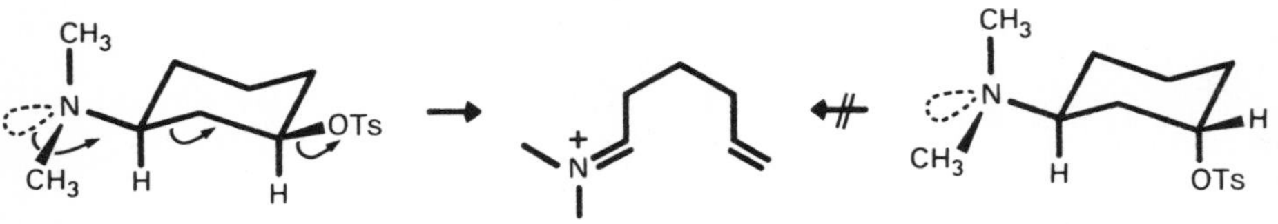

310   **311**   **312**

The cis bicyclic chloroamine **313** reacted at a faster rate than the corresponding decalin system **313**, (NCH$_3$ replaced by CH$_2$) and gave exclusively the cyclohexene iminium salt **314** while the chloroamine **315** reacted to give a mixture of other products.

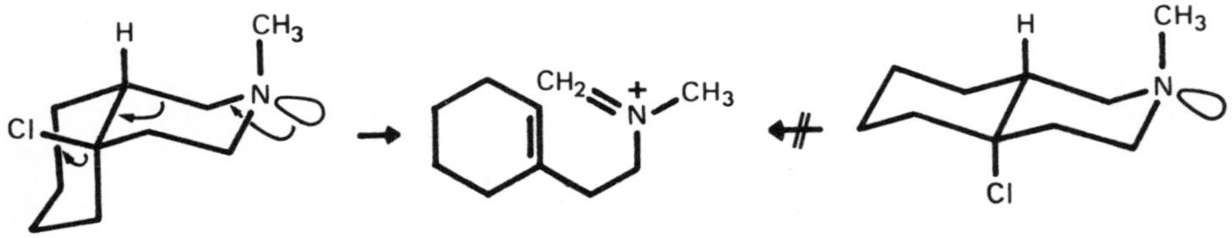

313   **314**   **315**

The bicyclic aminotosylates **316** and **317** gave exclusively the trans and the cis iminium salts **318** and **319** respectively. On the other hand, the other three isomers **320**, **321**, and **322** gave only substitution and elimination reactions.

316   **318**

317   **319**

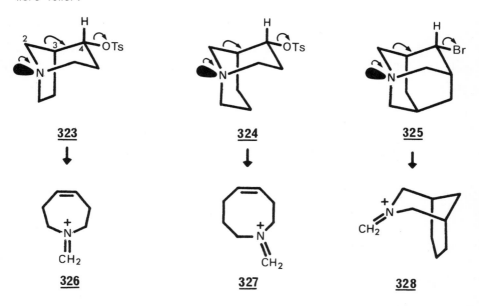

320   321   322

The aminotosylates **323** and **324** and the equatorial 4-bromo-adamantane **325** undergo a quantitative fragmentation to give the iminium salts **326**, **327**, and **328** respectively (91). Moreover, they react some 10^3, 10^4, and 10^5 times faster than their respective nitrogen-free parent compounds. These azabicyclic systems thus react by a concerted mechanism because the nitrogen electron pair and the $C_2 - C_3$ bond and the leaving group are respectively oriented antiperiplanar. Unexpectedly, fragmentation also took place with the C-4 isomeric products of **323**, **324**, and **325**, but the rates of the reactions were lower.

Fragmentation can also occur when the heteroatom is an oxygen atom and again the concerted mechanism is restricted to the molecular geometry **329** where the $C_2 - C_3$ bond is antiperiplanar to the leaving group. There are several examples of this fragmentation in the literature and only a few typical examples will be reported here.

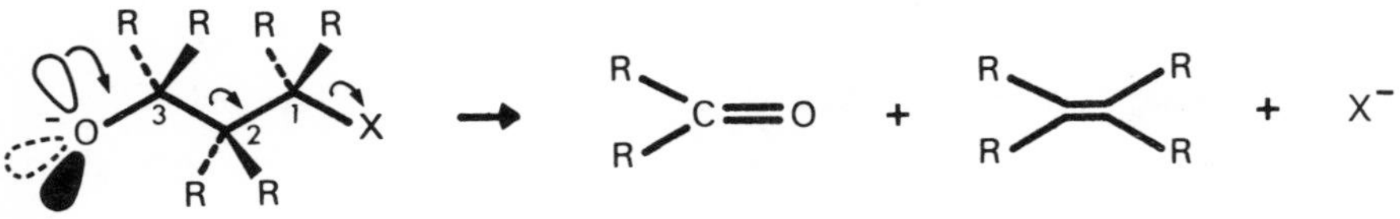

329

Corey, Mitra, and Uda (92) in their synthesis of caryophylene have found that the tricyclic hydroxy-mesylate **330** was converted under basic conditions into the bicyclic ketone **331** having the double-bond with the desired E geometry.

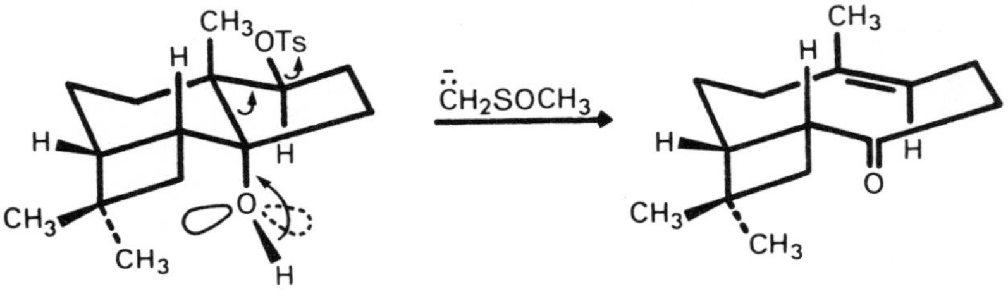

330   **331**

Reduction of the bicyclic ketotosylate **332** which gave the exo-methylene primary alcohol product **335** via the intermediates **333** and **334** was observed by Kraus (93).

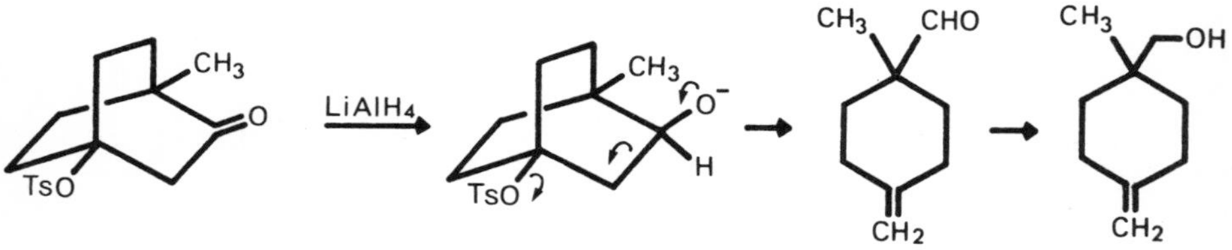

332   **333**   **334**   **335**

Siddall and co-workers (94) in their synthesis of juvenile hormones observed the clean base-catalyzed conversion of **336** into **337** while Marshall and Brady (95) utilized the fragmentation **338** → **339** in their synthesis of hinesol.

Wharton and Hiegel (96) found that upon treatment under basic conditions, cis and trans hydroxytosylates **340** and **341** yielded the trans olefin **342** while the cis compound **343** gave the cis olefin **344**. On the other hand, **345** gave neither **342** nor **344**.

336 **337**

338 **339**

340 **342** **341**

343 **344** **345**

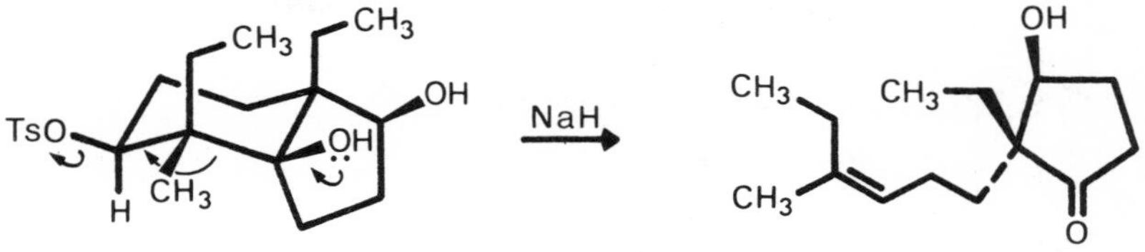

Fischli, Branca, and Daly (97) have reported that epoxysulfone 346 gave stereospecifically the bicyclic alcohol 347 upon treatment with sodium amide in refluxing toluene. Treatment of 347 with potassium t-butoxide gave the enlarged ring ketone 348 (for a similar fragmentation, see ref. 98).

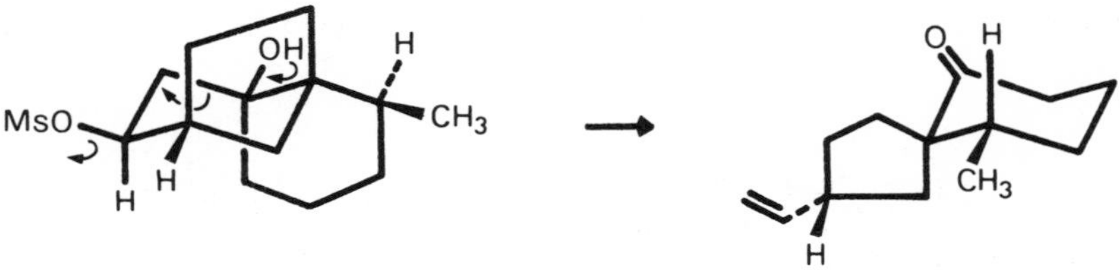

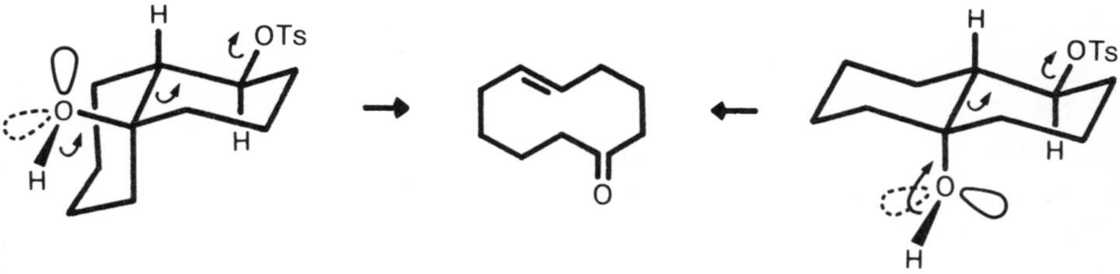

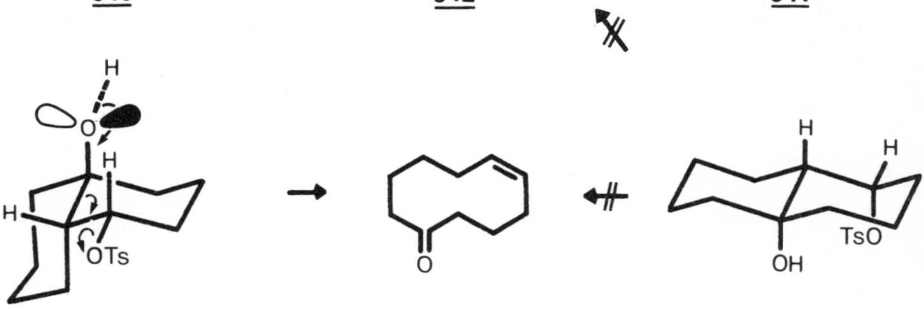

346 347 348

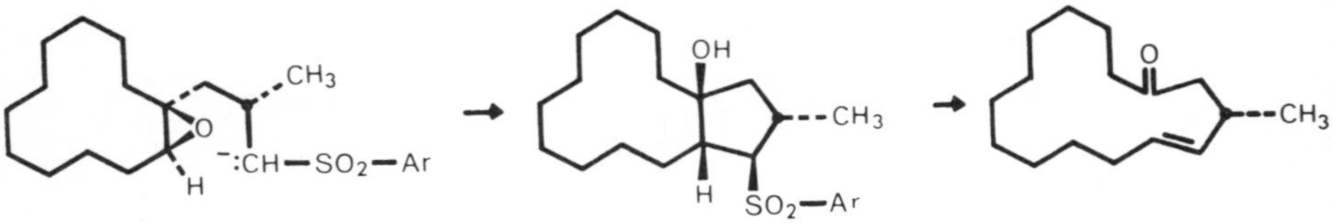

Treatment of the chlorobenzoate 349 with potassium t-butoxide gave the cyclopentane aldehyde 350 (99) while the reduction ($LiAlH_4$) of chloroketone 351 gave the cyclohexene alcohol 353 via the fragmentation of the intermediate 352 (100).

349 350

351 352 353

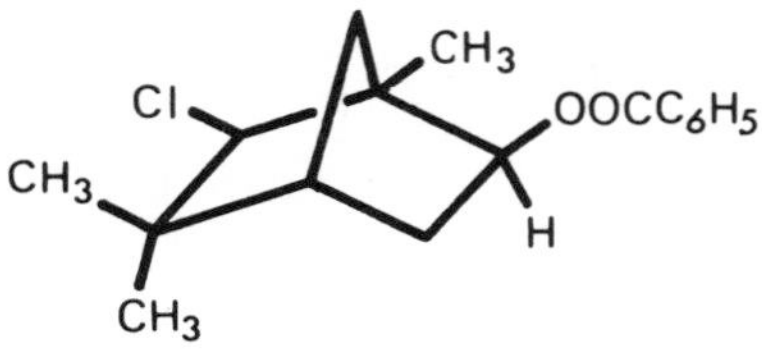

Interestingly, Marshall and Bundy (101) have found that the hydroboration of unsaturated mesylate 354 followed by usual basic treatment gave smoothly the diene 356 via the fragmentation of the intermediate 355. Note that the fragmentation can take place with the two possible stereoisomers 357 and 358.

354 355 356

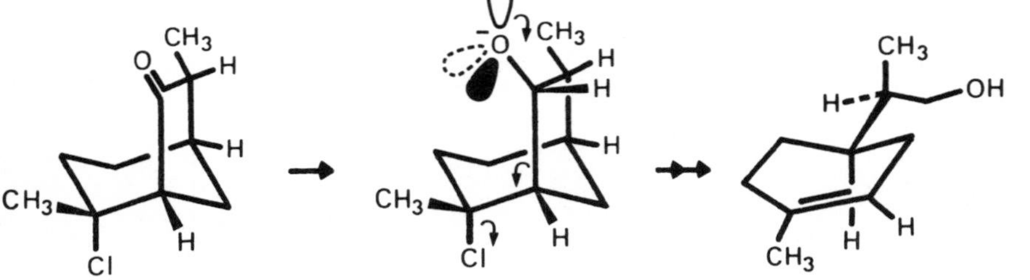

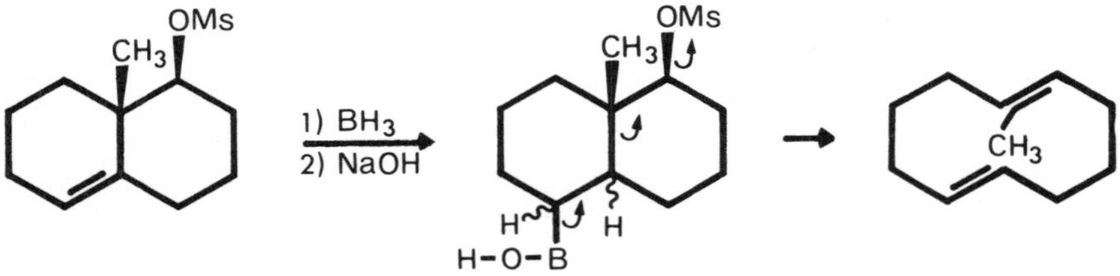

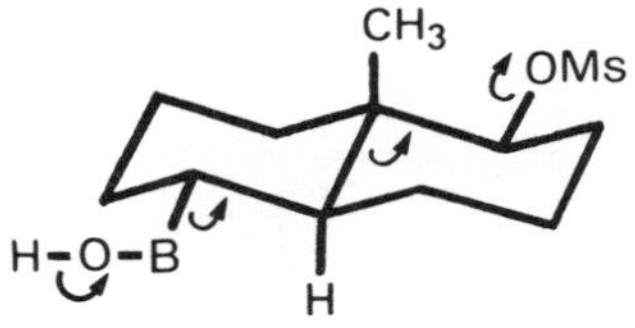

357 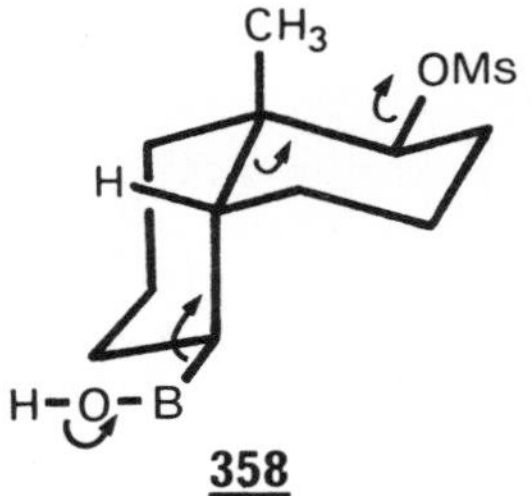   358

When two oxygen atoms can participate as in **359**, the fragmentation process should occur readily.

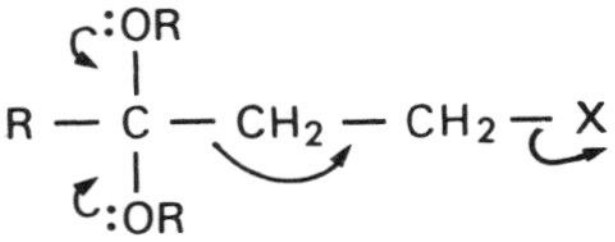

359

It has been reported (102) that the equatorial bicyclic ketotosylate **360** (R=CH₃) is transformed into monocyclic olefin ester **362** (R=CH₃) by treatment with sodium ethoxide while the axial isomer **363** yielded the bicyclic olefin **364**. Similarly, compound **360** (R=COOCH₃) gave **362** (R=COOCH₃) (103). Thus, reactions with the equatorial tosylates take place via the stereoelectronically controlled fragmentation of intermediate **361** (R=CH₃ or COOC₂H₅).

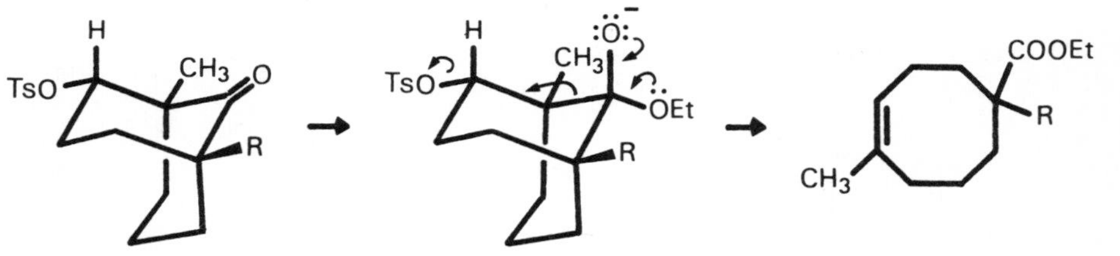

360   **361**   **362**

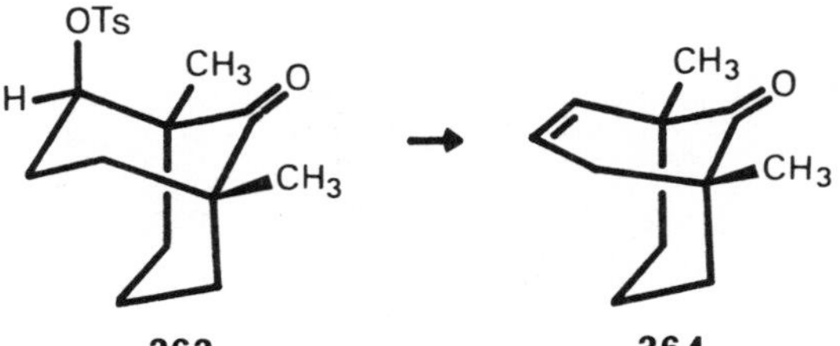

363   **364**

In another study, Buchanan and McLay (104) found that the equatorial bicyclic ketotosylate **365** gave the seven-membered olefinic ester **367** while the
axial isomeric tosylate **368** gave the seven-membered olefinic ester **371**.
Again, the reaction of the equatorial tosylate occurs through the fragmentation of intermediate **366**. The transformation **368** → **371** can be explained by
the stereoelectronically controlled retro-Dieckmann fragmentation of **369**
followed by the elimination of the tosylate group from **370**.

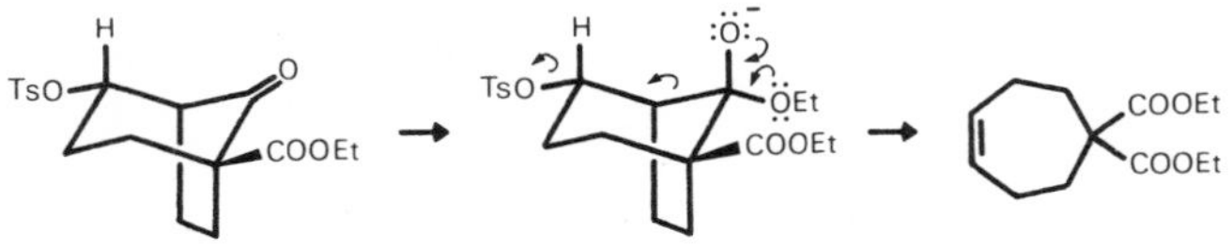

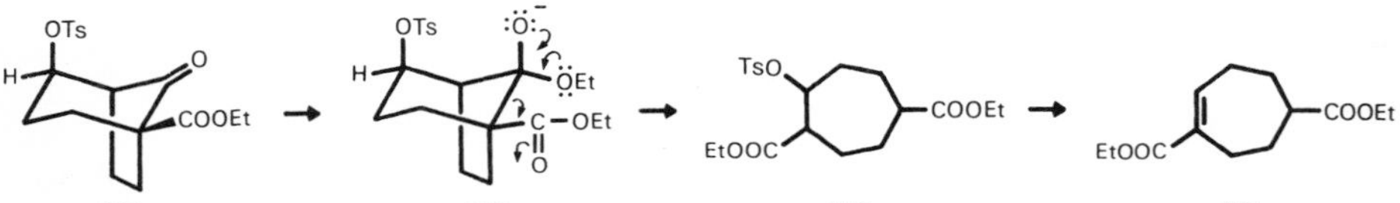

The facile acid-catalyzed conversion of ryanodol (**372**) into anhydroryanodol
(**373**) is another example of this process (105, 106). The fragmentation also
occurs under basic conditions (NaH, THF), indicating that when one of two
oxygen atoms is negatively charged (**372**, $O - H_1 = O^-$), the reaction can proceed even if the leaving group is a hydroxyl ion (107).

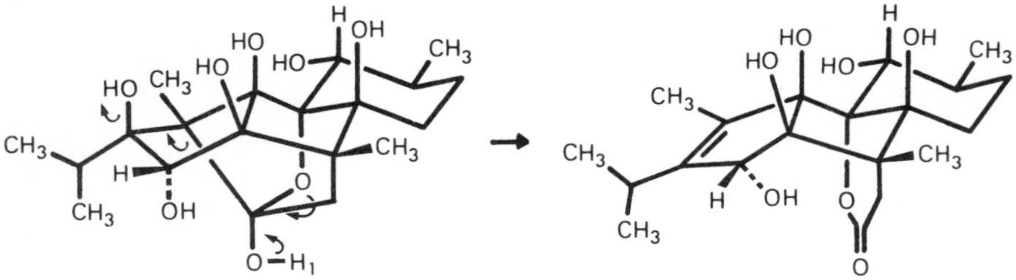

A similar fragmentation was also observed in the course of the total synthesis of ryanodol (108). Intermediate **374** was cleanly transformed into the
lactone olefin **375**. This fragmentation must however be the result of a syn
elimination with ring C of **374** in the boat form.

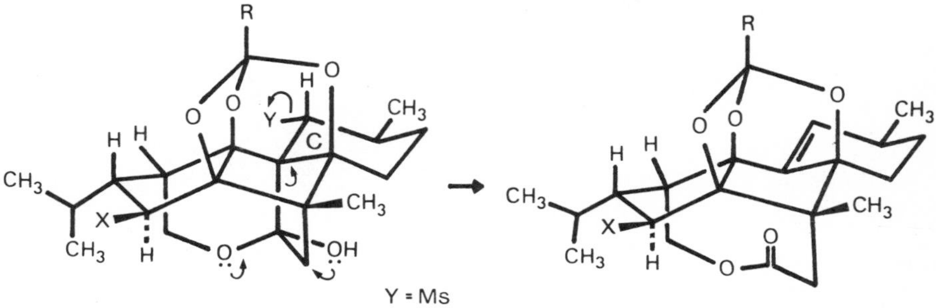

Y = Ms
X = CH(OCH$_3$)$_2$

374   **375**

Interestingly, the reduction (LiAlH$_4$), or the Grignard reaction (CH$_3$MgI), of the ketal tosylate **376** gave ketal **378** (R=H or CH$_3$) (109). The two oxygens of the ketal function in **376** have each an electron pair oriented antiperiplanar to ease the fragmentation process to give the dioxolenium ion intermediate **377**.

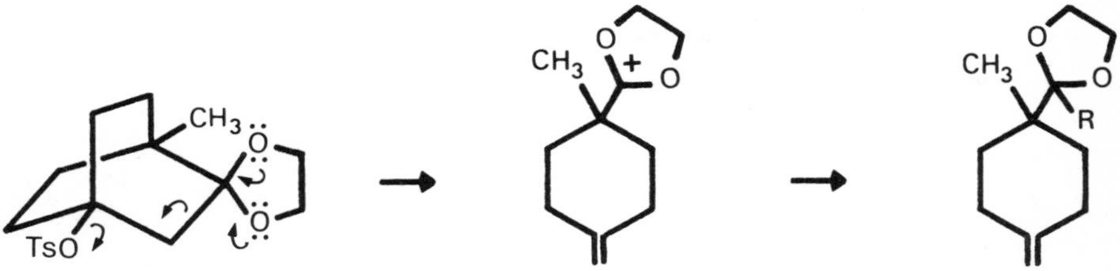

376   **377**   **378**

Eschenmoser and co-workers (110, 111) have recently carried out a series of decarboxylative double fragmentations which are completely stereoelectronically controlled. This work led to a new synthesis of unsaturated macrolides.

Heating the carboxylic acid salts **379** and **380** (X$^+$= amidinium ion) at their melting points (180 and 220°C respectively) gave the 5-E, 8-Z macrolide **382** in high yield. Eschenmoser has pointed out that these fragmentations probably take place in two consecutive steps. In the first step, the two oxygen atoms have each an electron pair properly oriented to eject the tosylate ion via the cleavage of the central C — C bond, producing the dipolar ions (**381** and **383**). Thence, in the second step, the C$_8$—C$_9$ Z double-bond is produced by a stereoelectronically controlled decarboxylation which occurs in an anti mode (C$_8$—COO$^-$ bond being antiperiplanar to the C$_9$—O bond).

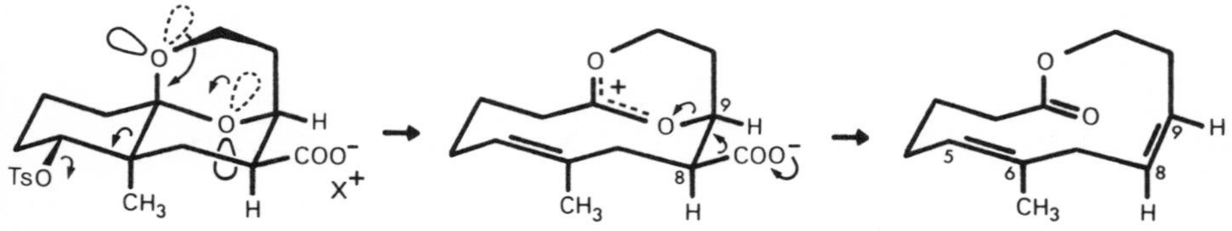

379   381   382

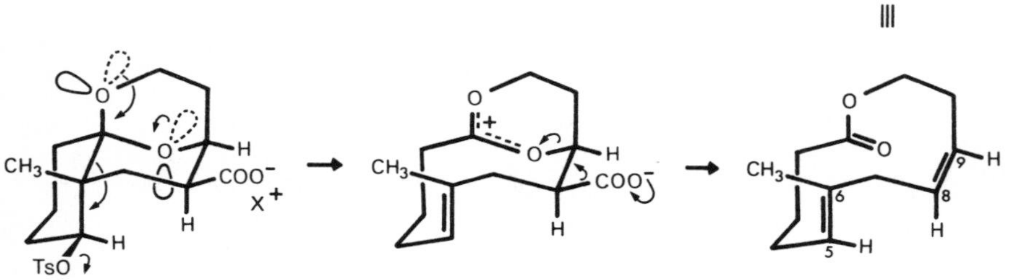

380   383   382

Confirming the above results, heating the carboxylic acid salts **384** and **385** (X=amidinium ion) for 1 min at their melting points (165° and 195°C respectively) gave a high yield of the 5-**E**, 9-**E** isomeric macrolide **386**. Note again that, in the dioxolenium intermediate **387**, the $C_{10}-COO^-$ bond becomes antiperiplanar to the C_9-O bond and consequently the decarboxylation produces an **E** double-bond.

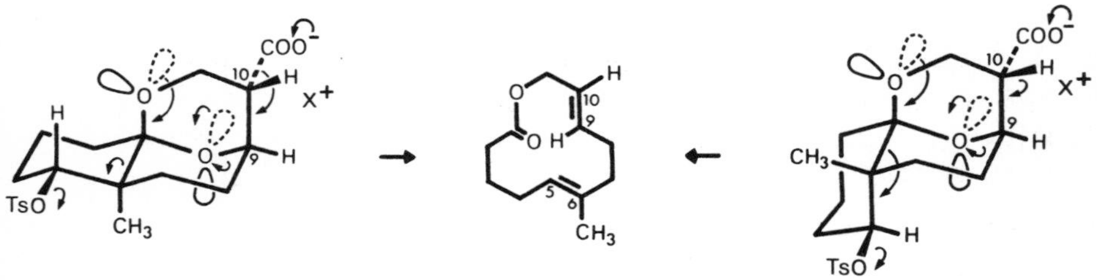

384   386   385

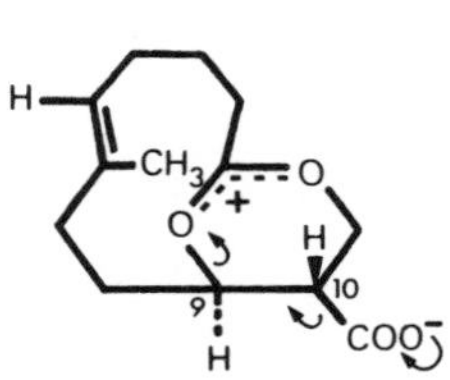

387

Retro-Claisen reaction

The retro-acylation reactions of β-ketoaldehydes (**388**, X=H), β-diketones (**388**, X=alkyl) and the retro-Claisen reaction of β-ketoesters (**388**, X=OR) occur through the formation of an intermediate **389** which gives an ester **390** and the enolate ion **391**. Protonation of **391** then gives the corresponding aldehyde (**392**, X=H), ketone (**392**, X=alkyl) or ester (**392**, X=OR).

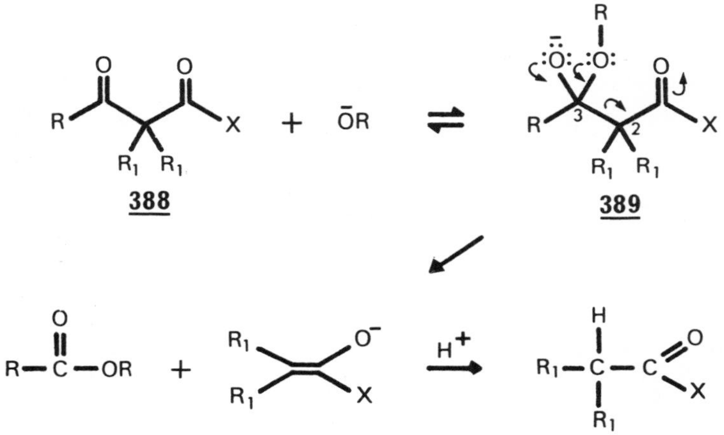

This process can take place with stereoelectronic control only if the two oxygen atoms at C$_3$ in **389** have each an electron pair oriented antiperiplanar to the C$_3$ —C$_2$ bond and if this bond is parallel to the π system of the carbonyl group. Thus a conformation such as **393** is required for this reaction. Consequently, in the reverse process **394** → **393**, the two reactants should approach each other as shown by **394**.

It is well known that bicyclic β-diketones are readily cleaved by hydroxide or alkoxide nucleophiles. For instance, diketone **395** gives readily the keto-acid **397** via intermediate **396** (112). This is a stereoelectronically allowed process.

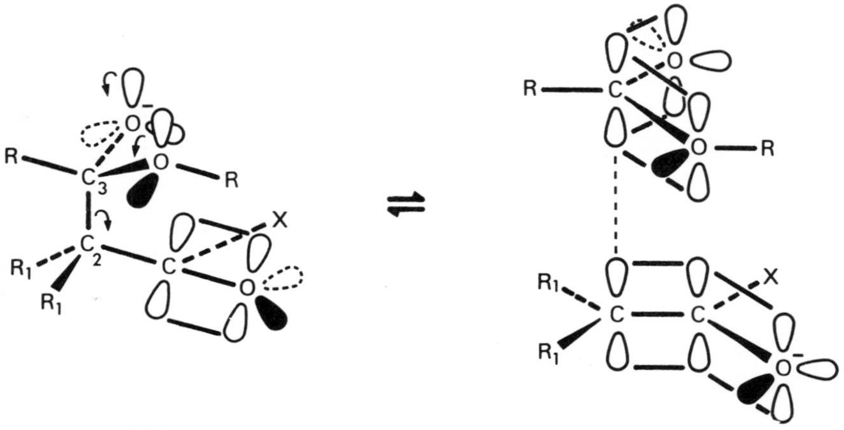

393   **394**

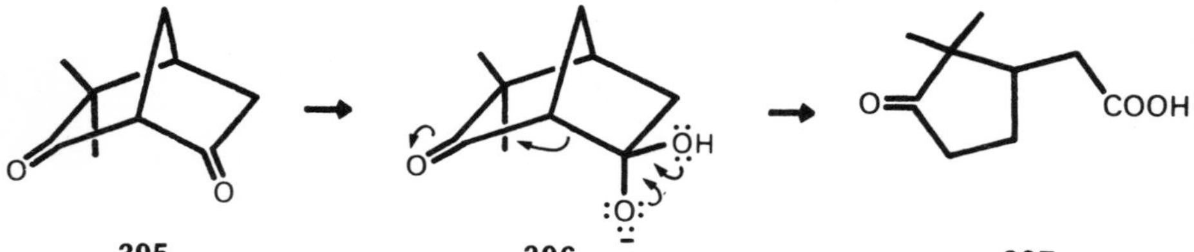

395   **396**   **397**

Several examples of retro-Claisen type fragmentations have been observed in the work toward the synthesis of ryanodol. Treatment of triketone acetate **398** with sodium methoxide gave specifically the cis-enedione ester **399**. On heating, **399** isomerized to the more stable trans-enedione **400** (113, 114). Methoxide ion reacts with **398** to give intermediate **401** which undergoes a retro-Claisen fragmentation to **402**. Loss of acetate ion from **402** yields the cis isomer **399**. The last step must be faster than the rotation of the $C_3 - C_4$ bond in **402** as this would have allowed the production of the trans isomer **400** directly from **402**. It is also possible that the formation of the cis isomer is the result of the direct fragmentation (Grob type) of **401** into **399** (dotted arrow).

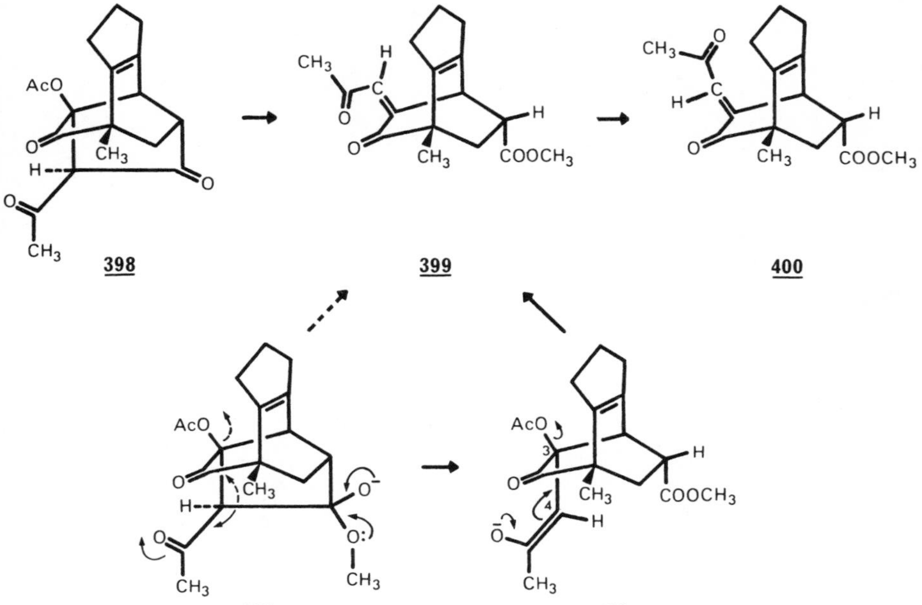

398   **399**   **400**

401   **402**

The reactivity of triketone orthocarbonate **403** is very interesting. The most hindered carbonyl group, i.e. $C_1 = O$, is the most reactive one towards nucleophilic addition. This behavior can be rationalized by the fact that the $C_1 - C_9$ bond is parallel to the π system of the carbonyl group at C-8. Indeed, this compound can be viewed as the hybrid of resonance structures **403** ↔ **404**.

Compound **403** is readily reduced with sodium borohydride at -78°C and yields the monoalcohol **405** (115). It also reacts with potassium t-butyl hydroperoxide at -20°C and gives the cis-enone-perester carbonate **406** in high yield (116). This last transformation can be explained by retro-Claisen fragmentation of intermediate **407** followed by the elimination of methoxide ion from **408**. It is also possible that **407** undergoes a direct stereoelectronically controlled Grob type fragmentation to compound **406.**

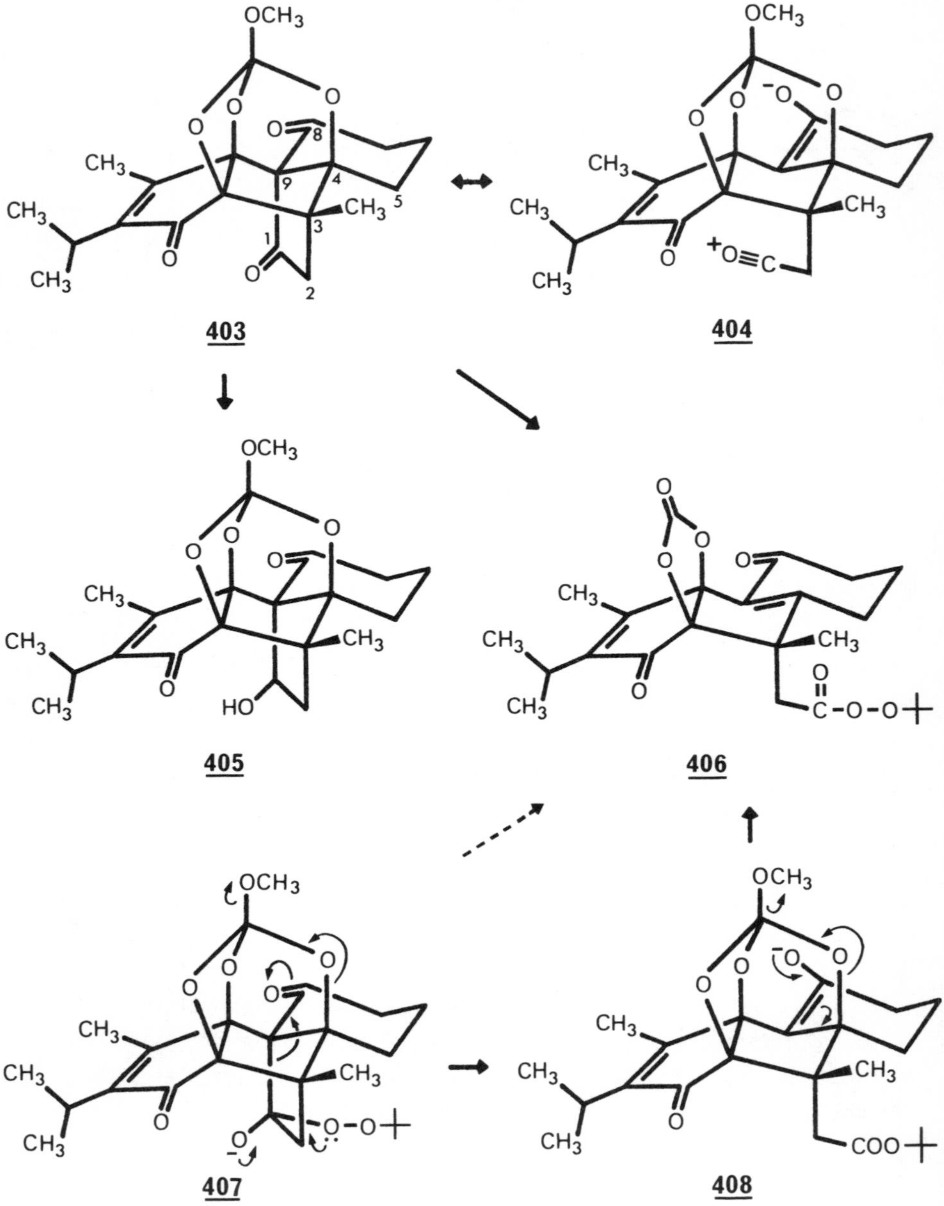

403   **404**

405   **406**

407   **408**

In another study (117), hydride reduction (LiBH(sec-Bu)$_3$) of the p-nitro-benzoate diketone **409** gave the carbonate enone lactone **410** in high yield. Reduction of p-nitrobenzoate ester must have produced the hemi-ketal intermediate **411** which underwent an internal retro-Dieckmann fragmentation to **412** which could then lose methoxide ion to form compound **410**.

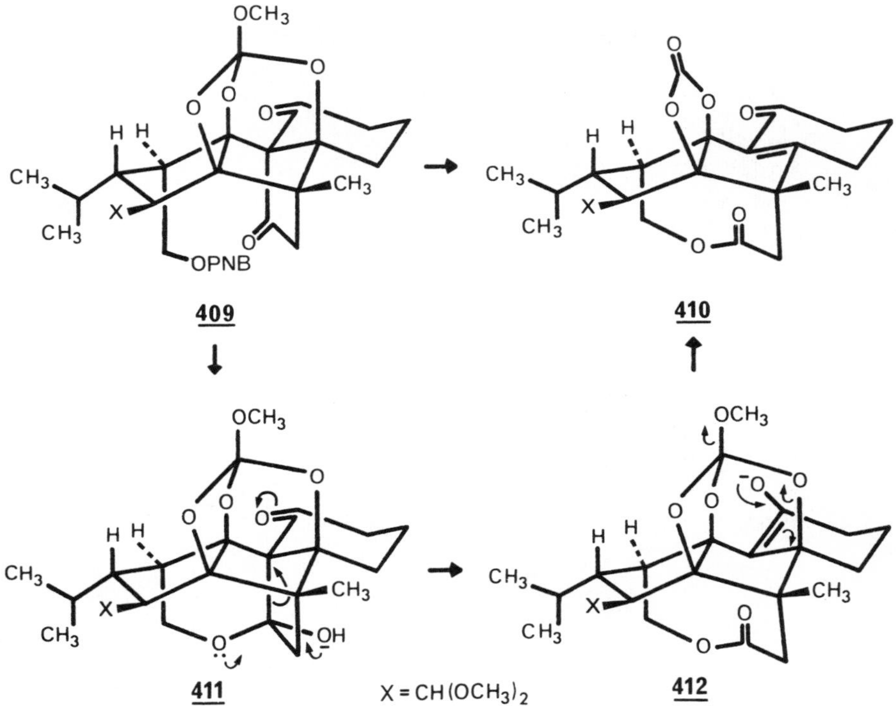

The enone lactone **410** could also be obtained in a single operation from the ozonolysis in methanol of hemi-ketal **413** (118). In this reaction, **413** produced first the diketone **414** which underwent an internal aldol condensation to **415** which is nicely set up to give **410** via the intermediate **416**.

Ozonolysis of diketone carbonate **417** in methanol afforded an almost quantitative yield of the bicyclic diene triketone hydroxy-ester **418** (119). This remarkable transformation can also be readily explained. Ozonolysis of **417** produces the tetraketone intermediate **419** followed by methanol addition to produce the hemi-ketal **420** which undergoes a retro-Claisen reaction to **421**. Then, loss of carbon dioxide from **421** yields **418**. Again, **420** could also undergo a Grob type fragmentation to yield **418** directly.

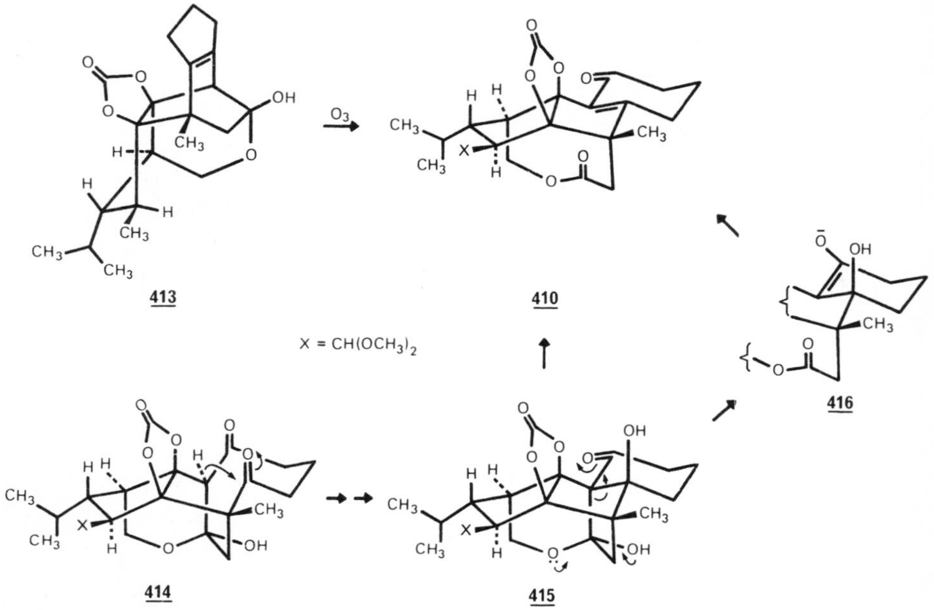

413   **410**

X = CH(OCH$_3$)$_2$

416

414   **415**

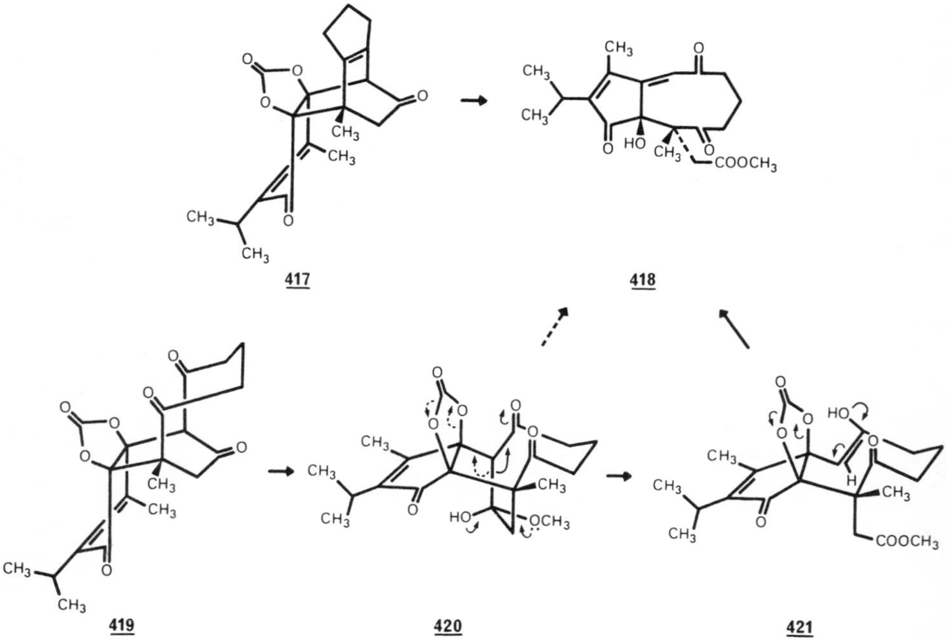

417   **418**

419   **420**   **421**

Buchanan and Young (120) have reported the methoxide-catalyzed conversion of tricyclic enedione 422 into a mixture of isomeric esters 425. This reaction can also be explained by the stereoelectronically allowed fragmentation 423 → 424.

422 423 424 425

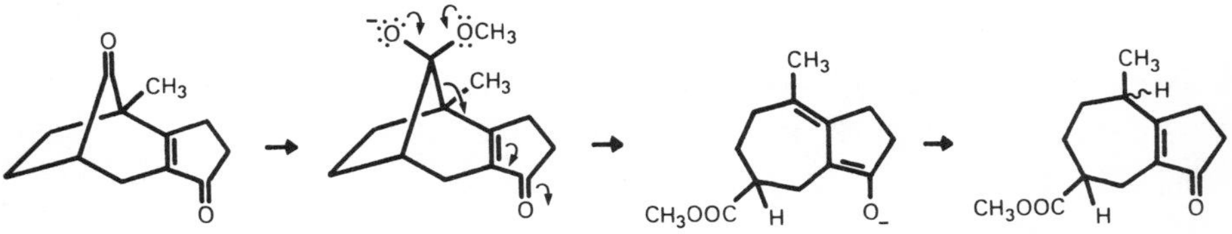

Trost and Frazee (121) have reported a very interesting stereospecific Grob type fragmentation where a cyclopropane and a cyclobutane ring are opened simultaneously. Treatment of 426 with sodium methoxide in refluxing methanol gave the E isomer 428 while the same reaction on 429 gave the Z isomer 431 stereospecifically. These fragmentations occur through the stereoelectronically allowed fragmentation of the intermediates 427 and 430 respectively. The same authors have also observed that treatment of keto-diester 426 (OCH_3 replaced by $OC(CH_3)_3$) with a methanolic solution of sodium borohydride in the presence of magnesium methoxide at 0°C and then reflux (to ensure complete transesterification) gave 434. This reaction can be visualized as 426 → 432 → 433 → 434 where the key step is equivalent to a 1,4-elimination reaction.

426 427 428

429 430 431

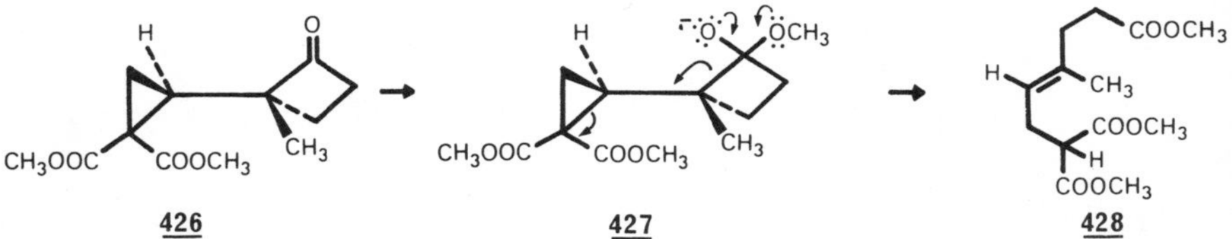

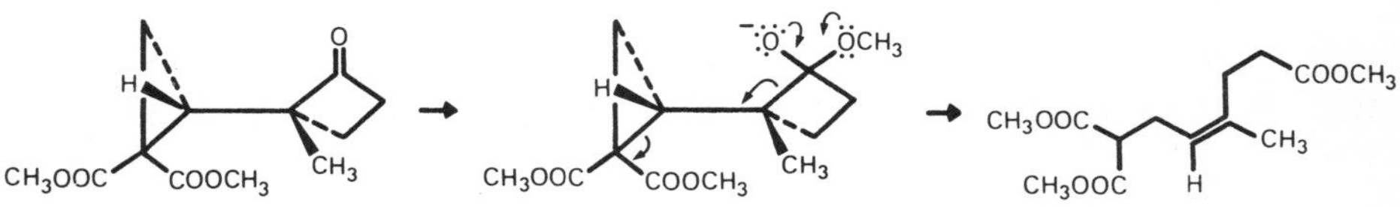

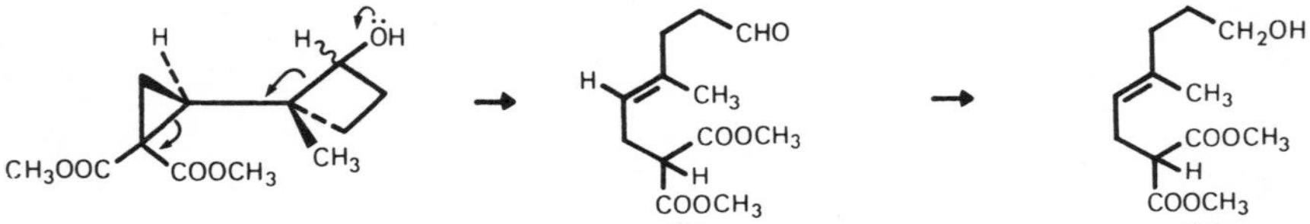

432   433   434

Eisele, Grob, Renk and Tschammer (122) have reported the quantitative transformation of the bicyclic ketone oximetosylate **435** into the cyanocarboxylic
acid **437**. This reaction occurs through the intermediate **436** and it is interesting to note that the formation of the nitrile function occurs through
an <u>anti</u> elimination process.

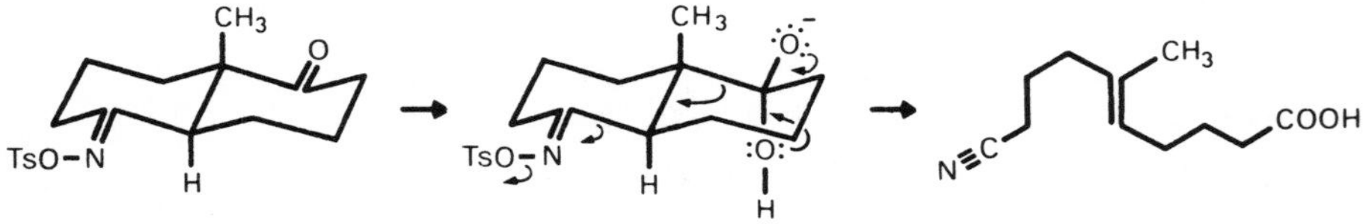

435   436   437

Enolate ion and related functions

The enolization process, <u>i.e.</u> conversion of a carbonyl compound such as **438**
into the intermediate enol **439** or enolate anion **440** is an important reaction
in organic chemistry because these intermediates can further react with
electrophiles to undergo either protonation, halogenation, alkylation, aldo-
lization, or acylation type reactions.

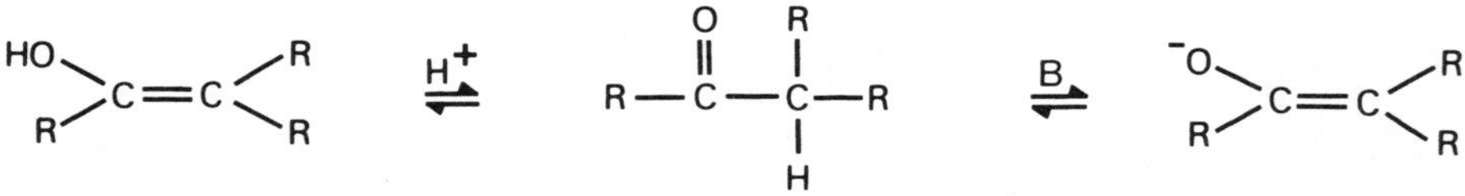

439   **438**   **440**

As early as 1953, Corey (123) observed that the kinetically controlled bromination of ketosteroids always gives the epimer in which bromide is "polar"
(<u>i.e.</u> axial) and in 1954 (124), he proposed that these results can be explained on the following theoretical basis:

> "Ketonization of an enol and the reverse reaction, enol-
> ization of a ketone proceed through the same transition
> state and hence the same geometrical requirements for
> minimizing the energy of the transition state hold
> for both reactions. The energy of the transition state
> for enolization will be at a minimum when there is

maximum opportunity for bond formation between the
sp^3 → p-orbital made available by the leaving hydrogen
and the p-orbital of the carbonyl carbon."

"In the case of a cyclohexanone this implies that in
enolization a "polar" (i.e. axial) α-hydrogen is lost
in preference to an equatorial α-hydrogen (cf. **441**
⇌ **442**). Furthermore, it follows that in the ketoniza-
tion of an enolized cyclohexanone (e.g. by bromination
or protonation) the incoming substituent should adopt
preferentially the polar (axial) orientation."

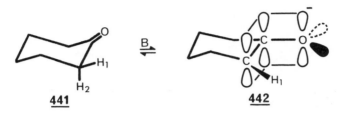

441   **442**

In a subsequent publication, Corey and Sneen (125) mentioned that this "non-
steric effect is stereochemical-electronic in nature" and on that basis
introduced the term "stereoelectronic". This work (123-125) must therefore
be considered one of the very first experimental and theoretical contributions to the principle of stereoelectronic control in organic chemistry.

Subsequently, certain authors (126-128) have supported this interpretation,
but alternative explanations have also been proposed (129-132). Another
early discussion of the stereochemistry of the enolization process was that
of Valls and Toromanoff (133). They proposed that if stereoelectronic effects are an important parameter, the cyclohexanone enolate should react by
two different pathways, one involving a chair-like transition state (**443** →
444) and the other a boat-like transition state (**443** → **445** → **446**). Thus, both
of these reactions proceed by perpendicular attack (134) of the electrophile. Their energy difference results from the difference in strain between
the chair (**444**) and the twist-boat (**445**) forms.

The strongest evidence against stereoelectronic control has been the lack
of selectivity in the base-catalyzed H — D exchange of 4-tert-butylcyclo-
hexanone, the axial-equatorial rate ratio for exchange being between 5:1
(128) and 3.5:1 (132). However, the low ratio does not hold up as a strong
argument, since this ketone can take a twist-boat conformation. Fraser and
Champagne (135) have observed a much greater selectivity in the exchange of
the conformationally fixed substrate **447**, a bridged biaryl ketone. The three-
dimensional representation **448** of the bridged biaryl ketone **447** shows that

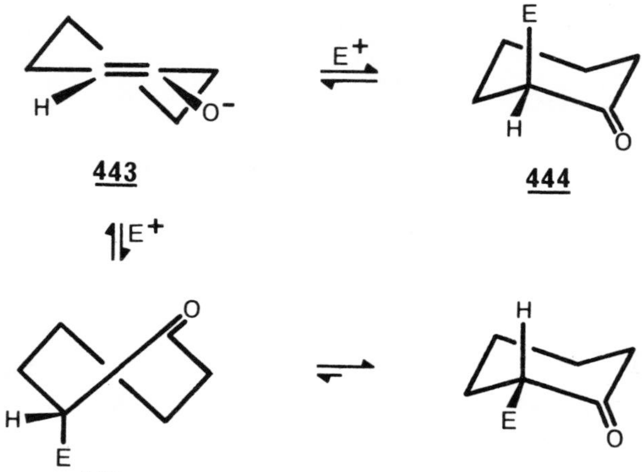

443   **444**

445   **446**

the $C-H_1$ bond is parallel to the p-orbital of the carbonyl group and the H_1/H_2 ratio of exchange was found to be 73 using sodium methoxide and 30 using sodium phenoxide in agreement with the principle of stereoelectronic control.

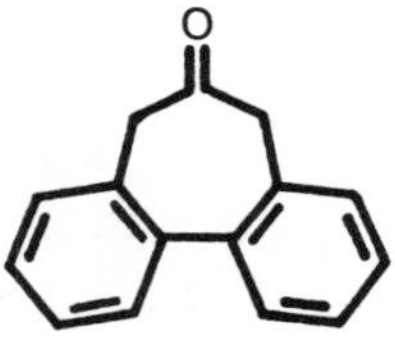

447 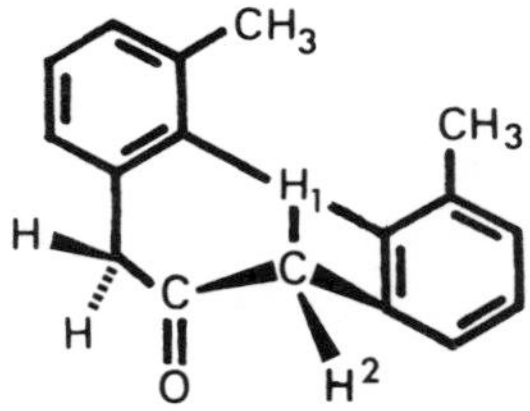   **448**

Stronger support was obtained also by Fraser and Champagne (136), who studied the base-catalyzed $H-D$ exchange reaction of 4-twistanone (**449**). A Dreiding model of **449** shows that the $CH_2 - CO$ fragment is oriented so that one $C-H$ bond alignment (H_1) gives maximum overlap with the adjacent π orbital of the carbonyl function whereas the other $C-H$ bond (H_2) is at a 30° angle to the plane of the π system (cf. the perspective formula **450**). Accordingly, when a sample of 4-twistanone was treated with sodium methoxide in methanol-0-d, the relative rates of exchange of the diastereotopic protons H_1 and H_2 were found to be in the ratio 290:1. When a weaker base was used (C_6H_5ONa), this ratio was essentially the same (280:1). The only reasonable explanation for this large difference in the rate of exchange is the effect of stereoelectronic control.

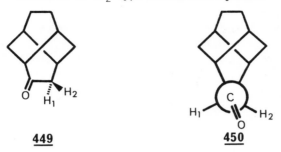

449   **450**

Interestingly, Wolfe, Schlegel, Csizmadia, and Bernardi (137) have predicted on the basis of <u>ab initio</u> MO calculations, that the enolate derived from the removal of H_1 should be more stable than that from H_2 by 18 kcal/mol when the geometry of the CH_2-CO fragment is similar to that in **450**.

Nickon and co-workers (138) have observed that brendan-2-one undergoes a remarkably easy bridgehead exchange at C-3 (**451** → **452**) under mild basic conditions (CH_3ONa, CH_3OD, 25°C). By comparison, using similar conditions, noradamantan-2-one (**453**) does not undergo C_3 exchange while bicyclo[3.2.1]-octan-2-one (**454**) exchanges only its two enolizable protons at C-3.

451   **452**   **455**

453   **454**

The cyclohexanone ring is locked in a boat form in **451**, in a chair form in **453** while it is flexible in **454**. As a result, the bridgehead C_3-H bond in **451** is appropriately aligned to overlap with the π orbital of the carbonyl group, whereas it is not in **453** and **454**. On that basis, the facile bridgehead enolization of **451** by comparison with **453** and **454** is readily explained by the principle of stereoelectronic control. This is also supported by the fact that **455** which is isomeric with **451** does not undergo exchange at the bridgehead position. Molecular models clearly indicate that the bridgehead C$_3$ — H bond in **455** is almost perpendicular to the π orbital of the carbonyl group. Consequently, no overlap is possible. These findings support the idea that prior to enolization of equatorial hydrogens, simple cyclohexanones must change from a chair to a boat-like conformation in order to improve initial stereoelectronic alignment as first suggested by Valls and Toromanoff (133) (cf. **446** → **445** → **443**).

Meerwein and Schürmann (139) found that the diketodiacid **456** was decarboxylated rapidly (→ **457**) in water under relatively mild conditions (≈180°C) for a compound of this structure, while Bootger (140) found that **458** is stable toward decarboxylation. These results together with the base-catalyzed deuterium exchange at the bridgehead position in bicyclo[3.3.1]nonan-2-one (**459** → **460**) led Schaeffer and Lark (141) to propose the following: the successful decarboxylation of **456** and the facile hydrogen exchange at the bridgehead in **459** are both due to the mobility of the cyclohexanone ring which can adopt a boat form in these compounds. In **458**, all rings are locked in the chair conformation and the decarboxylation is thus not favored stereoelectronically.

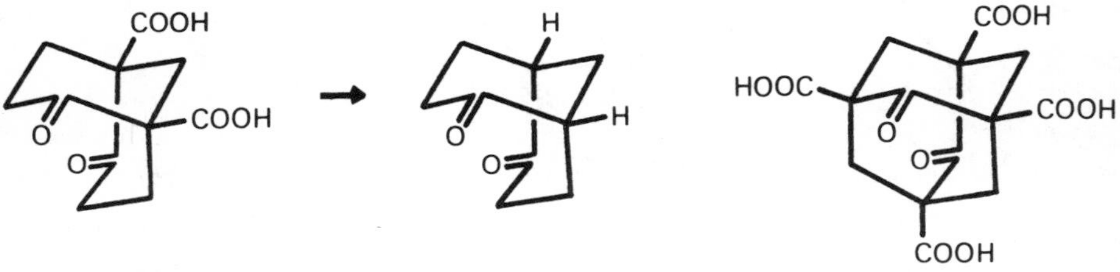

456   **457**   **458**

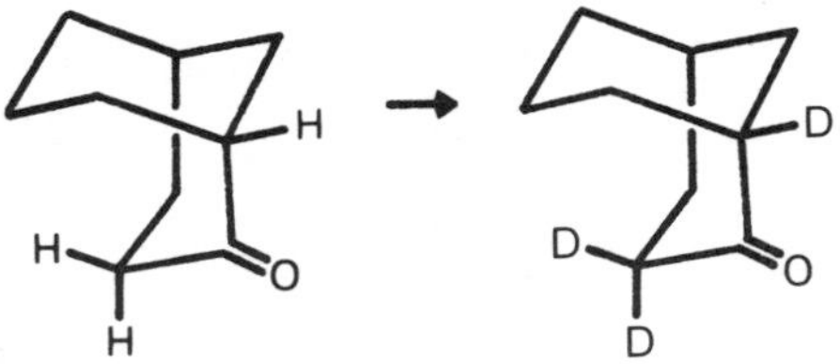

459   **460**

A more recent investigation by Yamada and collaborators (142) confirms the above results and conclusion. They have studied the base-catalyzed hydrogen-deuterium exchange at the bridgehead position of the bicyclo[3.3.1]nonan-2-one system using the five compounds 460-464. The result on the deuteration óf ketone 461 was virtually the same as that of the ketone 460, i.e. ≈45% deuteration of the bridgehead position C-1. There was no significant incor-poration of deuterium (≈3%) in ketone 462 under the same conditions whereas no deuterium was incorporated into the bridgehead of the ketone 463 which has the gem-dimethyl group at C-3. In contrast, the ketone 464 was found to be easily and exclusively monodeuterated at position C-1 (95% deuterium incorporation).

461

462 R = H
463 R = CH$_3$

464

Deuterium atom was neatly incorporated at the bridgehead position C-1 in ketone 464, the only compound in which the cyclohexane ring is locked in a boat conformation. Examination of molecular models indicates that the cyclohexanone ring can easily adopt a boat form in 460 and 461. It appears to be more difficult with ketone 462 and almost impossible with ketone 463.

These results strongly support the fact that the enolization of a cyclo-hexanone via the loss of an equatorial hydrogen must occur through a boat (or a twist-boat) conformation, a consequence of the principle of stereo-electronic control.

Using Δ^4-3-ketosteroids 465, Ringold and collaborators (143) have found that the rate ratio for axial and equatorial proton loss at C-6 is 53. The profound preference for axial proton loss (465 → 466) despite greater steric hindrance from the ß face (due to C-19 methyl group) strikingly emphasizes the importance of stereoelectronic effects. In view of this preference, these authors have essentially concluded that cases of equatorial deprotona-

tion must be considered in terms of axial (perpendicular) deprotonation
via a non-chair conformation of ring B.

465   **466**

In principle, reaction of enolate ions with electrophiles should be influenced by stereoelectronic effects. For instance, protonation of enolate
ions in acidic medium could provide information about the stereochemistry
of the forming of a C—H bond.

House, Tefertiller, and Olmstead (127) have found that the treatment of
the enolate ion **467** derived from 4-t-butylcyclohexanone with deuterium oxide
in deuterioacetic acid yielded a 7:3 mixture of the axial deuterioketone
468 and the equatorial deuterioketone **469**.

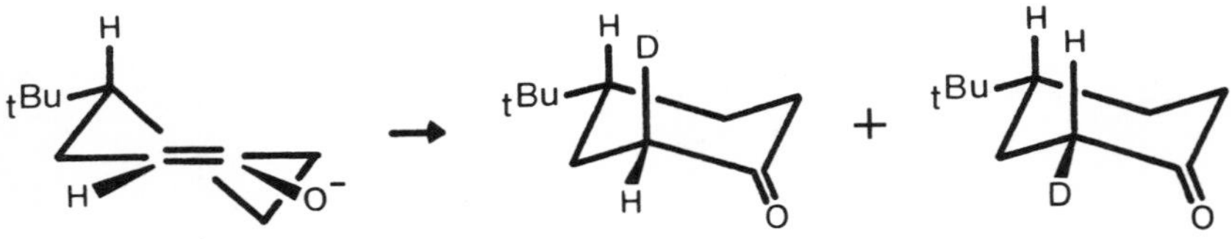

467   **468**   **469**

Axial protonation is not strongly favored. They concluded that in practice
this type of experiment is complicated by the fact that protonation of an
enolate anion can occur either at the carbon (to give **468** or **469**) or at
the oxygen atom (to yield the enol). Further reaction of the enol with aqueous acid also yields the two possible ketones **468** and **469**. Furthermore,
since the protonation steps of this strongly basic anion (either at C or O)
are diffusion-controlled (144), it is possible that the transition state
geometries for both reactions resemble the geometry of the enolate anion,
so the energy difference between the direction of attack on the enolate
is small.

The same authors studied the stereochemistry of alkylation of 4-t-butyl-
cyclohexanone. Alkylation of enolate ion **467** with triethyloxonium fluoroborate yielded a mixture of O-alkyl product and approximately equal amounts of

the stereoisomeric 2-ethyl-4-t-butylcyclohexanones (**468** and **469**, D replaced by C$_2$H$_5$). A comparable mixture of C-alkylated products was obtained using ethyl iodide as electrophile. These results were interpreted as evidence that relatively little new C − C bond formation has occurred at the transition state for the alkylation of enolate anions. For this reason, they proposed that the geometries of the six-membered rings in the transition states resemble much more closely the geometry of the planar enolate anion (e.g. **470** and **471**) than the geometry of chair (i.e. **472**) or twist-boat (i.e. **473**) cyclohexane rings. Huff, Tuller, and Caine (145) have also arrived at a similar conclusion. A more detailed discussion on the stereochemistry of alkylation of enolate ion has been published by House (146).

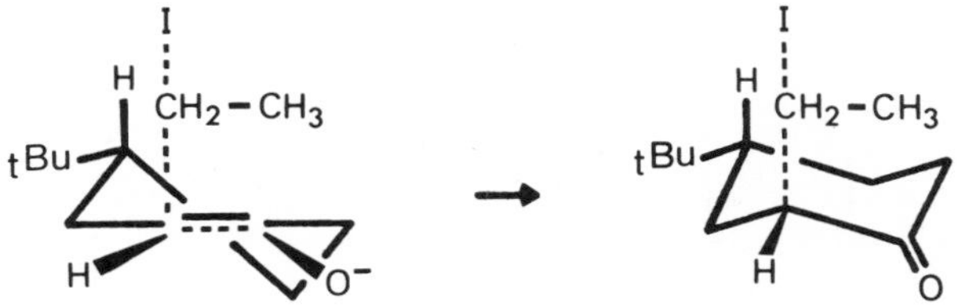

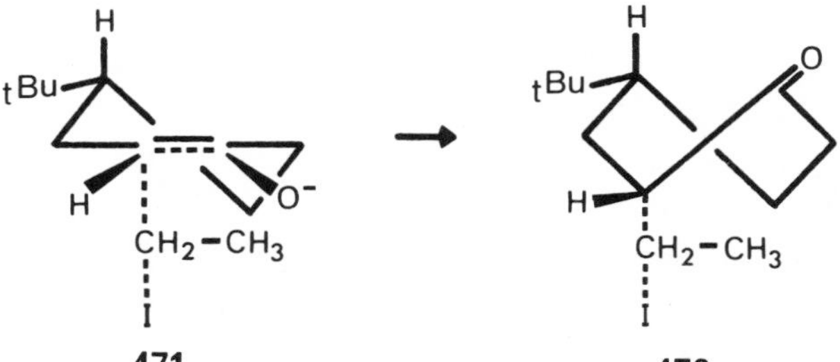

Grieco and co-workers (147) have carried out the kinetic protonation at -78° of the lactone enolate **474** and obtained a 3.5:1 ratio of **475** and **476**. Axial protonation is again not highly favored. This low selectivity may be due to competing C and O protonation.

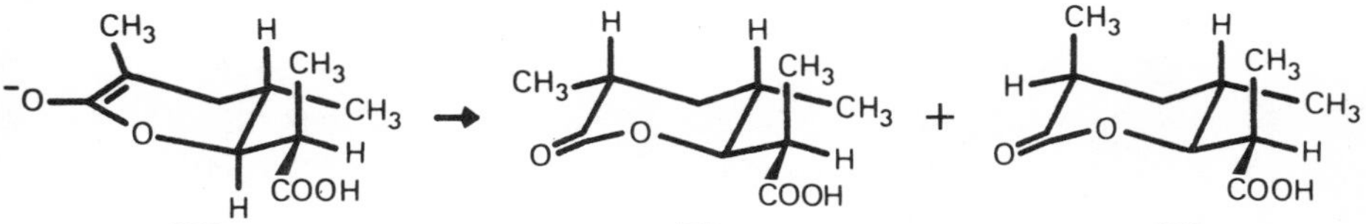

Schaefer and Weinberg have reported (126) that the deuterolysis of 4-t-butyl-1-morpholino cyclohexene (477) produced specifically 2-β-deuterio-4-t-butyl-cyclohexanone (478). Malhotra and Johnson (148) have shown that the enamine of 2-methylcyclohexanone gives on hydrolysis the cis-deuteromethylcyclo-hexanone 481. This enamine is known to exist in conformation 479 with the methyl group axial. Consequently, axial protonation must have taken place on 479 to give first the iminium ion 480 which was then hydrolyzed to 481.

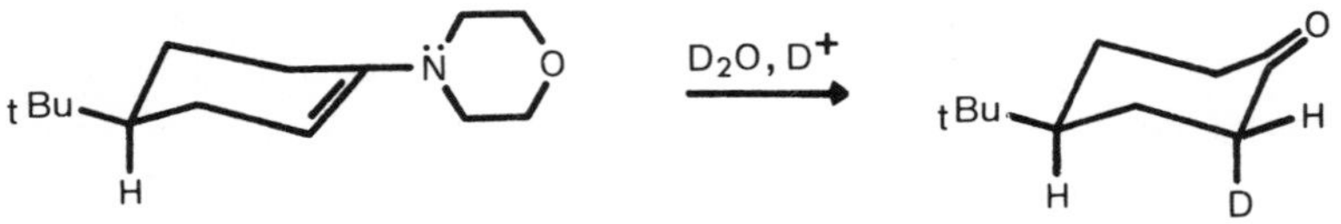

$$D_2O, D^+$$

477   478

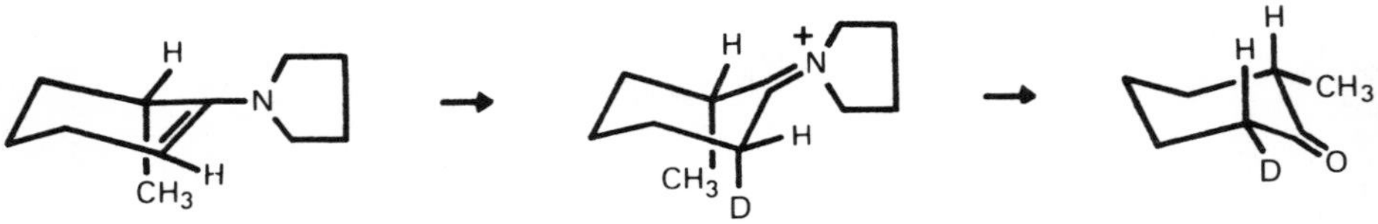

479   480   481

The stereochemistry of the alkylation of enamine 482 has been reported by Karady, Lenfant, and Wolff (149) to give mainly the axial alkylated product

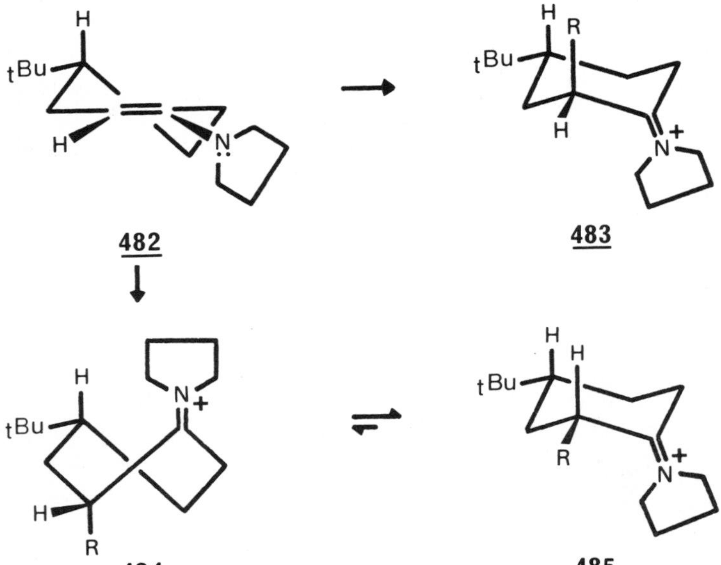

482

483

484   485

483 [R=CH$_3$ (70%), CH$_3$CH$_2$CH$_2$ (90%), and CH$_2$ CH—CH$_2$ (93%)]. The alkylation from the top face (**482** → **483**) of the enamine is thus preferred over that from the bottom face which would have produced the equatorial iminium ion **485** via the twist-boat **484**.

Spencer and co-workers (150) have provided the first evidence for stereo-electronic control in α-deprotonation of iminium ions. Treatment of the β-hydroxy-ketone **486** (R=H) and the β-acetoxy-ketone **486** (R=CH$_3$CO) with weakly basic non-tertiary amines (CF$_3$CH$_2$NH$_2$ or NH$_2$CH$_2$CN) gives the bicyclic enone **489** via the formation of imines **487** and **488**. Using appropriate deuterium labeling, they found that the axial proton (H$_1$) is preferentially removed and the stereoelectronic factor was estimated to be 18 (when OR=OH) and 110 (when OR=CH$_3$COO) in **486**. They also found that in the hydroxide ion catalyzed conversion of **486** (R=CH$_3$CO) into bicyclic enone **489**, the axial hydrogen was preferentially abstracted by a factor of 130.

Banks (151) has reported the decarbalkoxylation (NaCl in wet DMSO at 148-153°C) of the geminal diesters **490** and **493**. He found that the reaction of **490** was highly stereoselective yielding the axial isomer **491** in preference to the more stable equatorial isomer **492** (ratio ≈9:1). On the other hand, the reaction of **493** was non-stereoselective (ratio ≈1:1 of **494** and **495**).

These reactions likely proceed via the formation of an intermediate carbanion. Indeed, the carbanion **496** generated by treatment of **491** and **492** with lithium diisopropylamide gave 81% of **491** (H=D) and 19% of **492** (H=D). The preferential formation of **491** can be explained on the basis of stereoelectronic effects which influence the reactivity of the intermediate carbanion

496. The sp^2 hybridized carbanion **496** can also be viewed as an sp^3 hybridized anion and can therefore look like **497** or **498**. In **497**, the electron pair is antiperiplanar to the two C $-$ O bonds of the dioxane ring, so that the carbanion orbital can be delocalized by an overlap with the antibonding orbitals of the two C $-$ O sigma bonds (n-σ* interaction). On that basis, carbanion **496** would be closer to **497** than **498**, and the equatorial approach of the electrophile is thus readily understood. Banks has however given a different explanation based on the work of Klein (152, 153).

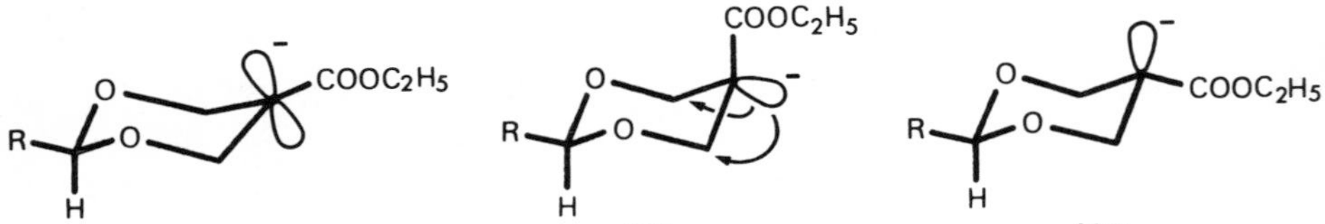

Similar experimental results were observed recently in our laboratory (154). Indeed, the lithium enolate of bicyclic spiro ester **499** gave almost exclusively the equatorial product **500** (E= C$_6$H$_5$S, CH$_3$S, C$_6$H$_5$Se, CH$_3$ or I) on reaction with various electrophiles.

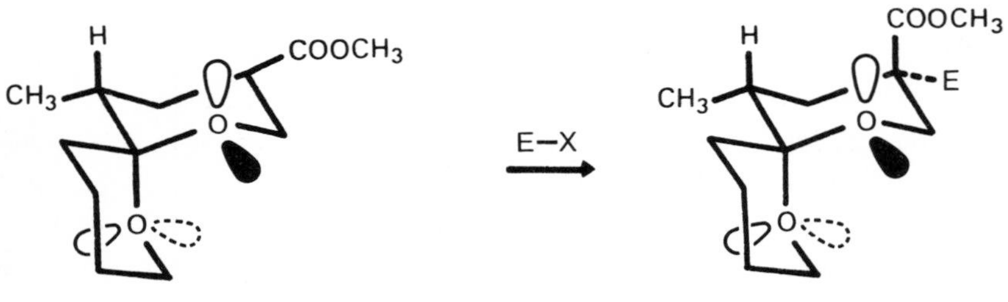

REFERENCES

(1) Baldwin, J.E. J. Chem. Soc., Chem. Commun. 1976, 738.

(2) Cram, D.J.; Abd Elhafez, F.A. J. Am. Chem. Soc. 1952, 74, 5828.

(3) Cram, D.J.; Greene, F.D. J. Am. Chem. Soc. 1953, 75, 6005.

(4) Cram, D.J.; Wilson, D.R. J. Am. Chem. Soc. 1963, 85, 1245.

(5) Chérest, M.; Felkin, H.; Prudent, N. Tetrahedron Lett. 1968, 2199.

(6) Karabatsos, G.J. J. Am. Chem. Soc. 1967, 89, 1367.

(7) Anh, N.T.; Eisenstein, O. Nouv. J. Chim. 1977, 1, 61.

(8) Chérest, M.; Felkin, H. Tetrahedron Lett. 1968, 2205.

(9) Cieplak, A.S. J. Am. Chem. Soc. 1981, 103, 4540.

(10) Woodward, R.B.; Bader, F.E.; Bickel, H.; Frey, A.J.; Kierstad, R.W. Tetrahedron 1958, 2, 1.

(11) Bohlmann, F.; Winterfeldt, E.; Studt, P.; Laurent, H.; Boroschewski, G.; Kleine, K.M. Chem. Ber. 1961, 94, 3151.

(12) Bohlmann, F.; Winterfeldt, E.; Boroschewski, G.; Mayer-Mader, R.; Gatscheff, B. Chem. Ber. 1963, 96, 1792.

(13) Toromanoff, E. Bull. Soc. Chim. Fr. 1966, 3357.

(14) Wenkert, E.; Dave, K.G.; Lewis, R.G.; Sprague, P.W. J. Am. Chem. Soc. 1967, 89, 6741.

(15) Stork, G.; Guthikonda, R.N. J. Am. Chem. Soc. 1972, 94, 5109.

(16) Stevens, R.V.; Lee, A.W.M. J. Chem. Soc., Chem. Commun. 1982, 102.

(17) Stevens, R.V.; Lee, A.W.M. J. Chem. Soc., Chem. Commun. 1982, 103.

(18) Kümin, A.; Maverick, E.; Seiler, P.; Vanier, N.; Damm, L.; Hobi, R.; Dunitz, J.D.; Eschenmoser, A. Helv. Chim. Acta 1980, 63, 1158.

(19) Petrzilka, M.; Felix, D.; Eschenmoser, A. Helv. Chim. Acta 1973, 56, 2950.

(20) Riediker, M.; Graf, W. Helv. Chim. Acta 1979, 62, 2053.

(21) Mueller, R.H.; Dipardo, R.M. J. Chem. Soc., Chem. Commun. 1975, 565.

(22) Wenkert, E.; Chang, C.-J.; Chawla, H.P.S.; Cochran, D.W.; Hagaman, E.W.; King, J.C.; Orito, K. J. Am. Chem. Soc. 1976, 98, 3645.

(23) Dean, R.T.; Rapoport, H. J. Org. Chem. 1978, 43, 4183.

(24) Ungemach, F.; DiPierro, M.; Weber, R.; Cook, J.M. J. Org. Chem. 1981, 46, 164.

(25) Overman, L.E.; Fukaya, C. J. Am. Chem. Soc. 1980, 102, 1454.

(26) Johnson, F. Chem. Rev. 1968, 68, 375.

(27) Overman, L.E.; Freerks, R.L. J. Org. Chem. 1981, 46, 2833.

(28) Stevens, R.V.; Lee, A.W.M. J. Am. Chem. Soc. 1979, 101, 7032.

(29) Robinson, R. J. Chem. Soc. 1917, 762, 876.

(30) Schopf, C. Angew. Chem. 1937, 50, 779, 876.

(31) Toromanoff, E. Bull. Soc. Chim. Fr. 1962, 708.

(32) Djerassi, C.; Schneider, R.A.; Vorbrueggen, H.; Allinger, N.L. J. Org. Chem. 1963, 28, 1632.

(33) Alexander, C.W.; Jackson, W.R. J. Chem. Soc., Perkin Trans 2 1972, 1601.

(34) Nagata, W.; Yoshioka, M. "Hydrocyanation of Conjugated Carbonyl Compounds in Organic Reactions"; Wiley: New York, N.Y., 1977, 25, 255.

(35) Agami, C.; Fadlallah, M.; Levisalles, J. Tetrahedron Lett. 1980, 59; Tetrahedron 1981, 37, 903.

(36) Abramovitch, R.A.; Struble, D.L. Tetrahedron 1968, 24, 357.

(37) Irie, H.; Katakawa, J.; Mizuno, Y.; Udaka, S.; Taga, T.; Osaki, K. J. Chem. Soc., Chem. Commun. 1978, 717.

(38) House, H.O.; Fischer, Jr., W.F. J. Org. Chem. 1968, 33, 949.

(39) Allinger, N.L.; Riew, C.K. Tetrahedron Lett. 1966, 1269.

(40) Heathcock, C.H.; Kleinman, E.; Binley, E.S. J. Am. Chem. Soc. 1978, 100, 8036.

(41) Luong-Thi, N.T.; Rivière, H. C. R. Hebd. Séances Acad. Sci., Ser. A 1968, 267, 776.

(42) Rivière, H.; Tostain, J. Bull. Soc. Chim. Fr. 1969, 568.

(43) Posner, G.H. Org. React. 1972, 19, 1.

(44) Piers, E.; Britton, R.W.; de Waal, W. Can. J. Chem. 1969, 47, 4299, 4307.

(45) Marshall, J.A.; Andersen, N.H. J. Org. Chem. 1966, 31, 667.

(46) Spencer, T.A.; Smith, R.A.J.; Storm, D.L.; Villarica, R.M. J. Am. Chem. Soc. 1971, 93, 4856.

(47) House, H.O.; Respess, W.L.; Whitesides, G.M. J. Org. Chem. 1966, 31, 3128.

(48) Marshall, J.A.; Roebke, H. J. Org. Chem. 1966, 31, 3109.

(49) Campbell, J.A.; Babcock, J.C. J. Am. Chem. Soc. 1959, 81, 4069.

(50) Yanagita, M.; Inayama, S.; Hirakura, M.; Seki, F. J. Org. Chem. 1958, 23, 690.

(51) Zimmerman, H.E.; Singer, L.; Thyagarajan, B.S. J. Am. Chem. Soc. 1959, 81, 108.

(52) Zimmerman, H.E.; Ahramjian, L. J. Am. Chem. Soc. 1959, 81, 2086.

(53) Zimmerman, H.E.; Ahramjian, L. J. Am. Chem. Soc. 1960, 82, 5459.

(54) House, H.O. "Modern Synthetic Reactions"; 2nd Ed.; W.A. Benjamin Inc.: Menlo Park, California, 1972; pp. 635-636.

(55) Baldwin, J.E. J. Chem. Soc., Chem. Commun. 1976, 734.

(56) Baldwin, J.E.; Cutting, J.; Dupont, W.; Kruse, L.; Silberman, L.; Thomas, R.C. J. Chem. Soc., Chem. Commun. 1976, 736.

(57) Baldwin, J.E.; Kruse, L.I. J. Chem. Soc., Chem. Commun. 1977, 233.

(58) Baldwin, J.E.; Reiss, J.A. J. Chem. Soc., Chem. Commun. 1977, 77.

(59) Seebach, D.; Golinski, J. Helv. Chim. Acta 1981, 64, 1413.

(60) Caine, D. Org. React. 1976, 23, 1.

(61) Stork, G.; Darling, S.D. J. Am. Chem. Soc. 1960, 82, 1512; 1964, 86, 1761.

(62) Barton, D.H.R.; Robinson, C.H. J. Chem. Soc. 1954, 3045.

(63) Robinson, M.J.T. Tetrahedron 1965, 21, 2475.

(64) Rosenfield, R.S.; Gallagher, T.F. J. Am. Chem. Soc. 1955, 77, 4367.

(65) Chapman, J.H.; Elks, J.; Phillipps, G.H.; Wyman, L.J. J. Chem. Soc. 1956, 4344.

(66) Fry, A.J.; Ginsburg, G.S. J. Am. Chem. Soc. 1979, 101, 3927.

(67) House, H.O. "Modern Synthetic Reactions"; 2nd Ed.; W.A. Benjamin Inc.: Menlo Park, California, 1972; pp. 158-160.

(68) Djerassi, C. "Steroid Reactions"; Holden-Day Inc.: San Francisco, 1963; pp. 319-322.

(69) Hallsworth, A.S.; Henbest, H.B.; Wrigley, T.I. J. Chem. Soc. 1957, 1969.

(70) Johnson, W.S.; Cohen, N.; Habicht, Jr., E.R.; Hamon, D.P.G.; Rizzi, G.P.; Faulkner, J.D. Tetrahedron Lett. 1968, 2829.

(71) Norin, T. Acta Chem. Scand. 1965, 19, 1289.

(72) Dauben, W.G.; Deviny, E.J. J. Org. Chem. 1966, 31, 3794.

(73) Dauben, W.G.; Wolf, R.E. J. Org. Chem. 1970, 35, 374.

(74) Dauben, W.G.; Wolf, R.E. J. Org. Chem. 1970, 35, 2361.

(75) Wenkert, E.; Yoder, J.E. J. Org. Chem. 1970, 35, 2986.

(76) Paquette, L.A.; Wyvratt, M.J.; Berk, H.C.; Moerck, R.E. J. Am. Chem. Soc. 1978, 100, 5845.

(77) Paquette, L.A.; Snow, R.A.; Muthard, J.L.; Cynkowski, T. J. Am. Chem. Soc. 1978, 100, 1600.

(78) Balogh, D.W.; Paquette, L.A.; Engel, P.; Blount, J.F. J. Am. Chem. Soc. 1981, 103, 226.

(79) Paquette, L.A.; Balogh, D.W.; Blount, J.F. J. Am. Chem. Soc. 1981, 103, 228.

(80) Eaton, P.E.; Mueller, R.H.; Carlson, G.R.; Cullison, D.A.; Cooper, G.F.; Chou, T.-C.; Krebs, E.-P. J. Am. Chem. Soc. 1977, 99, 2751.

(81) Baker, K.M.; Davis, B.R. Tetrahedron 1968, 24, 1655.

(82) Paquette, L.A.; Wyvratt, M.J. J. Am. Chem. Soc. 1974, 96, 4671.

(83) Bartsch, R.A.; Závada, J. Chem. Rev. 1980, 80, 453.

(84) Ingold, C.K. Proc. Chem. Soc., London 1962, 265.

(85) Sicher, J.; Závada, J. Collect. Czech. Chem. Commun. 1968, 33, 1278.

(86) Lafontaine, J.; Mongrain, M.; Sergent-Guay, M.; Ruest, L.; Deslong-champs, P. Can. J. Chem. 1980, 58, 2460.

(87) Julia, M.; Julia, S.; Guégan, R. Bull. Soc. Chim. Fr. 1960, 1072.

(88) Julia, M.; Julia, S.; Tchen, S.-Y. Bull. Soc. Chim. Fr. 1961, 1849.

(89) Brady, S.F.; Ilton, M.A.; Johnson, W.S. J. Am. Chem. Soc. 1968, 90, 2882.

(90) Grob, C.A. Angew. Chem. Int. Ed. 1969, 8, 535.

(91) Grob, C.A.; Bolleter, M.; Kunz, W. Angew. Chem. Int. Ed. 1980, 19, 708.

(92) Corey, E.J.; Mitra, R.B.; Uda, H. J. Am. Chem. Soc. 1964, 86, 485.

(93) Kraus, W. Angew. Chem. Int. Ed. 1966, 5, 316.

(94) Zurflüh, R.; Wall, E.N.; Siddall, J.B.; Edwards, J.A. J. Am. Chem. Soc. 1968, 90, 6224.

(95) Marshall, J.A.; Brady, S.F. J. Org. Chem. 1970, 35, 4068.

(96) Wharton, P.S.; Hiegel, G.A. J. Org. Chem. 1965, 30, 3254.

(97) Fischli, A.; Branca, Q.; Daly, J. Helv. Chim. Acta 1976, 59, 2443.

(98) Ohloff, G.; Becker, J.; Schulte-Elte, K.H. Helv. Chim. Acta 1967, 50, 705.

(99) Schmidt, H.; Mühlstädt, M.; Son, P. Chem. Ber. 1966, 99, 2736.

(100) Larsen, S.D.; Monti, S.A. Synth. Commun. 1979, 9, 141.

(101) Marshall, J.A.; Bundy, G.L. J. Am. Chem. Soc. 1966, 88, 4291.

(102) Martin, J.; Parker, W.; Raphael, R.A. J. Chem. Soc. 1964, 289.

(103) Buchanan, G.L.; McKillop, A.; Raphael, R.A. J. Chem. Soc. 1965, 833.

(104) Buchanan, G.L.; McLay, G.W. Tetrahedron 1966, 22, 1521.

(105) Wiesner, K. Adv. Org. Chem. 1972, 8, 295.

(106) Wiesner, K.; Valenta, Z.; Findlay, J.A. Tetrahedron Lett. 1967, 221.

(107) Ruest, L.; Deslongchamps, P. Unpublished results.

(108) Bélanger, A.; Berney, D.J.F.; Borschberg, H.-J.; Brousseau, R.; Doutheau, A.; Durand, R.; Katayama, H.; Lapalme, R.; Leturc, D.M.; Liao, C.-C.; MacLachlan, F.N.; Maffrand, J.-P.; Marazza, F.; Martino, R.; Moreau, C.; Saint-Laurent, L.; Saintonge, R.; Soucy, P.; Ruest, L.; Deslongchamps, P. Can. J. Chem. 1979, 57, 3348.

(109) Kraus, W.; Chassin, C. Tetrahedron Lett. 1970, 1003.

(110) Sternbach, D.; Shibuya, M.; Jaisli, F.; Bonetti, M.; Eschenmoser, A. Angew. Chem. Int. Ed. Engl. 1979, 18, 634.

(111) Shibuya, M.; Jaisli, F.; Eschenmoser, A. Angew. Chem. Int. Ed. Engl. 1979, 18, 636.

(112) Beereboom, J.J. J. Am. Chem. Soc. 1963, 85, 3525.

(113) Brousseau, R.; Deslongchamps, P. Unpublished results.

(114) Brousseau, R. Ph.D. Thesis, Université de Sherbrooke, 1972.

(115) Borschberg, H.-J.; Deslongchamps, P. Unpublished results.

(116) Marazza, F.; Ruest, L.; Deslongchamps, P. Unpublished results.

(117) Doutheau, A.; Deslongchamps, P. Unpublished results.

(118) Soucy, P.; Deslongchamps, P. Unpublished results.

(119) Borschberg, H.-J.; Ruest, L.; Deslongchamps, P. Unpublished results.

(120) Buchanan, G.L.; Young, G.A.R. J. Chem. Soc., Chem. Commun. 1971, 643.

(121) Trost, B.M.; Frazee, W.J. J. Am. Chem. Soc. 1977, 99, 6124.

(122) Eisele, W.; Grob, C.A.; Renk, E.; von Tschammer, W. Helv. Chim. Acta 1968, 51, 816.

(123) Corey, E.J. Experientia 1953, 9, 329.

(124) Corey, E.J. J. Am. Chem. Soc. 1954, 76, 175.

(125) Corey, E.J.; Sneen, R.A. J. Am. Chem. Soc. 1956, 78, 6229.

(126) Schaefer, J.P.; Weinberg, D.S. Tetrahedron Lett. 1965, 1801.

(127) House, H.; Tefertiller, B.A.; Olmstead, H.D. J. Org. Chem. 1968, 33, 935.

(128) Trimitsis, G.B.; Van Dam, E.M. J. Chem. Soc., Chem. Commun. 1974, 610.

(129) Feather, J.A.; Gold, V. J. Chem. Soc. 1965, 1752.

(130) Fishman, J. J. Org. Chem. 1966, 31, 520.

(131) Bordwell, F.G.; Scamehorn, R.G. J. Am. Chem. Soc. 1968, 90, 6749.

(132) Lamaty, G. "Isotope Effects in Organic Chemistry"; Vol. 2; Buncel, E. and Lee, C.C., Eds; Elsevier: Amsterdam, 1976; p. 71.

(133) Valls, J.; Toromanoff, E. Bull. Soc. Chim. Fr. 1961, 758.

(134) Velluz, L.; Valls, J.; Nominé, G. Angew. Chem. Int. Ed. Engl. 1965, 4, 181.

(135) Fraser, R.R.; Champagne, P.J. Can. J. Chem. 1976, 54, 3809.

(136) Fraser, R.R.; Champagne, P.J. J. Am. Chem. Soc. 1978, 100, 657.

(137) Wolfe, S.; Schlegel, H.B.; Csizmadia, I.G.; Bernardi, F. Can. J. Chem. 1975, 53, 3365.

(138) Nickon, A.; Covey, D.F.; Huang, F.-C.; Kuo, Y.-N. J. Am. Chem. Soc. 1975, 97, 904.

(139) Meerwein, H.; Schürmann, W. Ann. Chem. 1913, 398, 196.

(140) Bootger, O. Chem. Ber. 1937, 70B, 314.

(141) Schaefer, J.P.; Lark, J.C. J. Org. Chem. 1965, 30, 1337.

(142) Yamada, K.; Manabe, S.; Kyotani, Y.; Suzuki, M.; Hirata, Y. Bull. Chem. Soc. Jpn. 1979, 52, 186.

(143) Subrahmanyam, G.; Malhotra, S.K.; Ringold, H.J. J. Am. Chem. Soc. 1966, 88, 1332.

(144) Eigen, M. Angew. Chem. Int. Ed. Engl. 1964, 3, 1.

(145) Huff, B.J.L.; Tuller, F.N.; Caine, D. J. Org. Chem. 1969, 34, 3070.

(146) House, H.O. "Modern Synthetic Organic Reactions"; 2nd Ed.; W.A. Benjamin Inc.: Menlo Park, California, 1972.

(147) Grieco, P.A.; Ohfune, Y.; Yokoyama, Y.; Owens, W. J. Am. Chem. Soc. 1979, 101, 4749.

(148) Malhotra, S.K.; Johnson, F. Tetrahedron Lett. 1965, 4027.

(149) Karady, S.; Lenfant, M.; Wolff, R.E. Bull. Soc. Chim. Fr. 1965, 2472.

(150) Ferran, H.E.; Roberts, R.D.; Jacob, J.N.; Spencer, T.A. J. Chem. Soc., Chem. Commun. 1978, 49.

(151) Banks, H.D. J. Org. Chem. 1981, 46, 1743.

(152) Klein, J. Tetrahedron Lett. 1973, 4307.

(153) Klein, J. Tetrahedron 1974, 30, 3349.

(154) Caron, M.; Deslongchamps, P. Unpublished results.

CHAPTER 7

REACTIONS ON TRIPLE-BONDS

The addition of a nucleophile Y⁻ to a triple-bond as in **1** can take place to give a product anion where the entering nucleophile is trans (**2**) or cis (**3**) to the non-bonded electron pair. Stereoelectronic effects should therefore affect product formation.

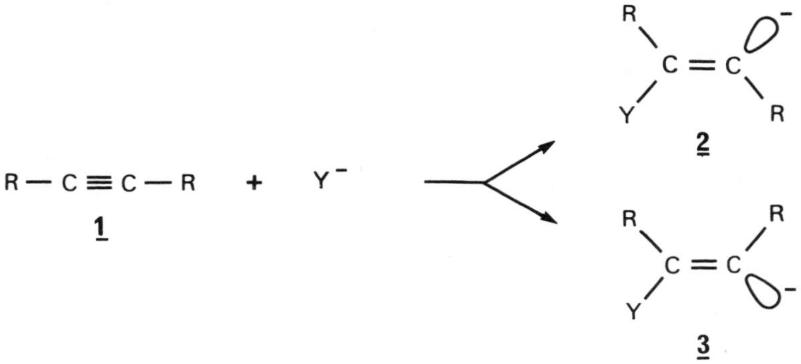

Ingold (1) has mentioned that the older literature contains several observations indicating the existence of stereochemically favorable and unfavorable situations for elimination reactions leading to acetylenic compounds. For instance, Michael (2) found that chlorofumaric acid (**4**) is converted by alkali about 50 times faster than in chloromaleic acid (**5**) into acetylene dicarboxylic acid (**6**). Chovanne (3) has observed that cis-dichloroethylene (**7**) is transformed by alkali about 20 times faster than is the trans-isomer **8** into chloroacetylene (**9**).

4 6 5

7 9 8

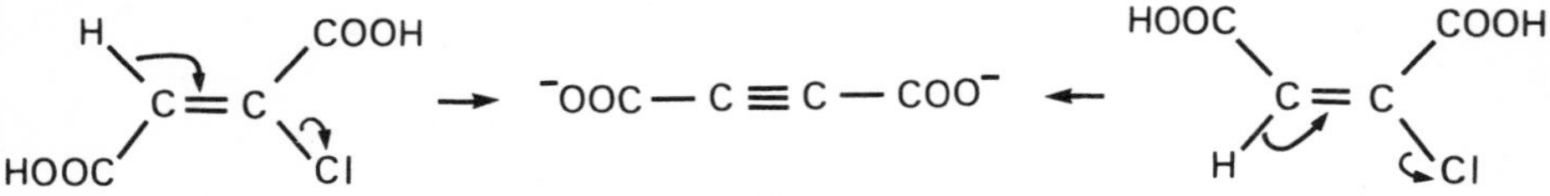

Cristol and co-workers (4) found that cis-p-nitrostyryl bromide 10, in the presence of ethanolic sodium ethoxide, undergoes elimination (yielding 11) 2300 times faster than the trans-isomer 12 which undergoes an alternative reaction of addition (→13).

10 11

12 13

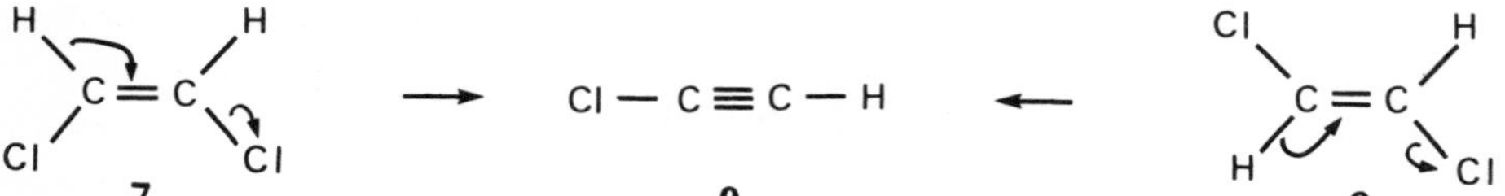

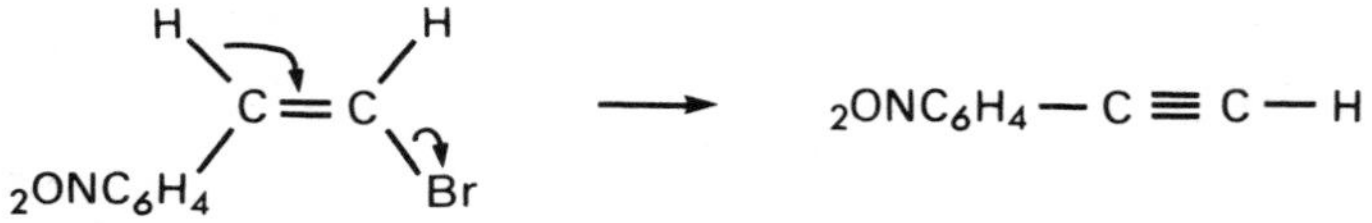

Modena and co-workers (5) have also reported that cis-β-arenesulfonylvinyl chlorides (14) undergo nucleophilic substitution of the halogen by methoxide ion by way of an E2 trans-elimination to an acetylene derivative (15) followed by a trans-addition to yield the cis-product 16.

14 15 16

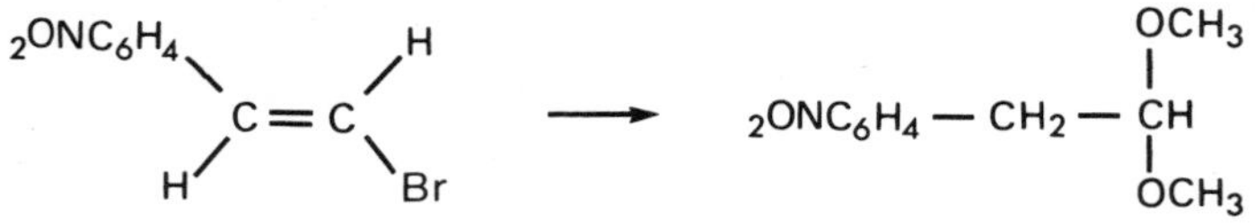

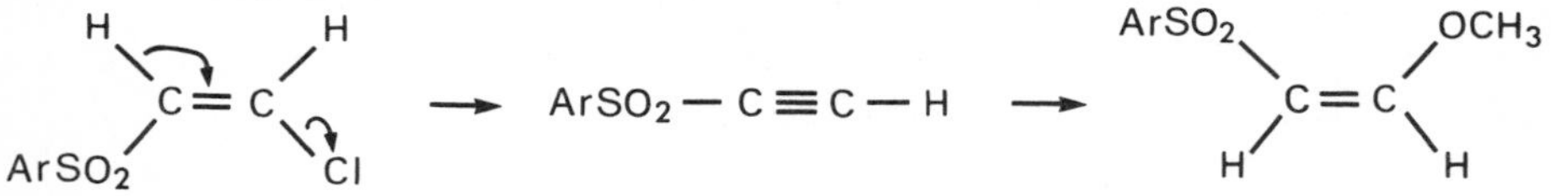

Miller (6) has reported that the methoxide-catalyzed addition of methanol to phenylacetylene (17) gave stereospecifically cis-β-methoxystyrene (19) via the intermediate 18.

$$C_6H_5 - C \equiv C - H \xrightarrow{\text{slow}} \underset{\textbf{18}}{\overset{C_2H_5}{\underset{O^-}{}}C=C\overset{OCH_3}{\underset{H}{}}} \xrightarrow{\text{fast}} \underset{\textbf{19}}{\overset{C_6H_5}{\underset{H}{}}C=C\overset{OCH_3}{\underset{H}{}}}$$

$$\underset{\textbf{17}}{C_6H_5 - C \equiv C - H}$$

Truce and collaborators (7-9) have shown that cis-dichloroethylene 20 reacts readily with sodium p-toluenethiolate in the presence of sodium ethoxide to give cis-1,2-bis-p-tolylmercapto-ethylene 22 while the trans-isomer 21, when subjected to the same conditions is recovered unchanged. Convincing evidence was obtained that the conversion 20 → 22 takes place via the intermediates 23, 24, and 25. Truce and Simms (9) have also observed that the base-catalyzed addition of p-toluenethiol to phenylacetylene and to 2-butyne yields cis-styryl p-tolylsulfide (26) and 2-p-tolylmercapto-trans-2-butene (27) respectively.

$$\underset{\textbf{20}}{\overset{H}{\underset{Cl}{}}C=C\overset{H}{\underset{Cl}{}}} \longrightarrow \underset{\textbf{22}}{\overset{H}{\underset{ArS}{}}C=C\overset{H}{\underset{SAr}{}}} \xleftarrow{\;\;/\!\!/\;\;} \underset{\textbf{21}}{\overset{H}{\underset{Cl}{}}C=C\overset{Cl}{\underset{H}{}}}$$

$$\downarrow$$

$$\underset{\textbf{23}}{H - C \equiv C - Cl} \longrightarrow \underset{\textbf{24}}{\overset{H}{\underset{ArS}{}}C=C\overset{H}{\underset{Cl}{}}} \longrightarrow \underset{\textbf{25}}{ArS - C \equiv C - H}$$

$$\underset{\textbf{26}}{\overset{H}{\underset{C_6H_5}{}}C=C\overset{H}{\underset{SAr}{}}} \qquad \underset{\textbf{27}}{\overset{H}{\underset{CH_3}{}}C=C\overset{CH_3}{\underset{SAr}{}}}$$

Trans-elimination is therefore clearly stereochemically favored over cis-elimination indicating that stereoelectronic effects must play a decisive role in these reactions. Ingold (1) has pointed out that:

> "... a heterolytic elimination involves an internal SN_2 type substitution at the α-carbon atom, which will be restricted by the exclusion principle to receiving the new electron pair on the side remote from that which the old-electron pair becomes expelled.

Concerted with this process, there is an SE2 type substitution at the β-carbon atom. Such a substitution is
not expected to be stereochemically restricted by any
adamantine principle. However, the changing electron-
pair in substitutions of this type is usually observed
to exchange nuclei with retention of configuration of
behavior which, in the concerted reaction, leads to
trans-elimination, as shown below (cf. 28). If however,
in the transition state, the changing electrons were
sufficiently released by the departing proton to allow
their passage through the β-carbon atom, then their
entry by substitution into the β-carbon atom would produce a cis-elimination as illustrated (cf. 29)."

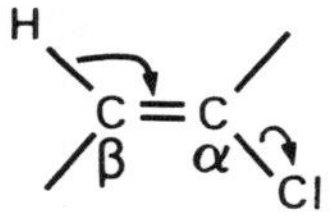

28 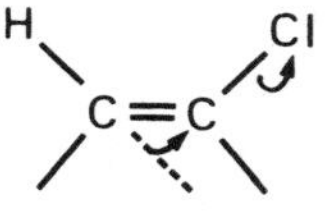   **29**

More recently, Hegarty has investigated the reactivity of nitrilium ion **30**
and found that in the product formed under kinetically controlled conditions,
the nitrogen electron pair is always trans to the incoming nucleophile, i.e.
31. In an excellent review (10), Hegarty states that in spite of careful
searching, they have unable to observe any detectable quantity of the cis
product **32** (or of a product which would be derived from **32** by further chemical transformation).

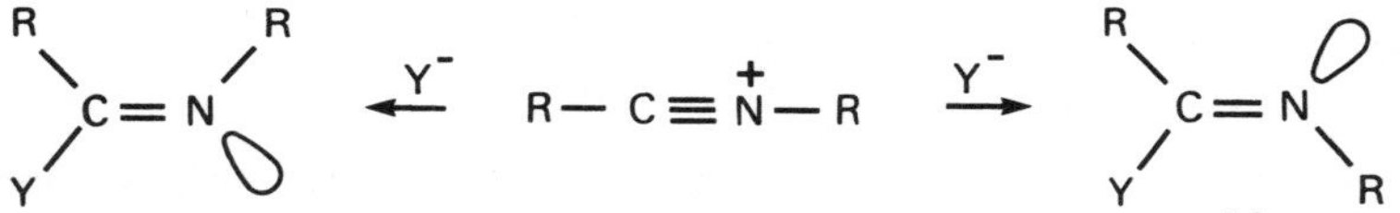

32   **30**   **31**

For instance, N-anilinonitrilium ion **34** formed in the solvolysis of hydraz-
onyl bromides **33** in the presence of sodium acetate at 30°C gives the Z-O-

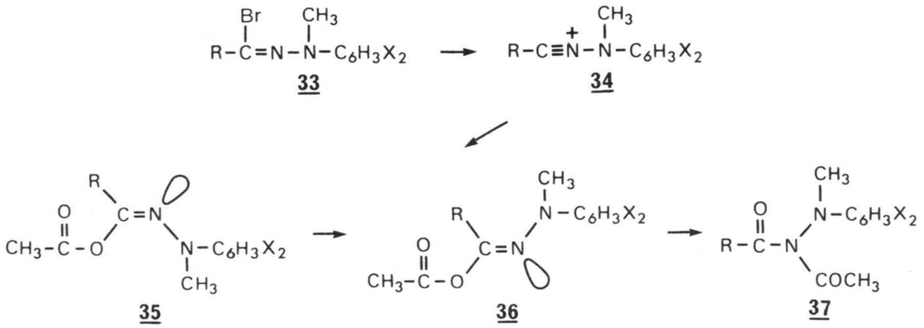

33   **34**

35   **36**   **37**

acylisoamide **35** in quantitative yield. On heating, the *Z* isomer **35** yields the *E* isomer **36** which is rapidly transformed into the corresponding amide **37** (11).

In another example (12), diazotization of amidoxime **38** gave in situ the nitrilium ion **39** which is trapped by acetate ion as the *Z* isomer **40**.

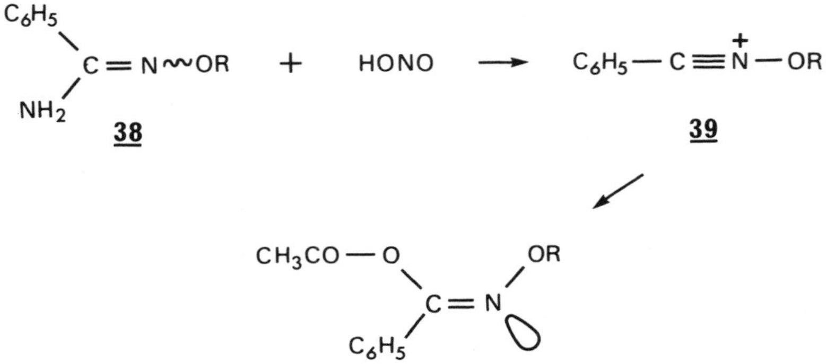

The observation that carboxylate ions react at a much faster rate than amines with nitrilium ions and that they give Z-O-acylisoamides only, has led to the development of a new method for the synthesis of peptides (13). Imide halide **41** on dissolution in a polar solvent undergoes rapid unimolecular ionization to the nitrilium ion **42** which reacts with the carboxylate ion to give the Z-O-acylisoamide **43** which in turn reacts with the amine to give the amide product **44**. The formation of the amide (or peptide) can be carried out by adding the halide **41** to a solution containing both the amine and the carboxylic acid. The initial reaction (**41** → **42** → **43**) is best carried out at pH=6 and when the pH is adjusted to ≈8, formation of the amide (**43** → **44**)

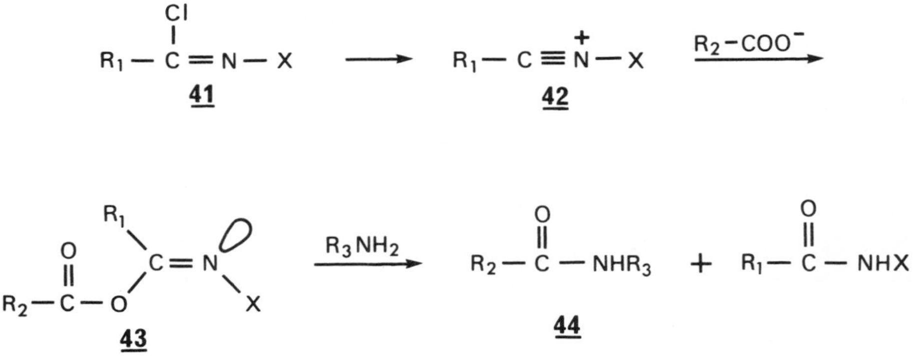

is rapid and complete. A vital feature is that, due to its geometry, the intermediate Z-0-acylisoamide **43** is stable to O−N acyl migration.

Hegarty and collaborators (14-16) have also found that the addition of a large variety of nucleophiles (Y= Cl⁻, N_3^-, CH_3O^-, R−C ≡ C⁻, $R_2\overset{..}{N}H$, CH_3COO^-, and AlH_4^-) on benzonitrile oxide **45** is completely stereospecific and yields the trans-adduct **46**.

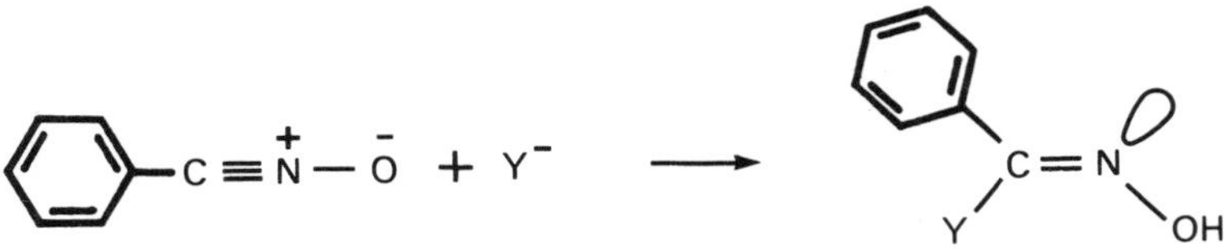

45   **46**

Hegarty and Chandler (17, 18) have found that secondary amines react with isonitriles **47** in the presence of AgCl at low temperature to give the iso-lable but thermodynamically unstable Z-amidines **50**. This reaction is believ-ed to occur via the formation of the metallated nitrilium intermediate **48** which adds the secondary amine stereospecifically to give **49**. Compound **49** is then converted into the Z isomer **50**.

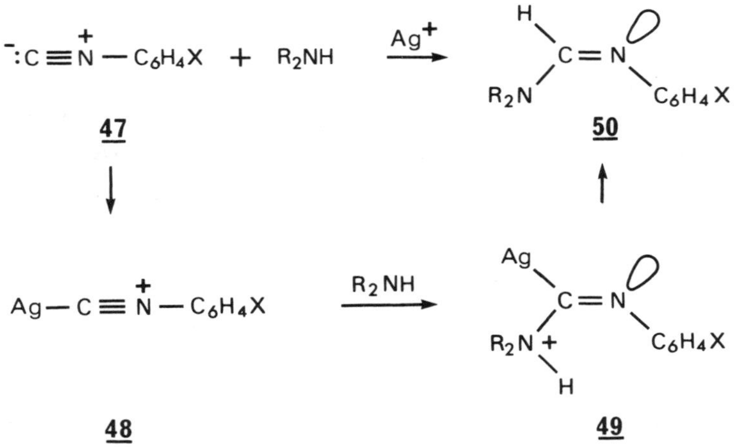

47   **50**

48   **49**

The marked stereospecificity observed in the addition of nucleophiles to nitrilium ions should also be apparent in the reverse process, i.e., in the loss of an atom or group leading to the nitrilium ion formation. Johnson and co-workers (19, 20) have prepared several pairs of isomeric O-alkylhy-droxamoyl chlorides **51** and **52**. These compounds undergo unimolecular loss of chloride ion under forcing conditions to give the nitrilium ion $\underline{53}$, and it was found that the $\underline{Z}$ isomer $\underline{51}$ is far more reactive than the $\underline{E}$ isomer $\underline{52}$ ($k_Z/k_E \approx 450$). Furthermore, both chlorides yield a single product, the $\underline{Z}$ isomer $\underline{54}$ on trapping the nitrilium ion $\underline{53}$ with methanol.

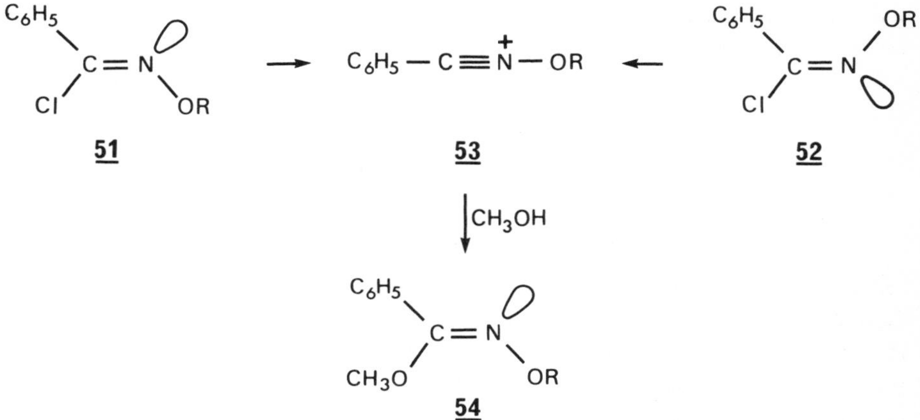

$$C_6H_5 - C \equiv \overset{+}{N} - OR$$

$$\underline{51} \qquad\qquad \underline{53} \qquad\qquad \underline{52}$$

$$\big\downarrow CH_3OH$$

$$\underline{54}$$

Broxton (21) has also observed that the formation of the arenediazonium ion $\underline{57}$ takes place at a much faster rate ($\approx 10^4$) with the $\underline{Z}$ isomer $\underline{55}$ than the $\underline{E}$ isomer $\underline{56}$.

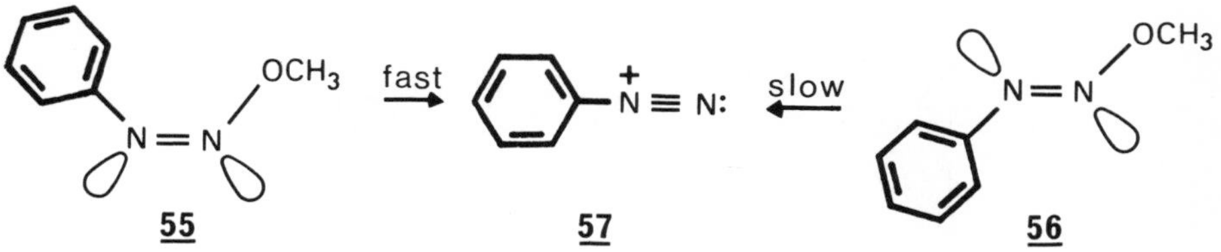

$$\underline{55} \qquad\qquad \underline{57} \qquad\qquad \underline{56}$$

A theoretical study on the addition of nucleophiles to fulminic acid ($\underline{58}$) and acetonitrile oxide ($\underline{59}$) has been carried out (22, 23). Ab initio calculations show that these compounds are linear in the ground state but are relatively easily deformed. It is markedly easier to bend the atoms in the nitrilium system in a trans fashion rather than in a cis mode. The transition state is reached relatively early when a nucleophile reacts with the nitrile oxide and its configuration corresponds to $\underline{60}$. Thus the configuration of the product is clearly determined at the transition state.

$$R-C\equiv\overset{+}{N}-\overset{-}{O} + Y^- \longrightarrow \overset{Y}{\underset{R}{\diagdown}}C\overline{\overline{\overline{}}}N\overset{O^-}{\diagup}$$

58 R = H

59 R = CH$_3$

60

Interestingly, the crystal structures of 8-methoxy-1-naphtonitrile and 8-nitro-1-naphtonitrile have been determined by X-ray analysis by Procter, Britton, and Dunitz (24). The structure of the methoxy derivative corresponds to **61** where the exocyclic C — O bond is bent inward (toward the nitrile group), the exocyclic C — CN bond is bent outward (away from the methoxy group). The C — C ≡ N bond angle is 174° instead of 180°. A similar observation has been made with 8-nitro-1-naphtonitrile. Crystals of this compound contain two symmetry independent molecules which differ in structure. Both show a bent C — CN bond and a short O⋯C ≡ N distance (cf. **62**), but the orientation of the nitro group is different with the result that in one molecule the O_1⋯C_{11} distance is 2.69 Å whereas in the other, it is 2.79 Å. This analysis is in complete agreement with the theoretical calculations and the experimental results presented above. Thus, it can be concluded that the nucleophilic addition on triple-bond (and the reverse process) is strongly influenced by stereoelectronic effects which favor the anti mode of addition.

61

62

The Beckman rearrangement converts oximes (and derivatives) to amides (25-27). In this rearrangement, the group which migrates is the one oriented antiperiplanar to the N — O bond and the stereochemical configuration of the migrating group is retained (cf. **63** → **66**). Thus, the R$_1$ group migrates in preference to the R$_2$ group forming first the nitrilium ion **65** via **64**. Hydration of nitrilium ion **65** followed by tautomerization gives the amide **66**. When the potential migrating group can form a relatively stable carbonium ion, fragmentation (**67** → **68**) takes place instead of migration yielding the corresponding nitrile (see also Chapter 6, p. 274).

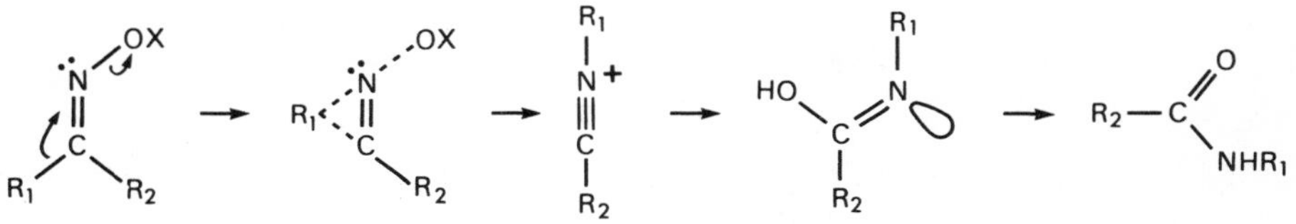

$$63 \rightarrow 64 \rightarrow 65 \rightarrow \rightarrow 66$$

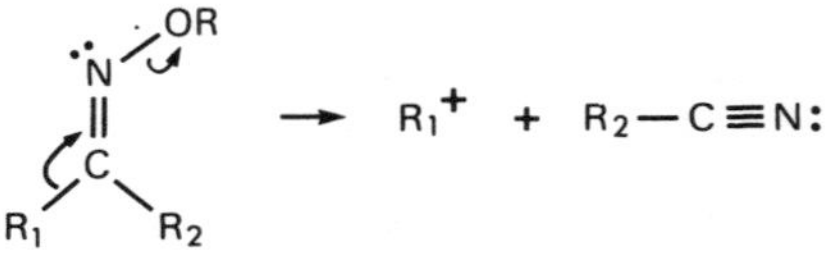

$$67 \rightarrow R_1{}^+ + R_2-C\equiv N:$$

$$68$$

Eschenmoser and collaborators (28) have reported the base induced fragmentation of α,β-epoxy-tosylhydrazones to produce acetylenic ketones (e.g. **69** →
71). Interestingly, in this fragmentation, both the triple-bond and molecular nitrogen are produced from the key intermediate **70** via an anti mode.
The decarboxylation of nitrobenzisoxazole carboxylate (**72**) into 2-cyano-
5-nitrophenol (**73**) can also be viewed as a trans-elimination (29, 30).

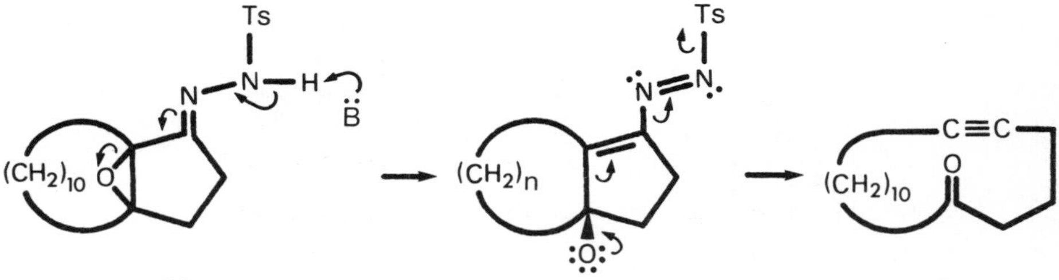

$$69 \rightarrow 70 \rightarrow 71$$

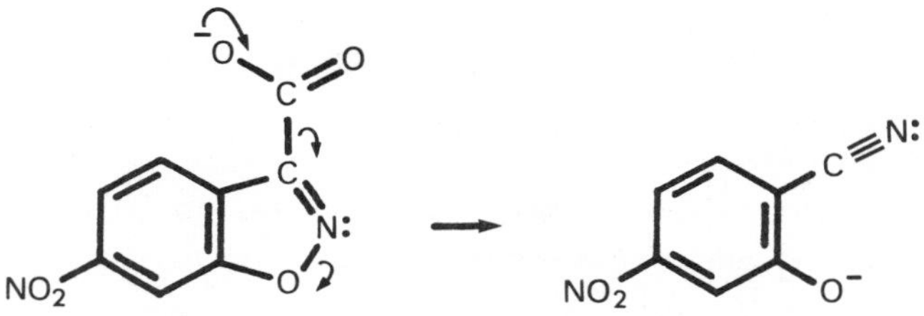

$$72 \rightarrow 73$$

In the Curtius rearrangement of an acylazide which yields as the initial
product an isocyanate ($\underline{74} \rightarrow \underline{75}$) and molecular nitrogen, the migrating group
retains its stereochemical configuration as in the Beckman rearrangement
(26, 27). It is therefore likely that stereoelectronic effects control this
rearrangement. The $R - C$ bond must therefore be antiperiplanar to the $N - N_2^+$
bond ($\underline{76} \rightarrow \underline{77}$). Note also that the oxygen atom in $\underline{76}$ has one electron pair
oriented antiperiplanar to the migrating R group.

REFERENCES

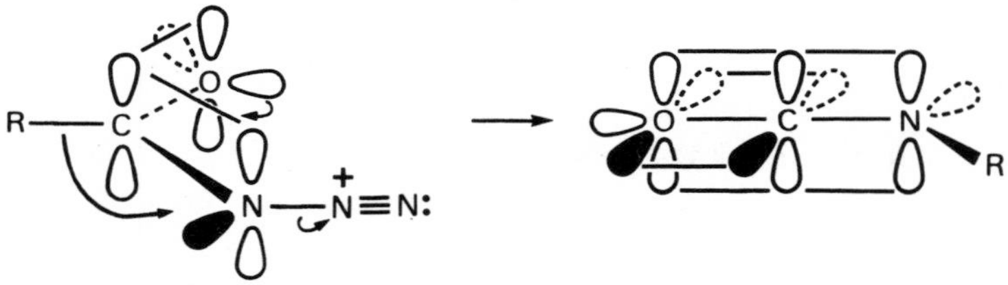

(1) Ingold, C.K. "Structure and Mechanism in Organic Chemistry"; 2nd Ed.;
     Cornell University Press: Ithaca, N.Y., 1969; p. 689.

(2) Michael, A. J. Prakt. Chem. 1895, 52, 308.

(3) Chovanne, G. Bull. Soc. Chim. Belge 1912, 26, 287.

(4) Cristol, S.J.; Begoon, A.; Norris, W.P.; Ramey, P.S. J. Am. Chem. Soc.
     1954, 76, 4558.

(5) Di Nunno, L.; Modena, G.; Scorrazo, G. J. Chem. Soc. B 1966, 1186.

(6) Miller, S.I. J. Am. Chem. Soc. 1956, 78, 6091.

(7) Truce, W.E.; Boudakian, M.M.; Heine, R.F.; McManimie, R.J. J. Am. Chem.
     Soc. 1956, 78, 2743.

(8) Truce, W.E.; McManimie, R.J. J. Am. Chem. Soc. 1954, 76, 5745.

(9) Truce, W.E.; Simms, J.A. J. Am. Chem. Soc. 1956, 78, 2756.

(10) Hegarty, A.F. Acc. Chem. Res. 1980, 13, 448.

(11) Hegarty, A.F.; McCormack, M.T. J. Chem. Soc., Chem. Commun. 1975, 168;
     J. Chem. Soc., Perkin Trans 2 1976, 2, 1701.

(12) McCarthy, D.G.; Hegarty, A.F. J. Chem. Soc., Perkin Trans 2 1977, 1080.

(13) Hegarty, A.F.; McCarthy, D.G. J. Am. Chem. Soc. 1980, 102, 4537.

(14) Dignam, K.J.; Hegarty, A.F.; Quain, P.L. J. Org. Chem. 1978, 43, 388.

(15) Dignam, K.J.; Hegarty, A.F.; Quain, P.L. J. Chem. Soc., Perkin Trans 2 1977, 1457.

(16) Dignam, K.J.; Hegarty, A.F. J. Chem. Soc., Chem. Commun. 1976, 862.

(17) Hegarty, A.F.; Chandler, A. Tetrahedron Lett. 1980, 885.

(18) Hegarty, A.F.; Chandler, A. J. Chem. Soc., Chem. Commun. 1980, 130.

(19) Johnson, J.E.; Nalley, E.A.; Kunz, Y.K.; Springfield, J.R. J. Org. Chem. 1976, 41, 252.

(20) Johnson, J.E.; Silk, N.M.; Nalley, E.A.; Arfan, M. J. Org. Chem. 1981, 46, 546.

(21) Broxton, T.J. Aust. J. Chem. 1979, 32, 1031; 1978, 31, 1519.

(22) Leroy, G.; Nguyen, M.T.; Sana, M.; Dignam, K.H.; Hegarty, A.F. J. Am. Chem. Soc. 1979, 101, 1988.

(23) Dignam, K.J.; Hegarty, A.F. J. Chem. Soc., Perkin Trans 2 1979, 1437.

(24) Procter, G.; Britton, D.; Dunitz, J. Helv. Chim. Acta 1981, 64, 471.

(25) Donaruma, I.G.; Heldt, W.Z. Org. React. 1960, 11, 1.

(26) Smith, P.A. "Open Chains Nitrogen Compounds"; Vol. II; W.A. Benjamin: New York, N.Y., 1966.

(27) Smith, P.A. "Molecular Rearrangements"; Vol. I; P. de Mayo Ed.; Interscience: New York, N.Y., 1963.

(28) Felix, D.; Schreicher, J.; Ohloff, G.; Eschenmoser, A. Helv. Chim. Acta 1971, 54, 2896.

(29) Kemp, D.S.; Paul, K.G. J. Am. Chem. Soc. 1970, 92, 2553; 1975, 97, 7305.

(30) Suh, J.; Scarpa, I.S.; Klotz, I.M. J. Am. Chem. Soc. 1976, 98, 7060.

CHAPTER 8

POT-POURRI

We have seen throughout the previous Chapters of this book that stereoelectronic effects combined with the usual steric interactions strongly influence the conformation and the reactivity of organic molecules. It is also
through the understanding of these effects that it becomes possible to consider the stereochemistry of the transition states of many organic reactions. As a result, better synthetic sequences can be elaborated as shown
by the several examples described in Chapters 5-7. More examples which cannot be appropriately placed in these Chapters are described here. This Chapter will also show how to use stereoelectronic principles either to imagine
and discover organic substances with rather unusual reactivities or simply
to develop new strategies in organic synthesis.

Miscellaneous organic transformations

The reaction of orthoesters with Grignard reagents provides a well known
route to acetals and ketals ($\underline{1} \rightarrow \underline{2}$). Eliel and Nader (1) have investigated
the stereochemistry of this reaction and have concluded that it is governed
by powerful stereoelectronic effects.

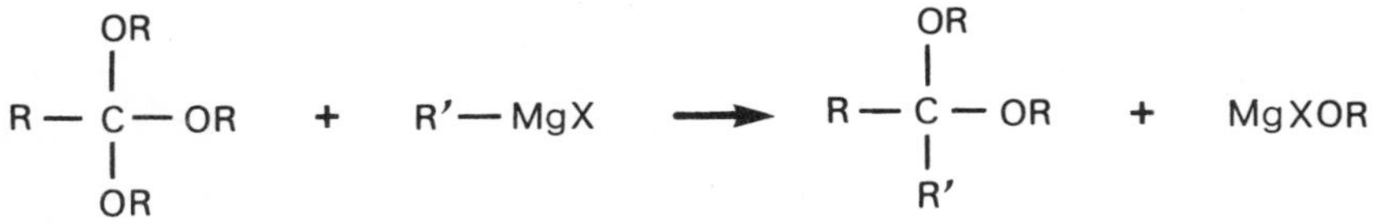

$$
R - \underset{\underset{\underset{\underline{1}}{OR}}{\overset{OR}{|}}}{\overset{OR}{\underset{|}{C}}} - OR \quad + \quad R' - MgX \quad \longrightarrow \quad R - \underset{\underset{\underset{\underline{2}}{R'}}{\overset{OR}{|}}}{\overset{OR}{\underset{|}{C}}} - OR \quad + \quad MgXOR
$$

They have studied the reactivity of axial and equatorial 2-alkoxy-1,3-diox-
anes **3** and **4** (R_1 =R_2 =H; R_1 =R_2 =CH_3 ; R_1 =CH_3 and R_2 =H). Reaction of axial
orthoesters **3** with methyl, ethyl, isopropyl and various p-substituted phenyl
Grignard reagents proceeded smoothly at room temperature to give largely
the corresponding 2-alkyl-1,3-dioxanes **5** having the 2-alkyl group axially
oriented. In contrast, the equatorial orthoesters **4** failed to react under
corresponding conditions.

Thus, both the departing alkoxy group and the incoming alkyl group prefer
the axial position. In compound **3**, the dioxane oxygens have each an electron
pair properly aligned to assist the ejection of the leaving group (probably
catalyzed by MgX_2 or RMgX, cf. **6**) and to form the dioxolenium ion **7**. Then,
the Grignard reagent attacks the ion **7** from the same side to yield the reaction product **5** having the 2-alkyl group in the axial orientation. Note that
the attack of the Grignard reagent on the opposite face of the dioxolenium
salt is not favored as it would require a boat-like transition state (**7** → **8**).
Similarly, the low reactivity of the equatorial alkoxy group in **4** is due
to the fact that its ejection cannot be stereoelectronically assisted unless
compound **4** takes first the unfavorable boat form **8** (R=OCH_3). This work and
analysis which appeared in 1969-70 (1) is one of the first contributions
to the principle of stereoelectronic control in systems derived from the
ester function. Bailey and Croteau have recently reported (2) that the products formed from the reaction of 2-methoxy-1,3-dioxane with Grignard reagents depend on reagent substrate complexation and stereoelectronic control.

Spencer and collaborators (3, 4) have observed the cyclization of **9**, **10**, and **11** under mild basic conditions which yielded the corresponding cis-decalin ketol **13**. On the other hand, cyclization of 12 afforded the trans-fused ketol **14** (R=H, X=H$_2$) exclusively (5). Spencer concluded that the cyclization process must be governed by the size of the incipient angular substituent: when R is small like a hydrogen atom, the transition state leading to the trans-product (cf. **15**, R=H) would be of lower energy than that leading to the cis-product (cf. **16** or **17**, R=H). However when R is larger than hydrogen, the opposite would be true. So, product formation depends on the steric effect caused by the angular R group and the enolate double-bond. It is however still possible that stereoelectronic effects play also an important role in these reactions. For instance, the approach of the enolate ion to the carbonyl group is different in **15** and **16** than in **17**. One approach could therefore be electronically favored over the others, but the experimental results described above give no information on that matter.

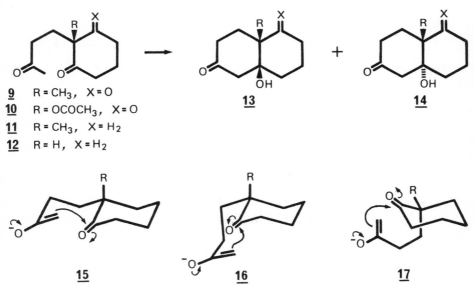

9   R = CH$_3$, X = O
10   R = OCOCH$_3$, X = O
11   R = CH$_3$, X = H$_2$
12   R = H, X = H$_2$

13

14

15   **16**   **17**

The first results which indicate that stereoelectronic effects play an important role in the aldol condensation were reported by Hajos and Parrish (6) who found that (a) triketone **18** cyclizes to the bicyclo[3.2.1]octane ketol **19** on treatment with piperidinium acetate in water and (b) ketol **19** undergoes an epimerization at C-4 yielding the more stable isomeric ketol **20** on treatment with piperidine. The authors concluded that the formation of ketol **19** from **18** under kinetically controlled conditions is the result of

"... maximum π-orbital overlap by an almost parallel
alignment of the enolic double-bond of the five-
membered ring and the carbonyl group in the butanone
side-chain."

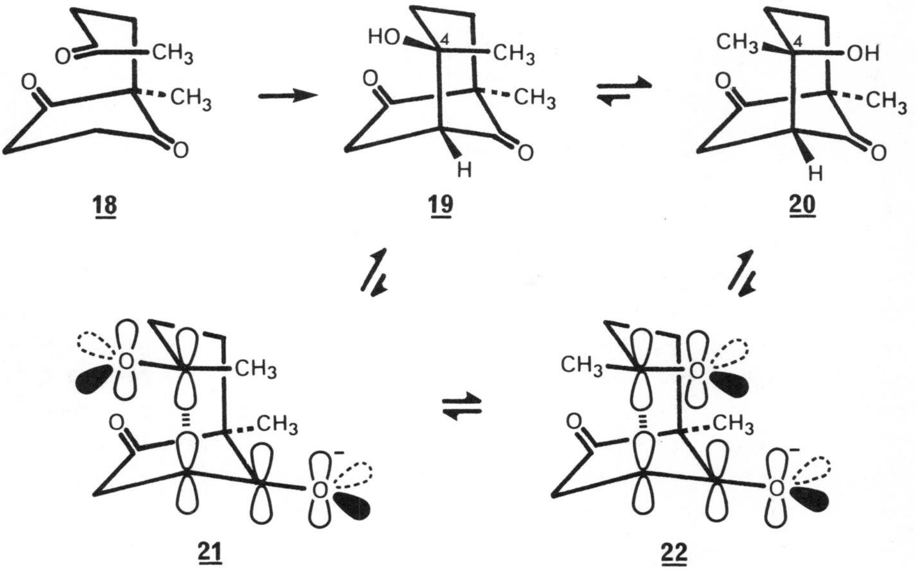

18   **19**   **20**

21   **22**

Indeed in the transition state leading to ketol **19** (cf. **21**), the enolate
double-bond is oriented antiperiplanar to the carbonyl group. This stereochemical approach must therefore be electronically favored over that leading
to the isomeric ketol **20**. In the last case, the carbonyl group of the side
chain is gauche (synclinal) with the enolate double-bond (cf. **22**).

In the course of the total synthesis of enmein, Fujita and co-workers (7)
have discovered that the intramolecular cyclization of the enolate **23** of
the corresponding tetracyclic keto-aldehyde at room temperature gave only
ketol **24**. However, when the same reaction is conducted at 60°C, thermodynamically controlled conditions prevail, and the epimeric product **25** is obtained. Inspection of molecular models indicates that the kinetically controlled
product **24** is again the result of an antiperiplanar arrangement of the enolate and the aldehyde double-bonds. Also, as in the previous examples, the
isomer **25** comes from a synclinal arrangement of the reacting functional
groups.

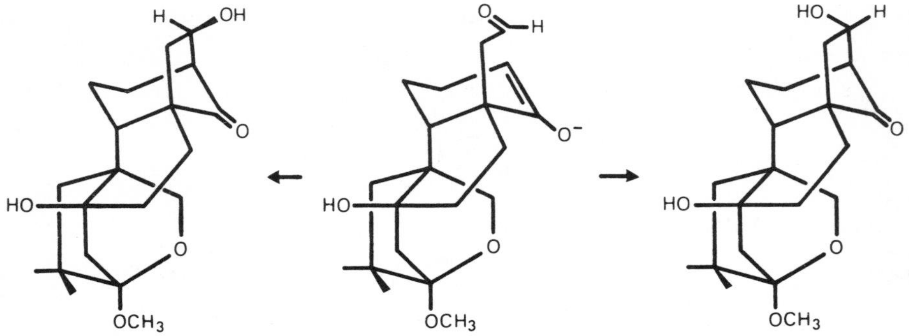

25   **23**   **24**

Schoemaker and Speckamp (8) have reported the quantitative conversion of
hydroxy-lactam **26** (n=1) (HCOOH, 18 h, r.t.) into the spirocyclic lactam
ester **27** (n=1). The other possible spiroisomer **28** (n=1) was not formed.
Hydroxy-lactam **26** (n=2) gave spiroisomer **27** (n=2), albeit in lower yield,
under similar conditions. The same authors (9) have also reported the successful cyclization of hydroxy-lactam **29** into spirolactam **30**. Analogous
results were obtained by Evans and Thomas (10) who found that the cyclization of a 9:1 mixture of enamides **31** and **32** in anhydrous formic acid gave
the spirocompound **30**. This compound is a key intermediate in Kishi's total
synthesis of perhydrohistrionicotoxin (11, 12).

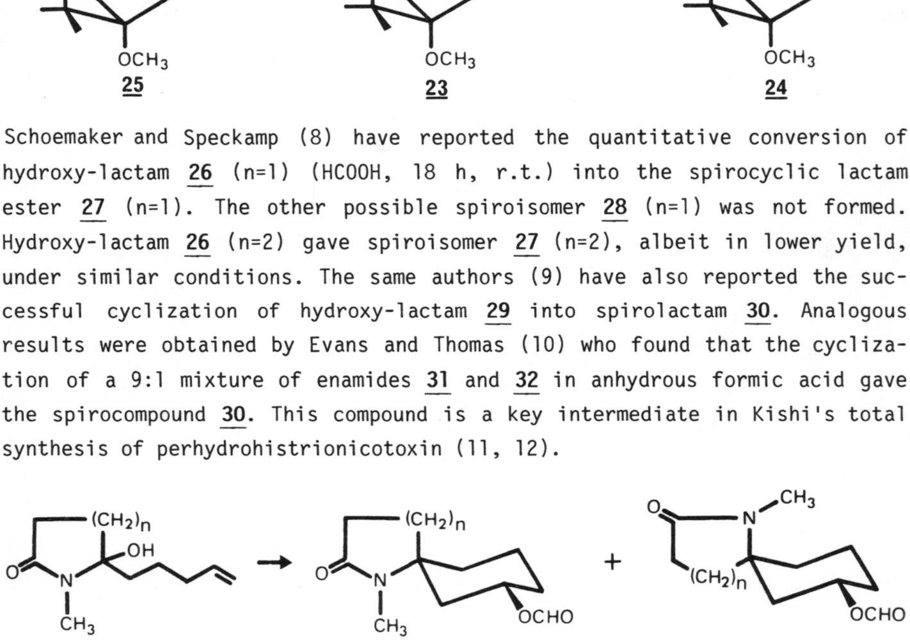

26   **27**   **28**

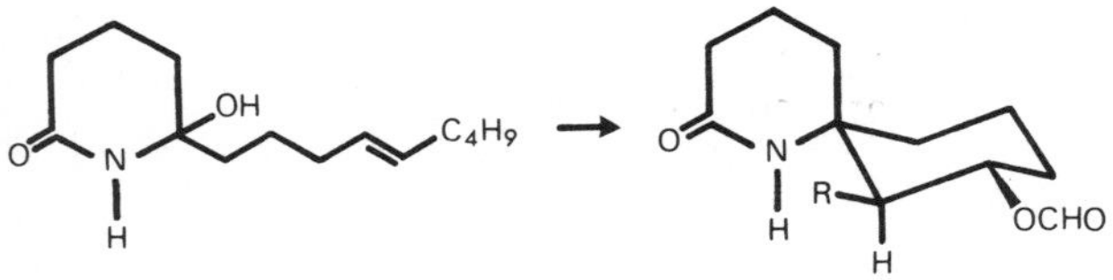

29   **30** (R = C$_4$H$_9$)

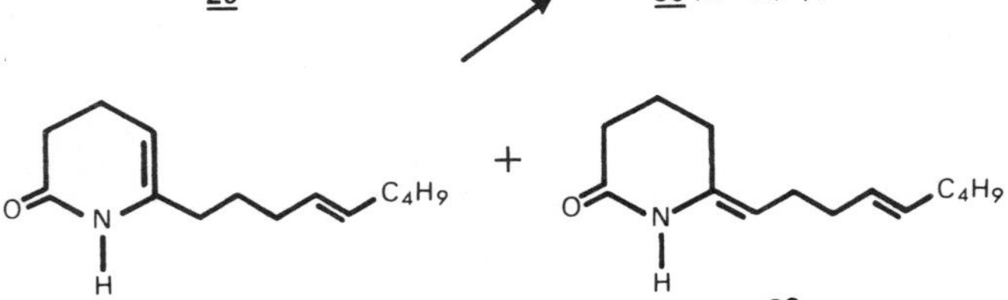

31   **32**

These results indicate that the cyclization via a chair-like transition state having the iminium cation in a pseudo-equatorial orientation (cf. 33) is more facile than that with an axial orientation (cf. 34). Note again the relative orientation of the iminium cation and the olefinic double-bond which is antiperiplanar in 33 and synclinal in 34.

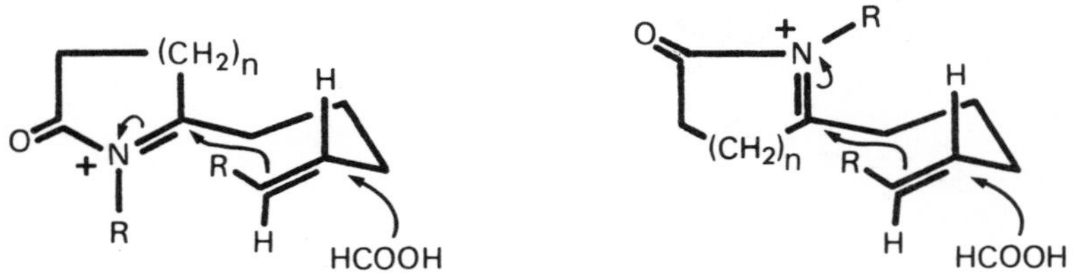

$$\underline{33} \qquad\qquad \underline{34}$$

Antiperiplanarity of double-bonds seems to be an important factor also in carbocyclic cases. Indeed, Harding and collaborators (13) have reported that the intramolecular cyclization of cyclohexenols 35 gave spiroisomer 36 in preference to 37 as the major product (ratio 4:1 when R=H and $\geqslant$9:1 when R=CH$_3$ or n-C$_4$H$_9$).

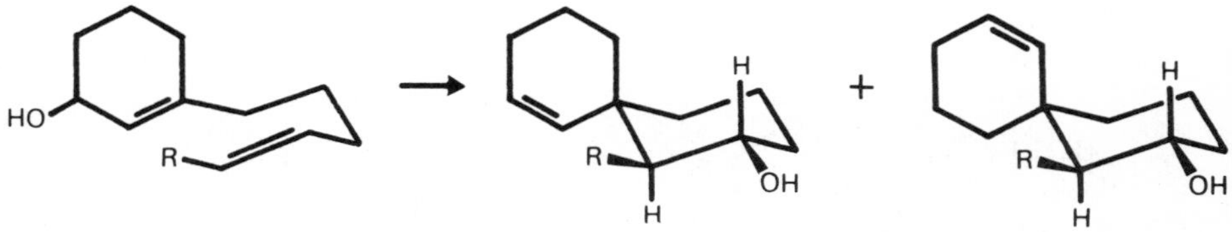

$$\underline{35} \qquad\qquad\qquad \underline{36} \qquad\qquad\qquad \underline{37}$$

Schoemaker, Kruk, and Speckamp (14) have discovered that the internal cyclization of cyclic α-acyliminium ions with double-bonds takes a completely different course when one methylene group of the ring is replaced by an oxygen atom: treatment of the E-olefin 38 (produced in situ) with formic acid gave the spiro[4.5]bicyclic compound 39 while the Z-olefin 40 yielded the isomeric product 41. The configuration at C_{11} in 39 and 41 depends on the configuration of the starting olefin; their formation can therefore be explained by a synchronous trans-coplanar attack of the α-acyliminium ion and formic acid on the double-bond. An anticlinal (cf. 42) is thus preferred over an antiperiplanar arrangement (cf. 43) in this particular series and it constitutes the first example of a 5-Exo-Trig mode of ring closure which takes place instead of the normally encountered 6-Endo-Trig ring closure.

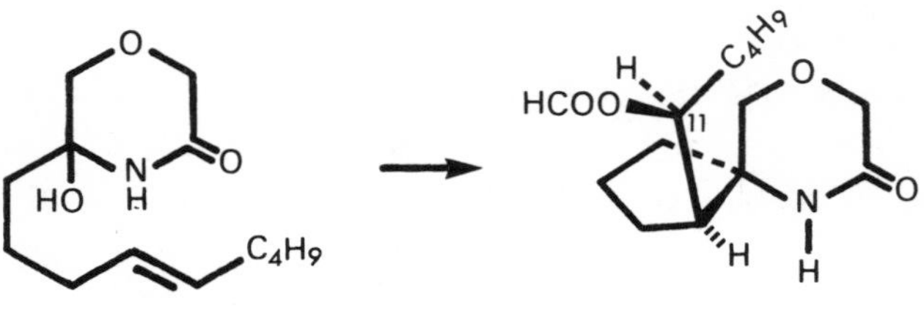

The synthesis of the eight possible diastereoisomeric racemates of dactyl-oxene-B (**44**) has been carried out recently by Ohloff and co-workers (15). In the course of this work, they have observed that the spiro bicyclic ethers **46** and **49** are formed more rapidly than their C-5 epimers **47** and **50** respectively on acid cyclization (p-TSA in CH_2Cl_2) of the hydroxy-diene precursors **45** and **48** (R_1= CH_3 and R_2= C≡CH or the opposite configuration).

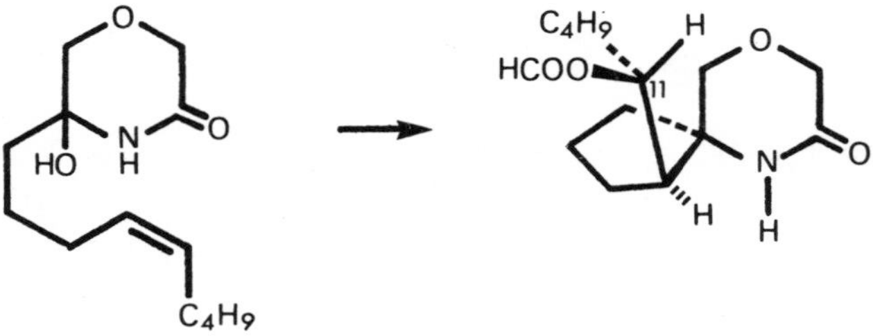

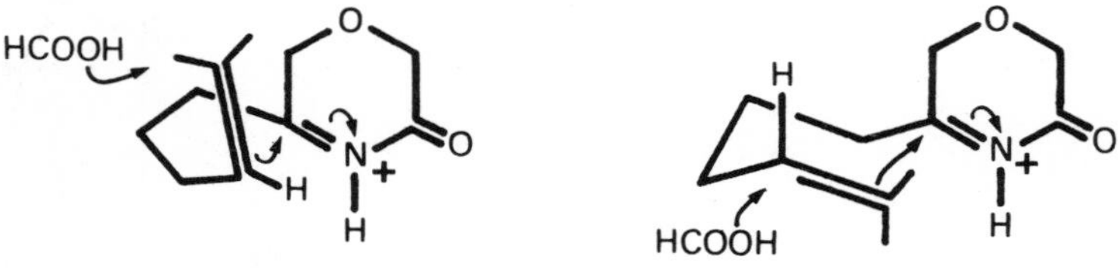

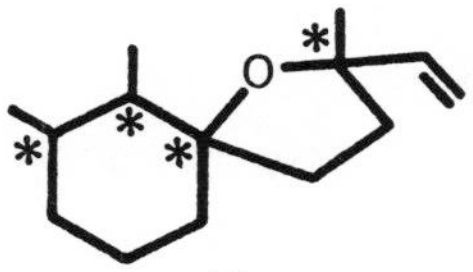

44

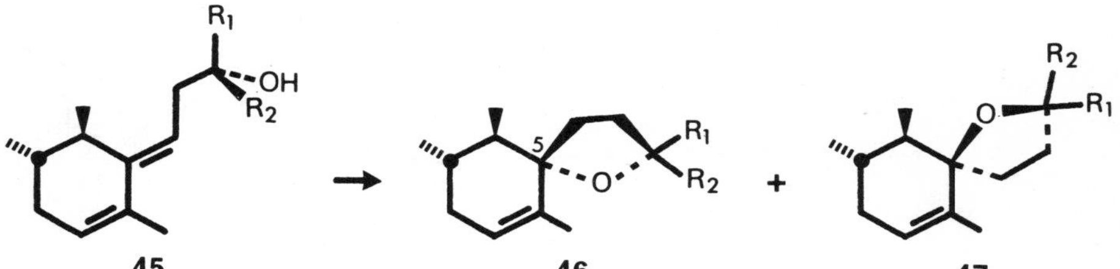

45 → **46** + **47**

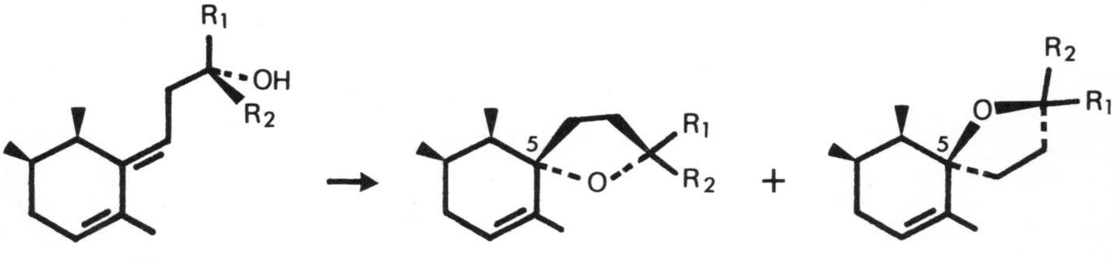

48 → **49** + **50**

The trans-dimethyl precursors (R₁= CH₃, R₂= C≡CH, or the opposite) are known to exist in conformation **45A** where both secondary methyl groups are axially oriented. The cis-dimethyl precursors **48** take conformation **48A** which has an equatorial methyl group at C-2 and a pseudo-axial one at C-1.

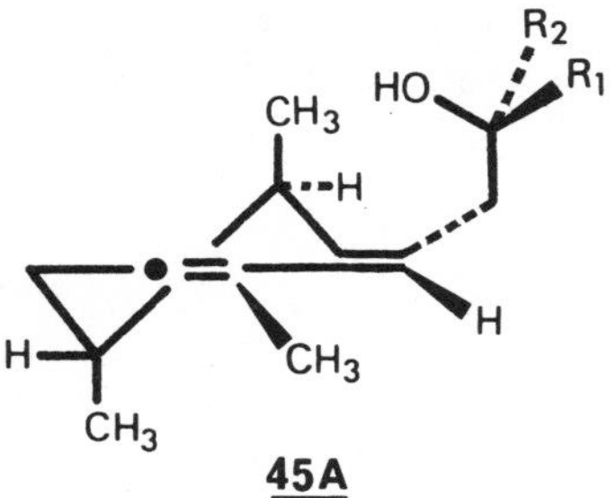

45A 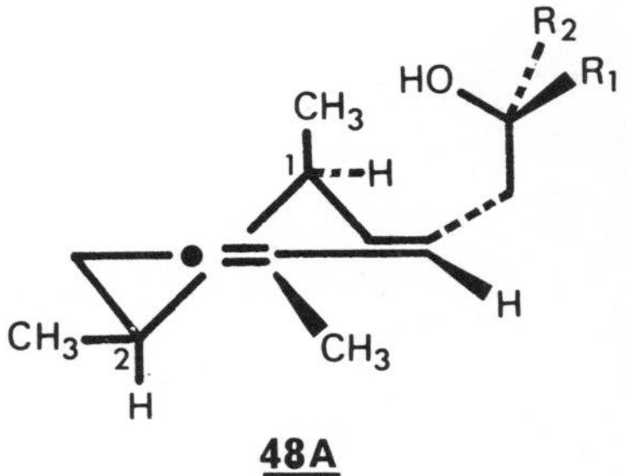   **48A**

A plausible explanation for the kinetically preferred formation of **46** and **49** can be obtained by taking the principle of stereoelectronic control into consideration: for instance, protonation of **45A** should give the allylic carbonium ion **51**. A nucleophilic attack on the bottom face of ion **51** (path A) should be favored over that on the top face (path B) because the former leads a half-chair transition state (**52**) while the latter leads to a less stable boat-like transition state (**53**). The preferred formation of **49** from **48A** can also be explained in a similar manner.

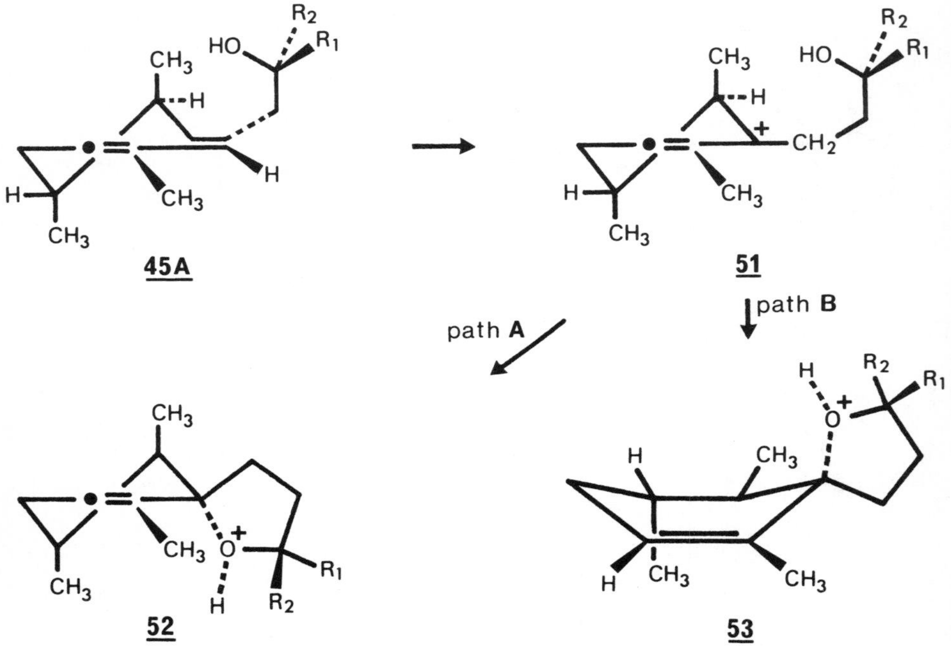

Ring opening of the β-lactam oxyanion **54** is now considered. Two different ring-openings are possible, a C — N bond cleavage to give **55** or a C — C bond cleavage to form **56**. Experimentally, the only process observed is **54** → **55** (16), although theoretical calculations (17) indicate that intermediate **56** is more stable than **55**.

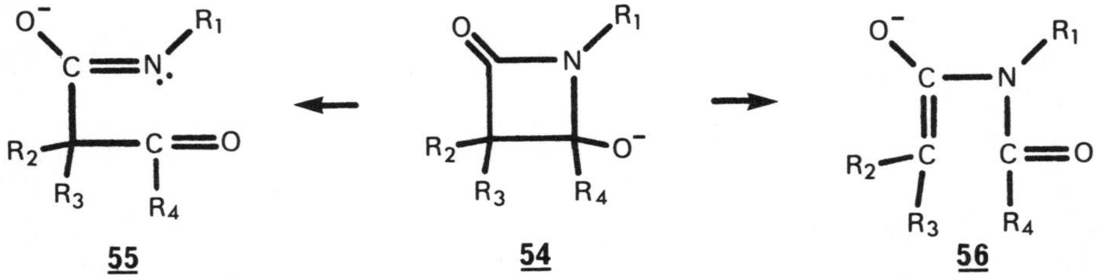

Kikuchi (16) has proposed that the experimental result can be readily explained because the process **54** → **56**, by comparison with the process **54** → **55**, is unfavorable from the stereoelectronic point of view. The C—C bond breaking process (**54** → **56**) may be expressed by a two-stage mechanism (**57** → **58** → **59**). In the first stage, the carbanion produced (**58**) is of high energy (no overlap) and the stabilization is achieved by the 90° rotation about the C—C bond (→ **59**). Thus even if the C—C bond breaking and the C—C bond rotation occur concertedly, the negative charge developing at the carbon atoms is not largely stabilized at the transition state as the overlap between the negative charge and the carbonyl π system is small.

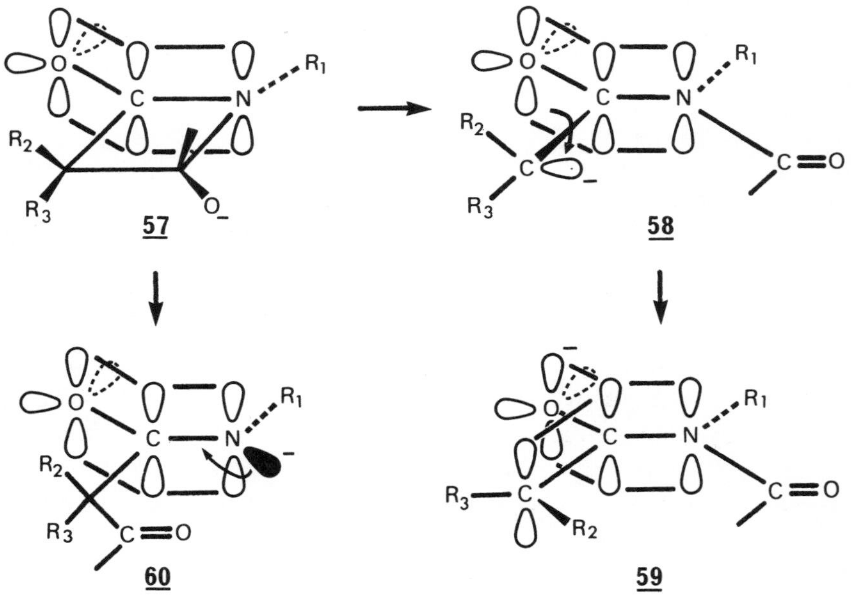

On the other hand, cleavage of the C—N bond gives directly the stabilized amide ion **60**. In the ion **60**, one nitrogen electron pair (p-orbital) is delocalized through a n-π* interaction (primary electronic effect) while the other is delocalized by an n-σ* interaction (secondary electronic effect; indeed, the newly generated electron pair is antiperiplanar to the C—O σ bond of the carbonyl function). Thus, both electron pairs of the nitrogen atom are delocalized.

Treatment of tricyclic β-lactam **61** which t-butyl hypochlorite in wet THF provided β-chloromethylpenam **63** (18). This 55% yield conversion takes place presumably through the intermediacy on the sulfenyl chloride $\underline{62}$. Thus, specific cleavage of the $C_A - S$ in preference to the $C_B - S$ bond is observed in $\underline{61}$. Contrary to this result, preferential cleavage of the $C_B - S$ bond takes place with the bicyclic β-lactam $\underline{64}$ (19). Similarly, neither of the tricyclic diastereoisomers $\underline{65}$ provided the selective $C_A - S$ bond cleavage (18).

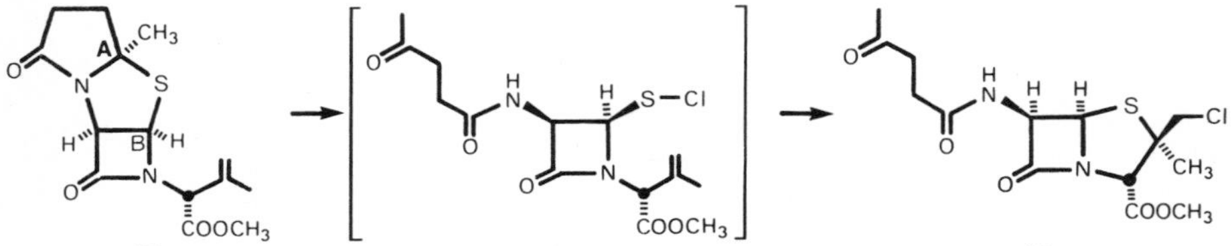

Baldwin and Christie (18) have proposed that the origin of this difference almost certainly derives from a stereoelectronic factor. In $\underline{64}$ and $\underline{65}$, the $C_B - S$ bond is more nearly orthogonal to the β-lactam amide plane than the $C_A - S$ bons is, with respect to the thiazolidine amide plane (cf. $\underline{66}$). The $C_B - S$ bond is therefore weaker than the $C_A - S$ bond and is preferentially cleaved. With tricyclic β-lactam $\underline{61}$, due to the five-membered ring, the $C_A - S$ bond becomes more nearly orthogonal to the thiazolidine amide plane (cf. $\underline{67}$). As a consequence, the ordering of bond lability is reversed and the $C_A - S$ bond is cleaved more readily. The stereoelectronically controlled step $\underline{61} \rightarrow \underline{63}$ has led to a stereospecific synthesis of a penicillin derivative from a peptide precursor.

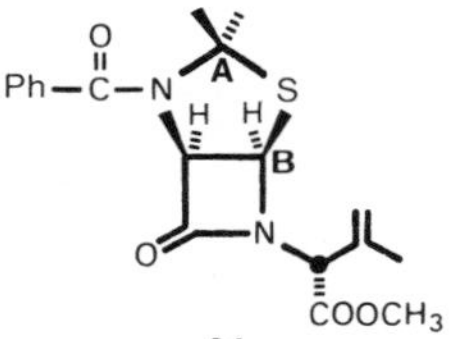

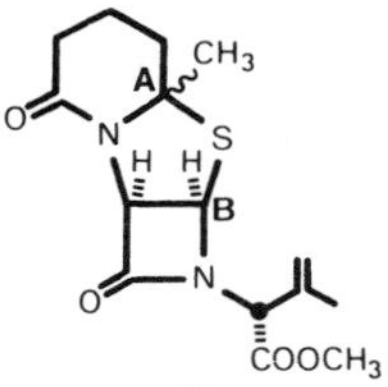

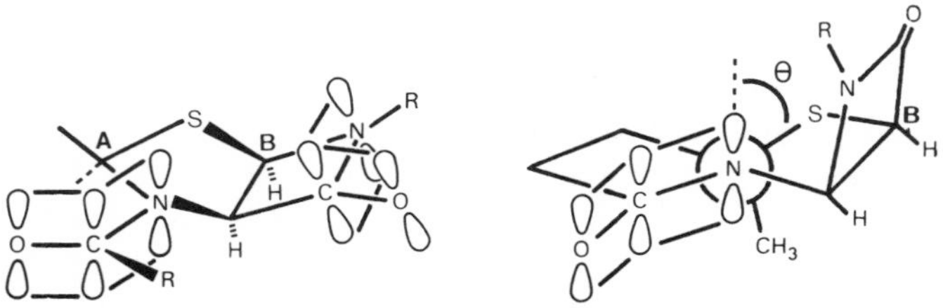

Ohuchida, Hamanaka, and Hayashi (20) have reported a synthesis of the dithia-
analogue $\underline{71}$ of tromboxane A. In one of the key steps, the stereoelectronically controlled axial conjugate addition of methyl 3-mercaptopropionate
to $\underline{68}$ was realized with stereoselectivity by using diisopropylethylamine
(0.2 equiv) in DMF. The product obtained $\underline{69}$, was then converted into $\underline{70}$
which was further transformed into the desired dithio derivative $\underline{71}$ [(a)
$\underline{t}$-BuO$^-$K$^+$, HMPA, 25°C; (b) NaOCH$_3$, CH$_3$OH; (c) NaOH 0.2 N, THF].

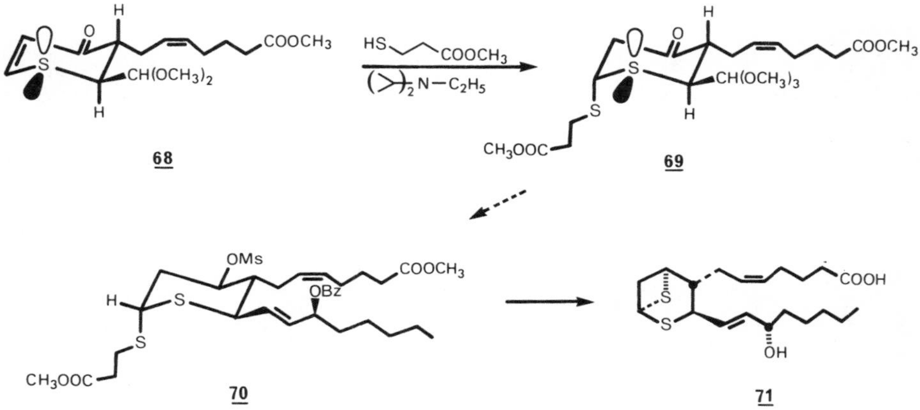

$\underline{68}$ $\underline{69}$

$\underline{70}$ $\underline{71}$

Noyori, Kobayashi, and Sato (21) have reported the Baeyer-Villiger oxidation
(22) of the substituted oxabicyclooctanones $\underline{72}$ (R= CH$_3$, $\underline{n}$-C$_5$H$_{11}$, C$_6$H$_5$ or
CH$_2$OR) with trifluoroacetic acid. They found that lactones $\underline{73}$ were produced
in a larger proportion ($\approx$2:1 to $\approx$3:1) than the isomeric lactones $\underline{74}$.

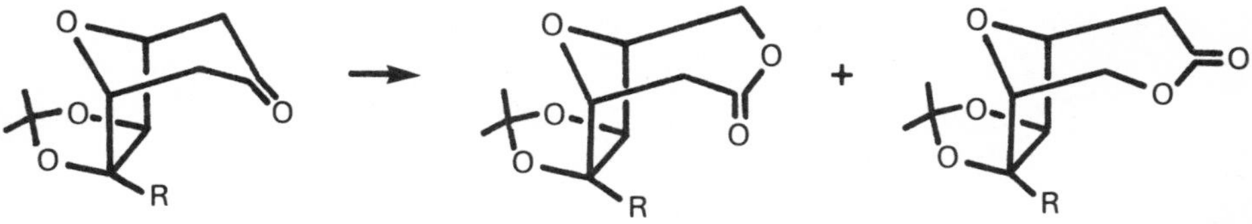

$\underline{72}$ $\underline{73}$ $\underline{74}$

The preferred formation of lactone $\underline{73}$ was explained by the orientation of
the non-bonded electron pairs of the hydroxyl group in the corresponding
tetrahedral intermediate and by assuming that the migrating C$-$C bond must
be antiperiplanar to the RCOO$-$O bond. They have also assumed that the per-
acid reacts on the least hindered face of the carbonyl group in $\underline{72}$.

Under such conditions, only four different conformations are possible (75, 76, 77, and 78) for the tetrahedral intermediate. Conformations 75 and 76 are eliminated on the basis of steric interactions between the hydroxyl hydrogen and the five-membered ring. Also, conformation 77 should be less stable than conformation 78 because the former has a steric interaction between the R group and the hydroxyl hydrogen. On that basis, the preferred formation of lactone 73 (from 78) over lactone 74 (from 77) would be due to a combination of steric and stereoelectronic effects. This rationalization is based on the reasonable assumption that an electron pair of an oxygen atom is a better electron donor than an O−H bond. This is the first rationalization which points out the importance of the stereochemistry of an hydroxyl group (relative orientation of the oxygen electron pair and the hydrogen atom with the remaining reacting bonds) in order to explain an experimental result.

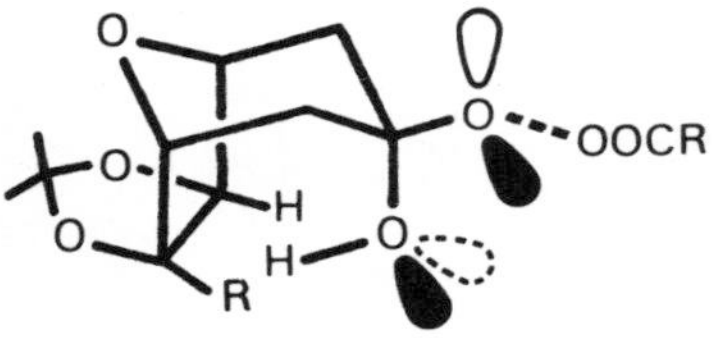

75

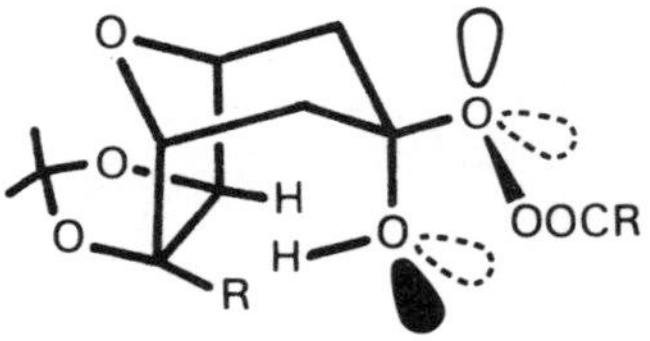

76

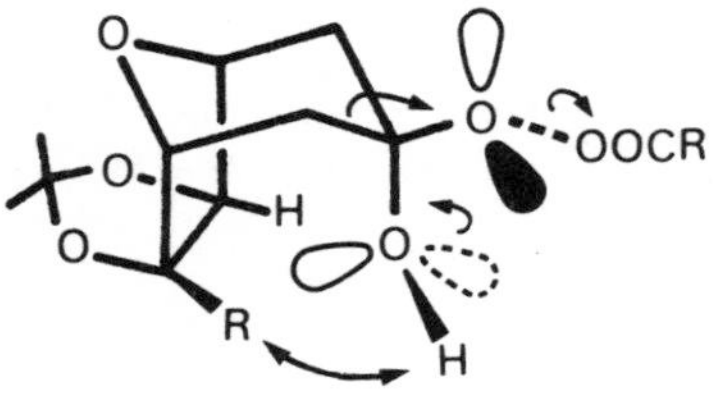

77

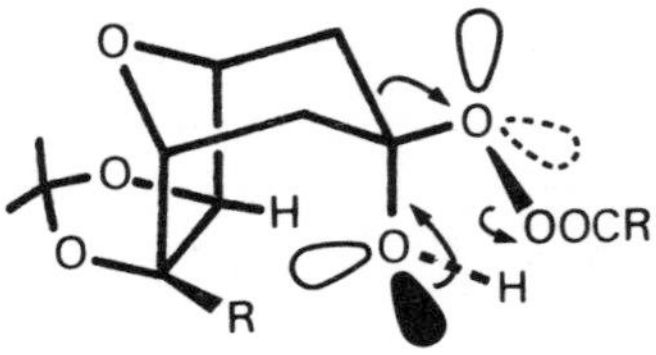

78

Lattes and co-workers (23) have studied the photochemical and thermal rearrangement of oxaziridines into amides (79 → 80).

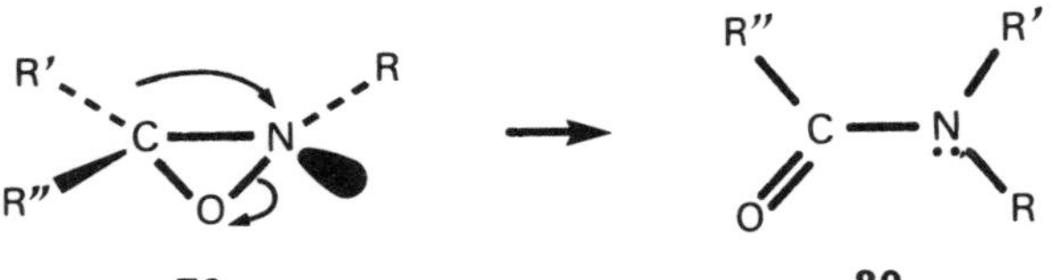

79

80

They have obtained unambiguous evidence that the photochemical process is completely regiospecific. They have further concluded that this process must be controlled by stereoelectronic factors because the bond (R' — C) which is anti to the nitrogen lone pair in oxaziridine 79 migrates more easily than the bond (R"— C) which is syn to the lone pair. Syn-anti isomerization by nitrogen inversion in oxaziridines 79 was rigorously excluded under the reaction conditions (r.t.).

Their key experiment is the following: optically active oxaziridine 81 (100% optical purity) was synthesized and its absolute configuration was established. The photolysis of 81 gave the optically active lactam 82 (having an S configuration at C-5) in 80% yield; none of the enantiomer 83 was observed. The driving force for the migration must therefore be the two electron pairs (one from the nitrogen and one from the oxygen atom) which are antiperiplanar to the C_3-C_8 bond.

81   82   83

They have further pointed out that the lower regioselectivity observed in the thermal rearrangement of oxaziridines does not rule out the above stereoelectronic requirement. In this case, the energy required for the reaction in sufficient to induce both nitrogen inversion in the starting oxaziridine and migration of the C-substituents.

Baldwin and Norris (24) have reported a study on the kinetically controlled addition of halogen (bromine or chlorine) or cyclopentadiene to 1,4-benzoquinone 4-(O-methyloxime) (84 → 86 and 84 → 85). The most remarkable result of the above addition reactions is the overwhelming predominance of products which have the configuration 87, where the methoxy group is anti to the remaining double-bond, rather than the alternative arrangement 88; in the case of either bromine and chlorine, over 70% of configuration 87 is produced and with cyclopentadiene, this arrangement is observed in excess of 93%.

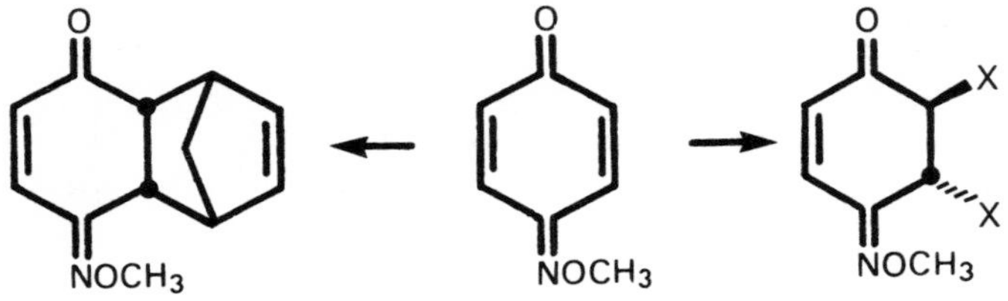

The authors have explained these results by invoking a stereoelectronic effect which involves the nonbonding electrons on the oxime nitrogen. They have proposed that the nonbonding electrons on the nitrogen atom of the oxime group exert a subtle but significant effect on the benzene sigma $C-C$ bond which is oriented antiperiplanar. This n → σ* interaction will produce some antibonding character in the antiperiplanar $C-C$ bond with concomitant lengthening of the bond. Thus, in structures **87** and **88**, the bonds being lengthened are those syn to the oxime substituent, i.e. C_4-C_5 and C_3-C_4 respectively. Lengthening of the C_3-C_4 bond in **88** will reduce the effective p-p orbital overlap between carbon atoms 3 and 4, reducing in turn the delocalization of the lone pair on the oxime oxygen atom, through the conjugated system, onto the carbonyl oxygen. The bond lengthening occurring in **87** does not lead to a reduction of electronic delocalization; this configuration is therefore more stable than **88**. It is also a logical consequence of this proposal that in quinonoid derivatives like **84**, the canonical structure **89** makes a greater contribution than structure **90**. This is in accord with the hypothesis suggested by Norris and Sternhell (25-28) to explain the ^{1}H nmr coupling constant and the position of syn-anti equilibrium in these derivatives.

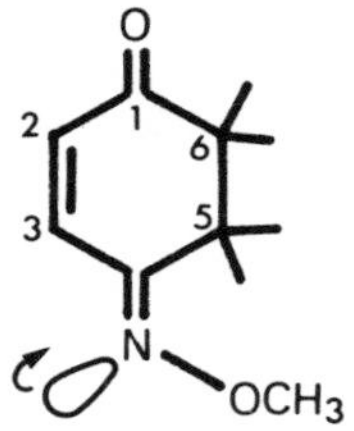

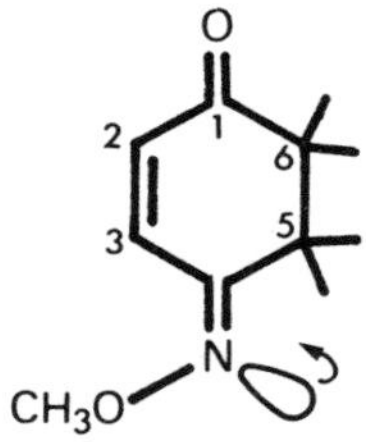

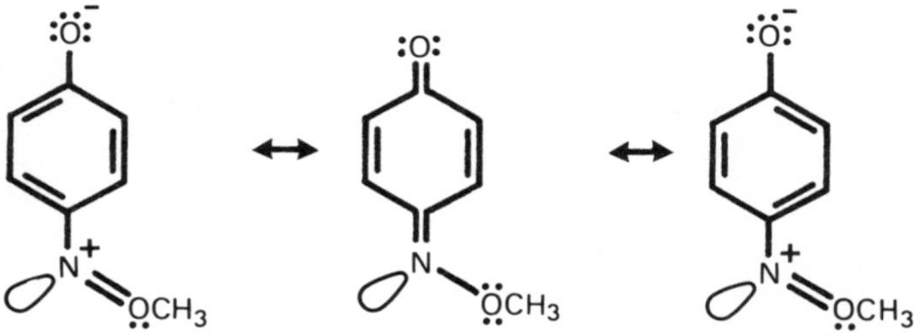

89   90

Vasella and collaborators (29-31) have reported that the cycloaddition of nitrones **91** ($R_1=R_2=H$; $R_1=CH_3$, $R_2=H$; $R_1=H$, $R_2=CH_3$) to methyl methacrylate yields diastereoisomeric adducts **92** and **93** which differ at C-5. For instance, the nitrone **91** ($R_1=R_2=CH_3$) gave a 95:5 mixture of **92** and **93** ($R_1=R_2=CH_3$). Vasella has suggested that the observed stereoselectivity is due to a stereoelectronic effect which strongly influences the stereochemistry of the transition state of the cycloaddition reaction. This stereoelectronic effect is caused by the nitrogen electron pair which is oriented antiperiplanar to the C—O bond of the furanose ring during the cycloaddition reaction (cf. **94** → **95**). When this additional electronic delocalization is operative, the energy of the transition state is lowered accordingly.

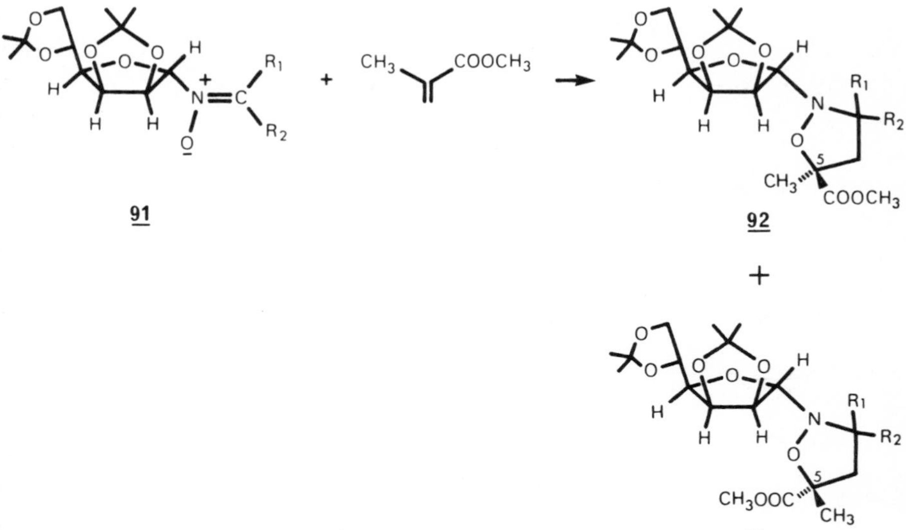

91   92

+

93

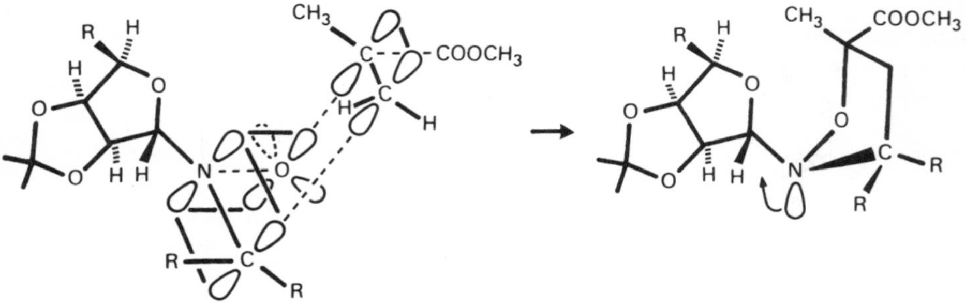

94   **95**

Dolson and Swenton (32) have found that the acid hydrolysis of bisketals **96A** and **96B** (R=CH₃) is highly regiospecific yielding monoketals **97A** and **97B** respectively. On the other hand, bisketal **96C** (R=CH₃) gave a ≈3:1 mixture of **97C** and **98C**. The selectivity observed in the hydrolysis of **96A** and **96B** was ascribed to the preferential formation of the linearly conjugated oxonium ion **101** relative to the cross-conjugated oxonium ions **99** and **100**. Also, as the transition state for conversion to the oxonium ion is approached, most effective stabilization of the positive charge requires the alkoxy group to be in the plane of the ring. For **99** and **100**, this planarity generates steric congestion while for **101**, this planar arrangement is quite acceptable sterically. For bisketal **96C**, the formation of a mixture of monoketals **97C** and **98C** is apparently due to additional stabilization of the p-methoxy group in cations **99** and **100**.

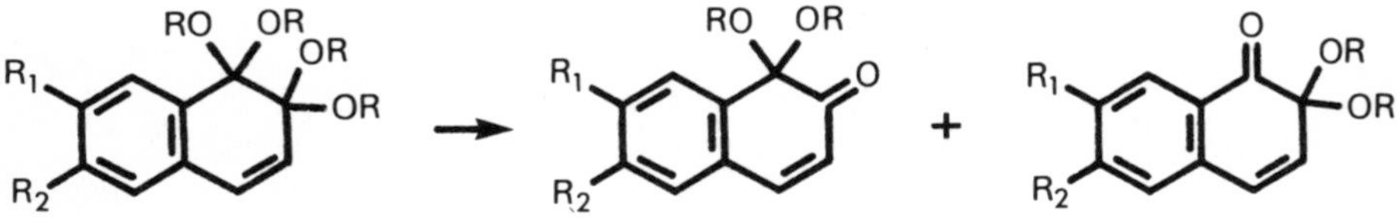

96A R₁ = R₂ = H   **97A**   **98A**
96B R₁ = OCH₃, R₂ = H   **97B**   **98B**
96C R₁ = H, R₂ = OCH₃   **97C**   **98C**

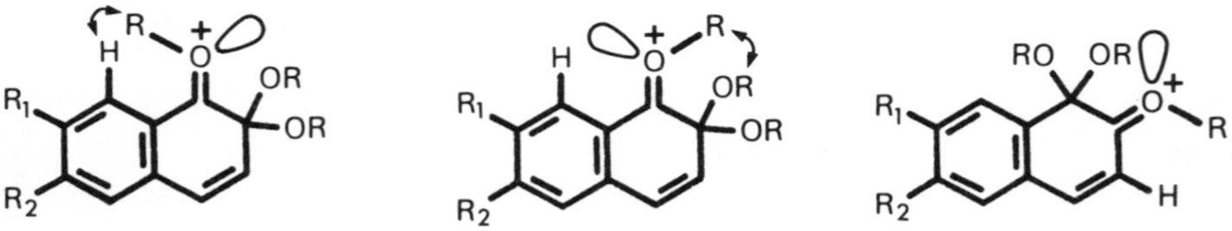

99   **100**   **101**

Burton, Le Page, Gabe, and Ingold (33) have recently proposed that the supe-
rior antioxidant activity of vitamin E (102) and the related phenol 103
by comparison with 4-methoxy-2,3,5,6-tetramethylphenol (104) is due to ster-

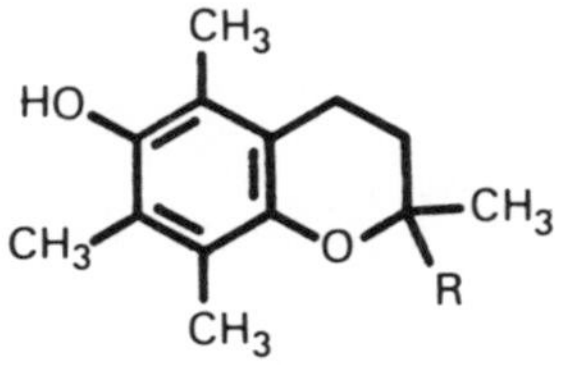

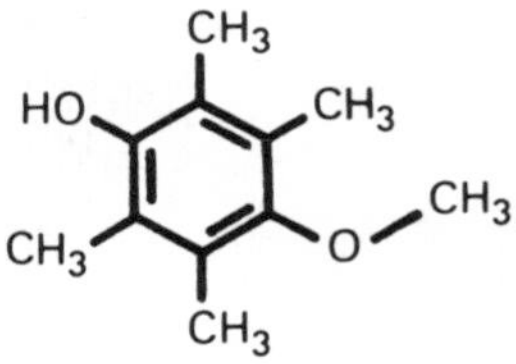

102 R = (CH$_2$CH$_2$CH$_2$CH(CH$_3$))$_3$CH$_3$ **104**
103 R = CH$_3$

eoelectronic factors. The antioxidant activity of phenols depends on their
capacity in trapping radicals as shown in the following equations.

$$ROO\cdot \; + \; ArOH \longrightarrow ROOH \; + \; ArO\cdot$$
$$ROO\cdot \; + \; ArO\cdot \longrightarrow \text{nonradical products}$$

With 4-methoxyphenols, the phenoxyl formed is further stabilized by delocalization of the unpaired electron to the p-type orbital of the methoxyl oxygen (105 ↔ 106). This interaction is allowed in vitamin E and in compound
103 but prohibited in compound 104 because the former two compounds exist
in conformation 107 and the latter in conformation 108. In conformation
108, due to the methyl groups in ortho position, the methoxyl group is twist-
ed out of the plane of the aromatic ring and the delocalization of the methoxyl oxygen electron pair is consequently prohibited.

105   **106**

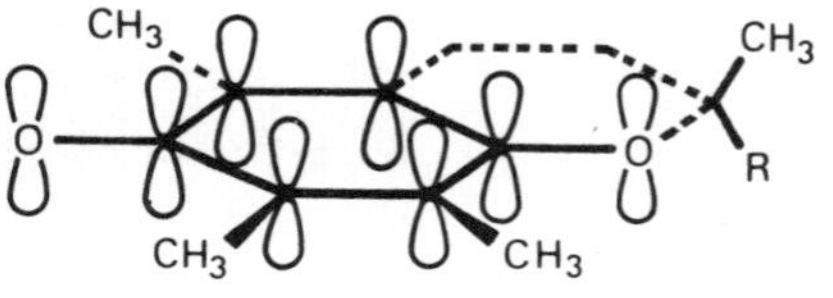

107 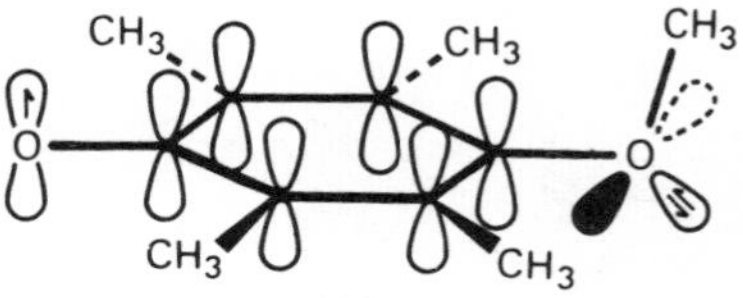   **108**

Hydrogen atoms at carbon next to sulfur tend to be acidic (34); carbanions
can therefore be generated under appropriate basic conditions with sulfides,
sulfoxides and sulfonium salts. It was also observed that the acidity depends

on the stereochemistry of the C—H bond. For instance, Eliel and collaborators (35-39) have shown that the equatorial carbanion 110 derived from 1,3-
dithiane 109 is thermodynamically more stable than the axial isomer 111.
Carbanion 110 is also preferentially generated from 109 under kinetically
controlled conditions and its reaction with electrophiles is stereocontrolled yielding the equatorial product 112 (E= H or alkyl). The stereoelectronic effect has also been demonstrated in 1,3-oxathianes (40) and in 1,3-
dioxanes (41).

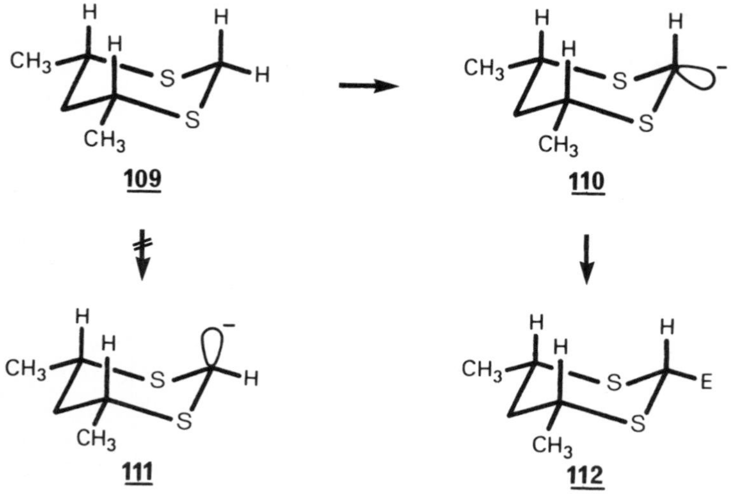

Wolfe and co-workers (42) were the first to observe the preferential exchange with D₂O/OD⁻, of one of the diastereotopic benzylic hydrogen in
benzyl methyl sulfoxide. Baldwin and co-workers (43) then showed that the
deuterium exchange of the kinetically more labile benzylic proton in (S)-
benzyl methyl sulfoxide (113) yields the (R)-configuration 114. The sulfoxide case has since been investigated extensively (44-47). Considerably
greater selectivity (rate ratio of over 1000) was found in the H/D exchange
and the alkylation of benzyl t-butyl sulfoxide (48, 49) and the exchange
of the α-hydrogens in the bridged biaryl sulfoxide 115 (50). There is also

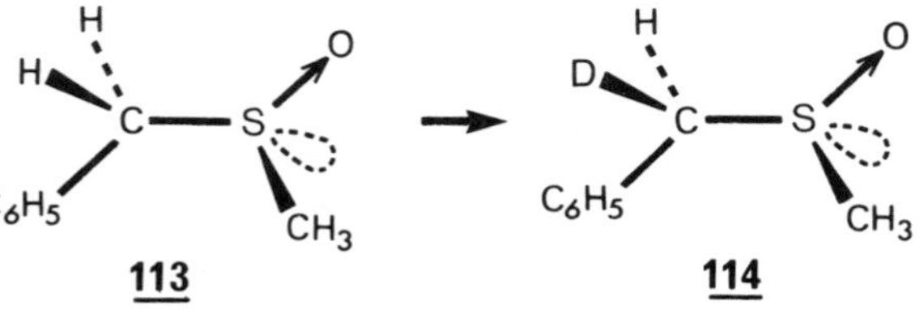

115

strong evidence that the acidity of the protons <u>alpha</u> to the sulfonium group depends upon the orientation of the C — H bond with respect to the substituent on the sulfur atom (51-53).

It can be concluded from these results that powerful stereoelectronic effects control the acidity in these compounds. There remains however to understand the nature of these effects. Several reports (54-61) indicate that the enhanced acidity of C—H <u>alpha</u> to sulfur is not due to a (d-p)π stabilization of the carbanion by the d orbitals on sulfur (62, 63). This electronic effect has been referred to (64) as the "gauche effect" which would be a tendency to adopt a structure which has the maximum number of gauche interactions between the adjacent electron pairs and/or polar bonds. Lehn and Wipff (61) have carried out an <u>ab initio</u> study which shows that carbanion stabilization by α-heteroatoms is subject to appreciable stereoelectronic effects. Their calculations have shown that equatorial type carbanions (<u>cf</u>. <u>110</u>) are much more stable than axial type carbanions (<u>cf</u>. <u>111</u>). According to these authors, these stereoelectronic effects may be interpreted in terms of a destabilizing interaction in <u>111</u> (mixing of two occupied orbitals) and a stabilizing interaction in <u>110</u> (C⁻ lone pair mixing with the antibonding α* C — S orbital, <u>cf</u>. <u>110A</u>). Theoretical work from Lehn, Wipff, and Demuynck (65) indicates that α-seleno carbanions are subject to appreciable stereoelectronic effects similar to those of α-thia carbanions.

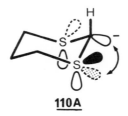

110A

Kishi and collaborators (66-68) have observed a regiospecific metallation (n-BuLi, THF, -78°C) and alkylation with ethyl iodide of anisylidenedithio-N,N'-dimethylpiperazine-2,5-dione (**116** → **117**). This result leads to the conclusion that H_3 is more acidic than the H_6 in **116** and that the relative acidity of these two bridgehead hydrogens must depend on stereoelectronic effects. An X-ray structure determination (69) shows that the dihedral angle $C_8 - S_9 - C_6 - H_6$ (154°) is close to that for $C_8 - S_7 - C_3 - Et$ (155.7°); it is therefore difficult to attribute the regiospecificity to a difference in overlap between the sulfur 3d orbitals and the sp^3 orbital at the bridgehead positions. Kishi has suggested that the regiospecificity could arise from the different environments around the lone pairs of the sulfur atoms. Indeed, the relative orientation of the two lone pairs of S_7 with the $C_3 - C_2$ and $C_3 - N_4$ bonds is completely different from that of S_9 with the $C_6 - C_5$ and $C_6 - N_1$ bonds. However, the nature of these stereoelectronic effects which make H_3 more acidic remains to be understood.

116   **117**

The reaction of thioacetals with two equivalents of peroxy acids generally affords epimeric gem-disulfoxides, the ratio of epimers being subject to kinetic and/or equilibrium control (70). However, Poje, Sikirica, Vicković, and Bruvo (71) have reported the first example of a stereospecific oxidation in the gem-disulfide series. They found that 2,2-bis(methylthio)-1,3-diphenylpropane is smoothly oxidized to the corresponding meso-disulfoxide with m-chloroperbenzoic acid. The study of conformational properties in solution by nmr indicated that the meso-disulfoxide exists in conformation **119** which

118   **119**

is the same as that found in the solid state by X-ray analysis. The above authors have suggested that the configuration and conformation of meso-disulfoxide 119 is the result of a stereospecific oxidation of conformation 118 of the gem-disulfide. They have further suggested that this particular stereochemical course cannot be explained by steric effects; it must have an electronic origin.

Lehn and Wipff (72) and Gorenstein and co-workers (73-80) have proposed on the basis of molecular orbital calculations that stereoelectronic effects similar to those observed in esters and amides play also an important role in the hydrolysis of phosphate esters. For instance, calculations suggest that the axial P — OR bond in the trigonal bipyramid conformation 120 is weaker than that in the conformation 121 because in the former, the oxygen atom of the equatorial OR group has an electron pair antiperiplanar to the axial P — OR bond. Experimental results tend to support this interesting proposal but additional experiments are needed before unambiguous conclusions can be reached (81).

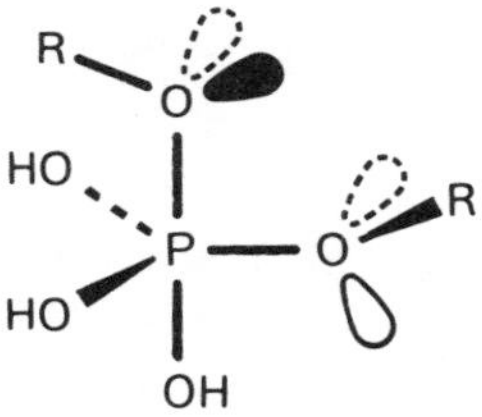

120 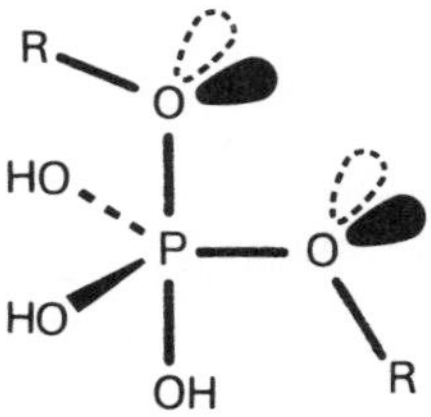   **121**

Unusual reactivity

Very unusual reactivity has been observed (82-84) with the tricyclic ortho-amides 122 and 123. This can be readily explained on the basis of the stereoelectronic effects due to the three nitrogen electron pairs.

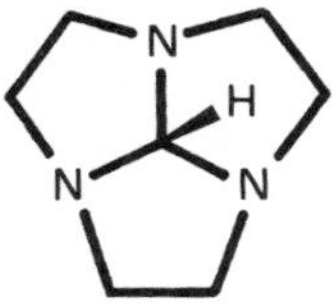

122 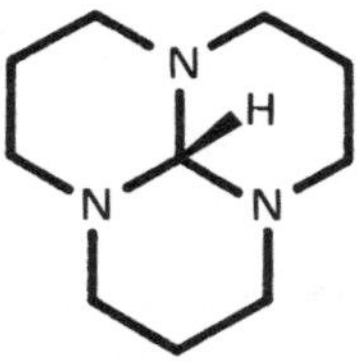   **123** 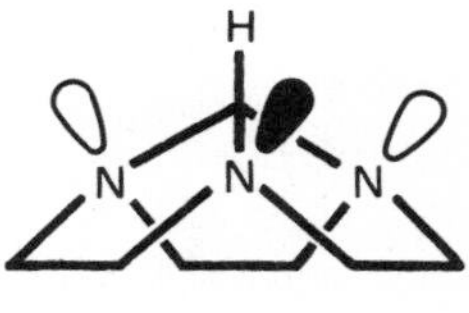   **122A**

123A   **123B**

Orthoamide <u>122</u> adopts a conformation (<u>122A</u>) in which the electron pairs are synperiplanar (or nearly) to the central carbon-hydrogen bond. This is confirmed by pronounced deshielding of the central hydrogen atom (δ=5.0 ppm) in the ^{1}H nmr spectrum and by the absence of Bohlmann bands (85, 86) in the infrared spectrum. The orthoamide <u>123</u> adopts conformation <u>123A</u> which has the nitrogen lone pairs antiperiplanar to the central C — H bond. Low temperature ^{13}C nmr analysis indicates that this conformer must predominate over conformer <u>123B</u> which is stabilized by two anomeric effects (two electron pairs antiperiplanar to a C — N bond) but destabilized by steric effects. The six equivalent axial carbon-hydrogen bonds adjacent to the nitrogens in conformer <u>123A</u> are responsible for strong Bohlmann bands between 2690 and 2800 cm^{-1} in the infrared spectrum. Furthermore, an additional Bohlmann band due to the central C — H bond appears at 2450 cm^{-1}, an extraordinarily low frequency. In the monodeuterated derivative <u>123</u> (H=D), this absorption is absent, and a new band appears near 1800 cm^{-1}. The positions of Bohlmann bands are therefore influenced by stereoelectronic effects.

The chemical shift of the methine proton in orthoamide <u>123</u> is 2.3 ppm. It is therefore at a much higher field than that of orthoamide <u>122</u>. This remarkable difference of 2.7 ppm can be ascribed to a dramatic stereoelectronic effect. The origin of the unusual spectroscopic properties of orthoamide <u>123</u> presumably is the antiperiplanar relationship of the central C — H bond to the three lone pairs. This arrangement permits mixing of the lone pair orbital with the antibonding orbital of the central C — H bond (σ*). As a result, the electron density at the methine hydrogen increases and the central C — H bond is weakened. Indeed, this hydrogen has a notably small chemical shift.

The orthoamides <u>122</u> and <u>123</u> are therefore completely different: one adopts a conformation in which the central C—H bond is synperiplanar to the adjacent lone pairs (<u>122A</u>) while the other takes a conformation in which the central C — H bond is antiperiplanar (<u>123A</u>). These two compounds are therefore expected to have quite a different reactivity.

Treatment of orthoamide **123** with equimolar amounts of aqueous hydrochloric acid yielded the salt **124** from which **123** can be regenerated by neutraliza- tion. NMR indicates that salt **124** and bicyclic formamidinium ion **126** inter- convert rapidly at 70°C. This is explained by a conformational change from **124** into **125** which permits elimination of the amine group with stereoelec- tronic control (**125** → **126**). Finally, addition of excess hydrochloric acid to orthoamide **123** precipitated the bicyclic dichloride **127**.

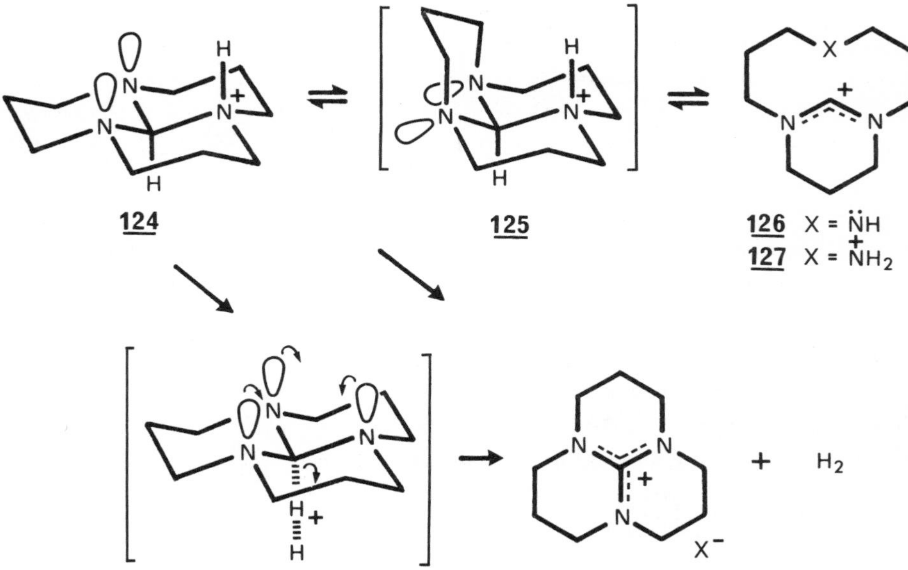

124   **125**   **126** X = $\overset{..}{N}H$
                                                                          127 X = $\overset{+}{N}H_2$

128   **129**

Pyrolysis of tetrafluoroborate orthoamide salt **124** under nitrogen cleanly gave guanidinium tetrafluoroborate **129** (X=BF$_4^-$) under remarkably mild condi- tions (110°C, 23 h). Molecular hydrogen, the necessary by-product was trapped in 76% yield. Erhardt and Wuest (84) have suggested that this reaction takes place in two steps: (1) dissociation of ammonium tetrafluoroborate **124** into orthoamide **123** and then (2) oxidation of orthoamide **123** by H$^+$ via a transi- tion state or an intermediate having the linear configuration **128**. This remarkable transformation can be rationalized only if the stereoelectronic effects of the three nitrogen lone pairs are taken into consideration (cf. **128**).

Orthoamide 123 cleanly reduced mercuric acetate in ethanol at 25°C to mercury or mercurous acetate. The organic product formed is guanidinium salt
129 (X= $^-$OAc). Similarly, iodine in methanolic potassium carbonate at 25°C
oxidized orthoamide 123 to guanidinium iodide 129 (X=$^-$I). On the other hand,
orthoamide 122 does not react with mercuric acetate even in boiling ethanol.
Syn-elimination of mercury and acetic acid from complex 130 must be slow
but anti-elimination from complex 131 (Y= I_2 or HgX_2) must occur readily.

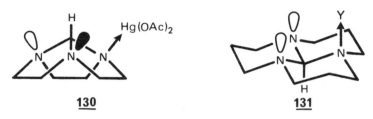

130   **131**

Oxidation of five-membered tricyclic orthoamide 122 to give the strained
guanidinium tetrafluoroborate 132 is however possible with triphenylcarbenium tetrafluoroborate. The six-membered tricyclic orthoamide 123 also reduced triphenylcarbenium tetrafluoroborate, but, very surprisingly, only
a very small amount (< 5%) of the expected guanidinium tetrafluoroborate
129 (X=BF_4^-) was formed. Thus, hydrogen is transferred not from the central
carbon atom but from one of the other six equivalent carbon atoms, giving
presumably iminium tetrafluoroborate 133 or the isomeric imine 134 as the
initial products of oxidation. These intermediates could not be isolated,
however, and the final products of oxidation appear to be oligomers of 134
due to a rapid cyclotrimerization characteristic of imines.

The low reactivity of the central carbon-hydrogen bond in orthoamide 123
was explained in two ways (82):

(1) steric factors make a methine hydrogen inaccessible to triphenylmethyl
     cations or radicals;
(2) the major products of oxidation by triphenylcarbenium may result from
     the cleavage of carbon-hydrogen bonds gauche or synperiplanar to lone
     pairs.

Thus, an intermediate similar to ammonium salt 135 would be formed first
and it would then undergo a syn-elimination by the loss of H_B to give only
iminium ion 133 since the triphenylmethyl group and the methine hydrogen
H_A in 135 lie on opposite faces of the molecule.

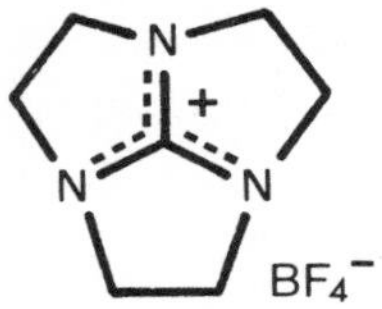

132 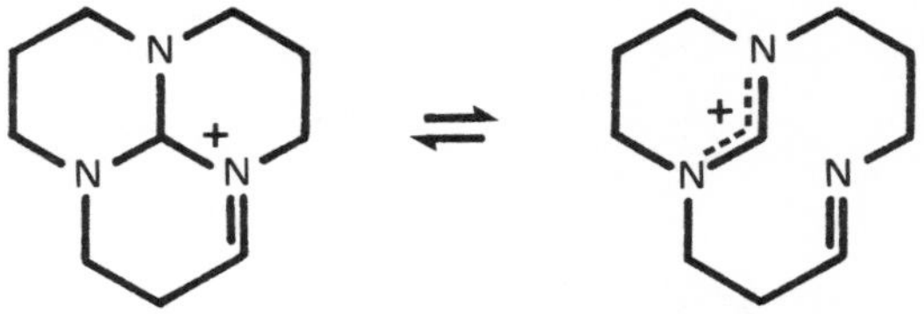   **133** ⇌ **134**

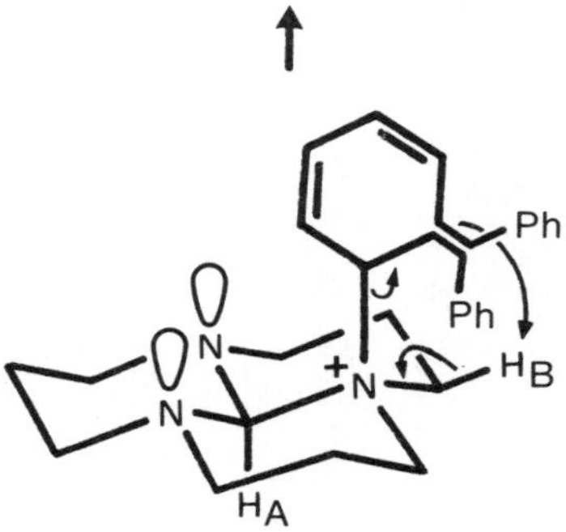

↑

135

Heating at 165°C for 12 h, equimolar amounts of magnesium perchlorate hemi-
hydrate, five-membered orthoamide **122** and methyl phenyl glyoxalate (**136**)
produced guanidinium perchlorate **132** (BF_4^- = ClO_4^-) and hydrolysis of the
crude product provided mandelic acid (**137**) in 45% yield. Six-membered orthoamide **123** also reduces methyl phenyl glyoxalate to mandelic acid. The predominant product of oxidation was again the perchlorate salt of the oligomer
obtained previously from imine **134**.

136 → **137**

The magnesium cation is essential in these oxidation reactions. Several
plausible mechanisms which provide a role for the magnesium cation and which
account for the absence of guanidinium ion **129** are possible. One of them
proposed by Wuest and collaborators (82) is the direct transfer of a proton
and two electrons within intermediates similar to the complexes **138** and
139.

138

139

New strategy in organic synthesis

In the total synthesis of optically active erythromycin A reported by Woodward and collaborators (87), the bicyclic compound **142** (Fig. 1) was used to produce the two segments C_9-C_{15} (**143**) and C_3-C_8 (**144**) of erythronolide A. These two segments were then combined (→**145**) and converted into **146**). Aldol condensation of a propionate ester derivative with **146** gave the erythronolide A secoacid derivative **147** (Fig. 2) which was successfully transformed into erythromycin A (**149**) through a series of chemical transformations where compound **148** was one of the key intermediates.

Interestingly, the optically active dithiadecalin **142** was prepared through several steps starting with the optically active (+)-tetrahydrothiopyran-thiol **141** which contains only one chiral center. As a first operation, racemic **141** was converted into a mixture of two diastereoisomeric thioesters by reaction with (-)-camphanylchloride and the desired optically active thioester of (+)-**141** was isolated by crystallization. Optically active (+)-**141** was then obtained from the optically active thioester by treatment with sodium methoxide in methanol. It is remarkable that (+)-**141** did not racemize under neutral and even strongly basic conditions. Indeed, the "normally anticipated" equilibrium **140** ⇌ **141** did not take place. This unexpected result is probably rooted in some stereoelectronic factor which is not yet understood. Compound (+)-**141** was used twice in the course of the synthesis (to produce **143** and **144**) and its only chiral center was destroyed twice in order to produce the two methyl groups at C-6 and C-12 (cf. **146**). Furthermore, it is interesting to point out that the chirality of (+)-**141** is responsible for the control of the relative and absolute stereochemistry of all the asymmetric centers of erythronolide A. This is indeed a very ingenious strategy.

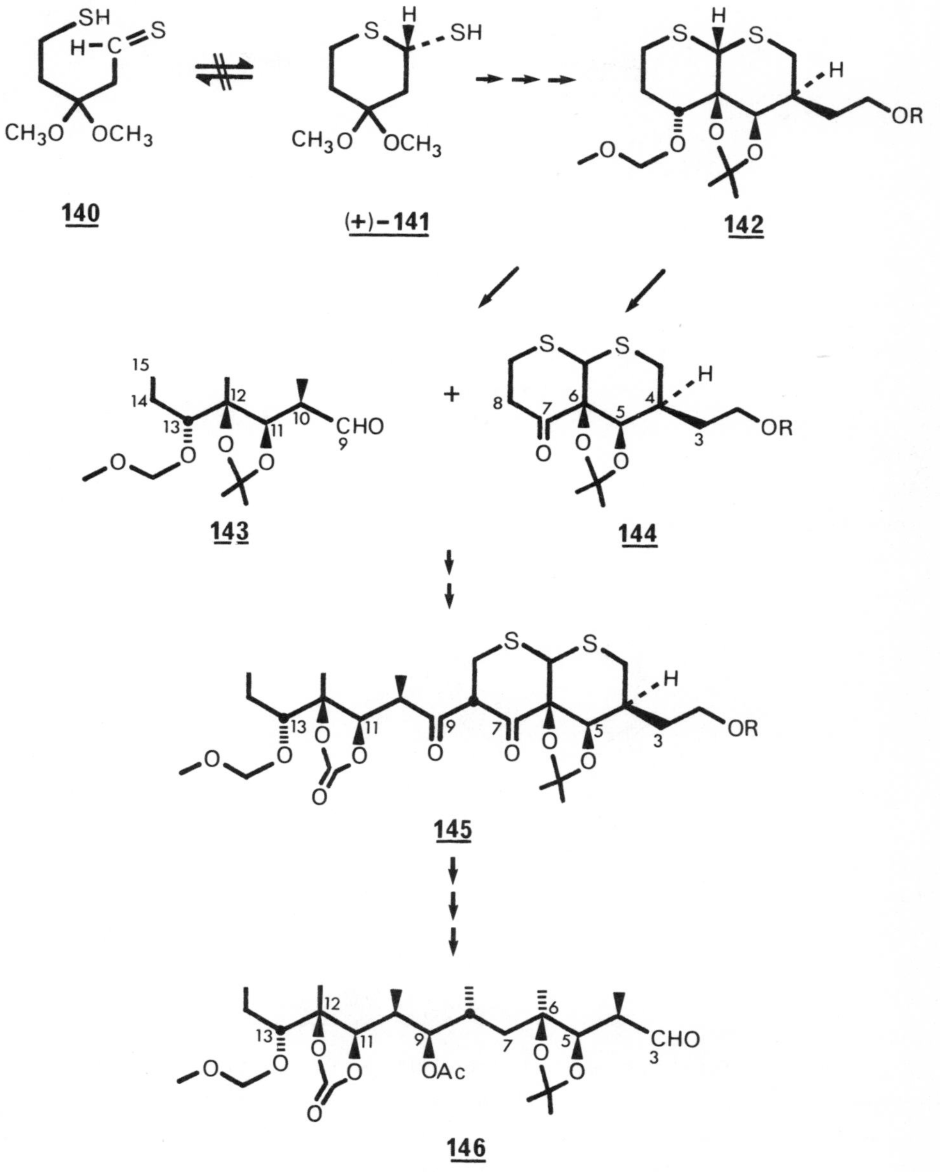

Fig. 1

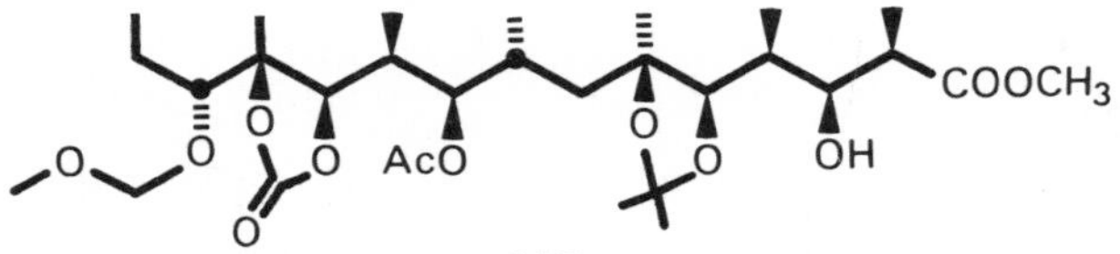

Fig. 2

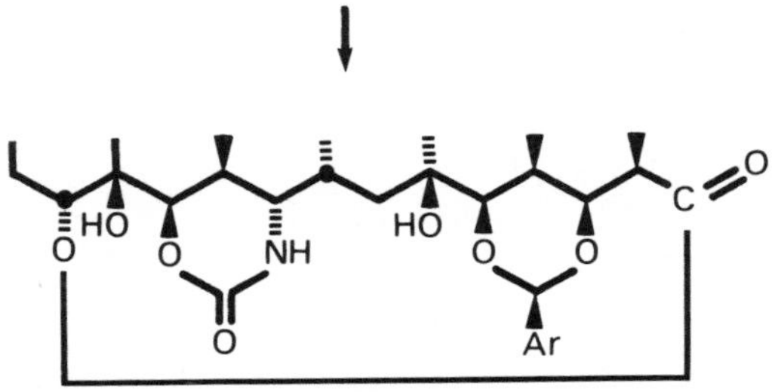

A new strategy for the synthesis of erythromycin A and closely related macrolide antibiotics was elaborated in our laboratory (88). This new approach
to synthesis is based on the knowledge that stereoelectronic effects control
the conformation of acetals. The strategy is based on the 1,7-dioxaspiro-
[5.5]undecane system which was found to be conformationally rigid, existing
in conformation **150** only (see Chapter 2). This is so because in this conformation, steric effects are at their minimum and the acetal function has

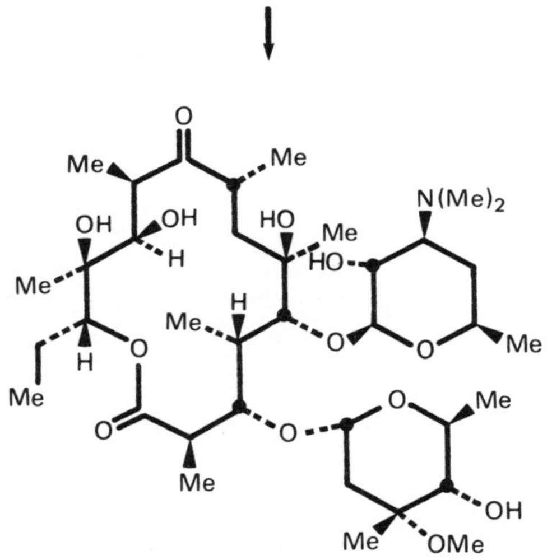

150   **151**

the maximum (stabilizing) stereoelectronic effects (two anomeric effects). Consequently, the spirosystem **150** can be viewed as an aliphatic chain of nine carbon atoms which is locked rigidly in space by the acetal function. The spiro system **150** can therefore be used as a template to introduce several chiral centers with a high degree of control. Indeed, this spirosystem has a potential for the creation of up to eight asymmetric centers (C*), and on hydrolysis, it should yield a dihydroxyketone aliphatic chain (cf. **151**), containing the same number of chiral centers.

Erythromycin A (**149**) was chosen as the target molecule because its structure fits admirably well with the above strategy. Indeed, the methyl ester of the ring-opened form of erythronolide A, the aglycone part of erythromycin A, can be drawn as shown by structure **152**. It can be readily seen using molecular models that **152** is an appropriate "dihydroxyketone" which can be converted into the substituted 1,7-dioxaspiro[5.5]undecane **153** (R=H), where the stereoelectronic effects of the spiroacetal are maximum. It is also important to note that seven (C_5, C_6, C_8, and $C_{10} - C_{13}$) of the ten chiral centers of erythronolide A are now part of the two spirorings. Furthermore, the substituents at C-5, C-8, C-10, C-11, and C-13 are all equatorially oriented and the tertiary hydroxyl group at C-6 is axial while that at C-12 is equatorial. Thus, the spirocompound **153** has no undesirable strong steric interactions and it should be readily made both from erythromycin A and by total synthesis. The reconversion of **153** into erythromycin A would then complete the synthesis. This was the plan.

The progress made using this approach follows. Erythromycin was successfully converted into the spirocompound **153** (R=$C_6H_5CH_2$) through a multistep sequence. Compound **153** obtained in this manner was used for testing the end of the synthetic plan. Degradation of the side chain of **153** (R=$C_6H_5CH_2$) gave the aldehyde **154**. Compound **154** was then reconverted into spiro **153** (R=

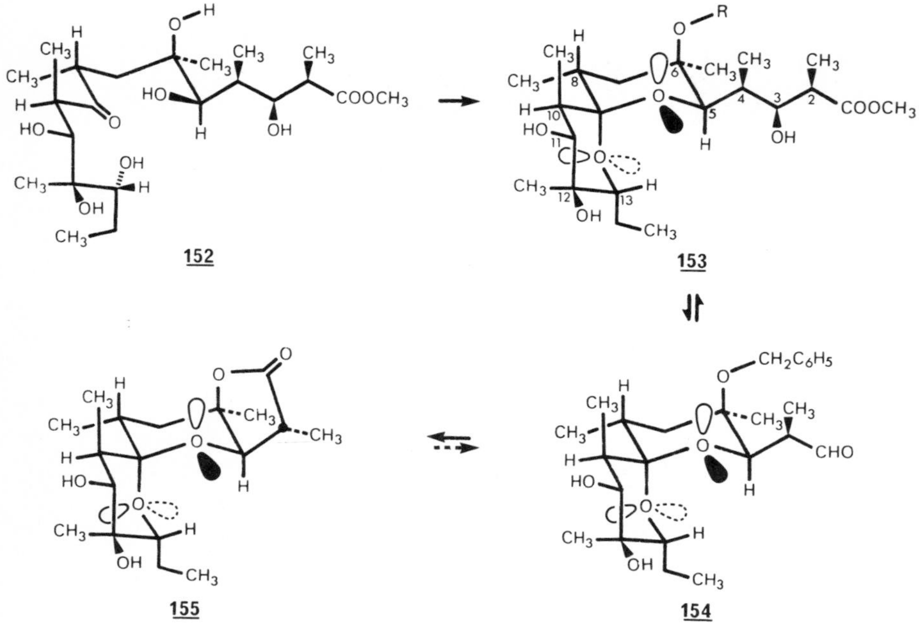

152   **153**

155   **154**

$C_6H_5CH_2$) using the methodology developed by Evans and McGee (89), i.e. via an aldol condensation using the zirconium enolate of methyl propionate. In this condensation, two diastereoisomers were formed in a 10:1 ratio and the major isomer was the desired **153** ($R=C_6H_5CH_2$). Therefore, the control of the stereochemistry of two (C-2 and C-3) of the three chiral centers of the side chain of the spiroketal **153** is solved. The spiroaldehyde **154** was also converted into the spirolactone **155** which has all the substituents fixed rigidly in space. The next successful operation was the total synthesis of spirolactone **155**.

Lactone **156** which is readily made was chosen as starting material. Condensation of **156** with the lithium acetylide **157** gave after treatment with trimethylsilyl chloride the adduct **158**. Synthons **156** and **157** have each one chiral center; their condensation leads to the formation of a diastereoisomeric mixture, but this is irrelevant as two (C-8 and C-9, using erythromycin numbering system) of the three chiral centers in **158** are epimerizable. Controlled catalytic hydrogenation of **158** gave the cis olefin **159**. Removal of the trimethylsilyl protecting groups and cyclization to a mixture of spiroisomers ((CH_3)$_3$SiOSO$_2$CF$_3$ in CH_2Cl_2) followed by equilibration (pyridinium tosylate in refluxing dichloroethane) gave one single spiroproduct having structure **160** in good yield.

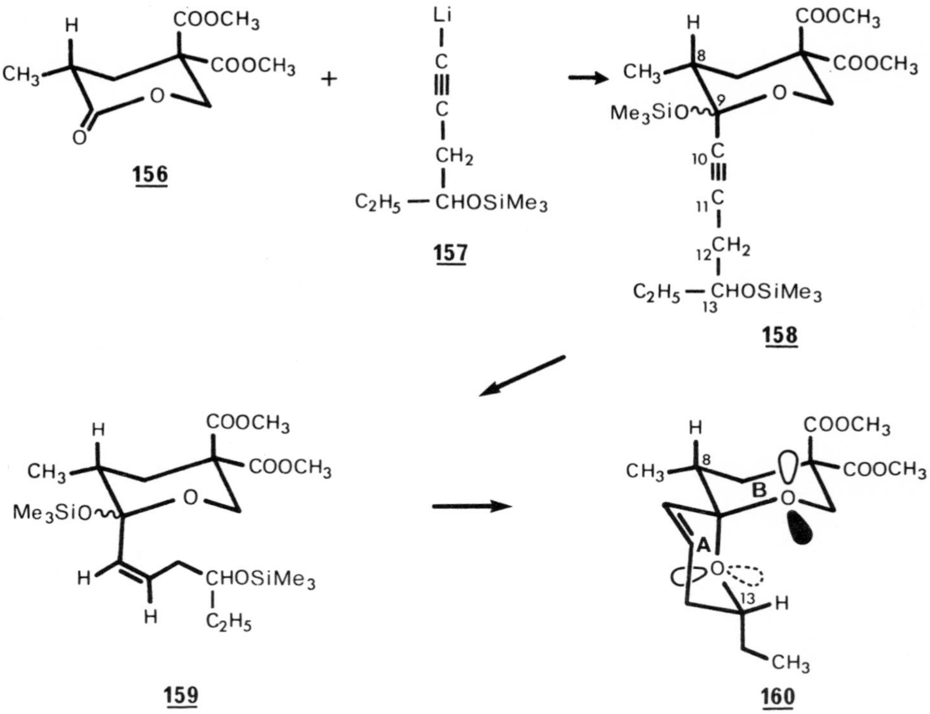

156 + **157** → **158**

159 → **160**

In compound **160**, the ethyl side chain at C-13 is in the equatorial orientation to avoid a strong 1,3-diaxial steric interaction with ring B. Then,
stereoelectronic effects of the acetal function adjust the stereochemistry
of the two oxygen atoms of the rings. The secondary methyl group at C-8 is
epimerizable under acid conditions; as a result, this group takes the
equatorial orientation in order to avoid a strong 1,3-diaxial steric interaction with the axial carbomethoxy group. Consequently, only isomer **160** is
obtained after equilibration and a complete control of the relative stereochemistry at C-8 and C-13 is achieved. Utilizing the appropriate optically
active synthon **157** should give optically active **160** having the absolute
configuration corresponding to that of erythromycin.

Compound **160** was then transformed into the conjugated enone **161**. Lithium
dimethylcopper 1,4-addition on enone **161** and reaction of the resulting
enolate with dibenzoyl peroxide gave compound **162** in good yield. Reaction
of **162** with methylmagnesium iodide at low temperatures furnished compound
163 again in good yield. Thus by this sequence, an excellent control of

stereochemistry at five chiral centers (C-8, C-10, C-11, C-12, and C-13) of
erythromycin A was achieved.

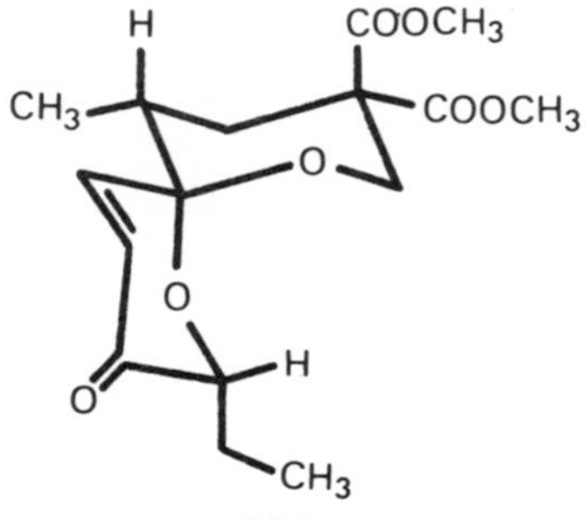

161 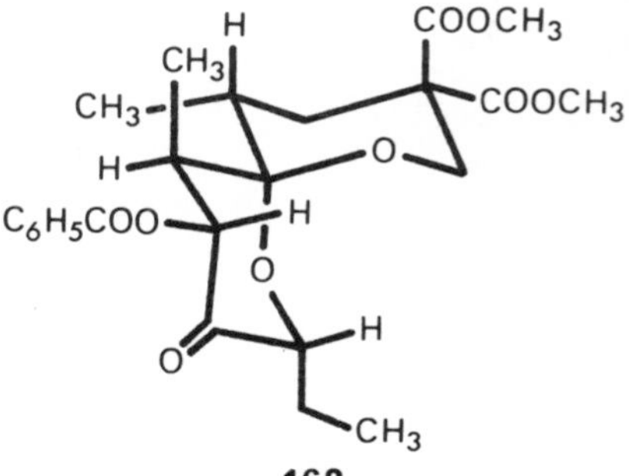   **162**

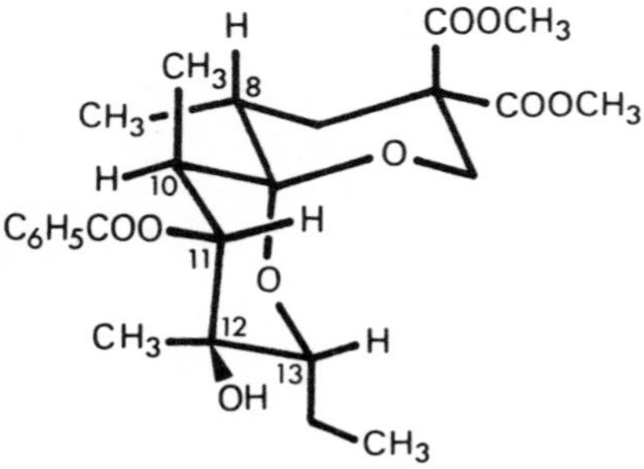

163 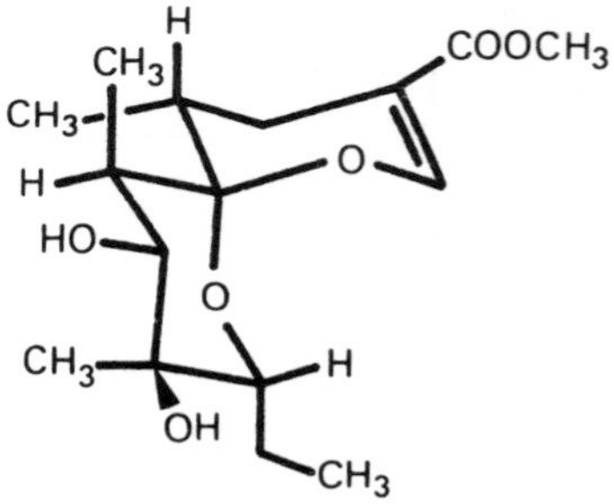   **164**

Through several steps, compound **163** was converted into enolpropionate ester
165 via the conjugated enol ester **164**. Using the Claisen rearrangement methodology developed by Ireland and co-workers (90), compound **165** gave a 4:1
mixture of two isomeric products (epimeric at C-4). The major epimer was
shown to have structure **166**. In this important transformation, two new
chiral centers are produced (C-4 and C-5) with the desired configuration.
The stereocontrolled introduction of the tertiary hydroxyl group at C-6
was carried out via iodolactonization and hydrogenolysis (**166** → **167** → **155**).
The product obtained was found to be identical in all respects other than
optical rotation with spirolactone **155** obtained from the degradation of
erythromycin. There now remains to convert spirolactone **155** into spiroaldehyde **154** in order to complete the total synthesis of spiroerythronolide A
153. Work is also progressing for the reconstruction of erythromycin A from
spiroerythronolide A **153**. This will complete the total synthesis. Interest-

ingly, this new strategy to organic synthesis, based on stereoelectronic principles, permits a control of stereochemistry of all the chiral centers of erythronolide A.

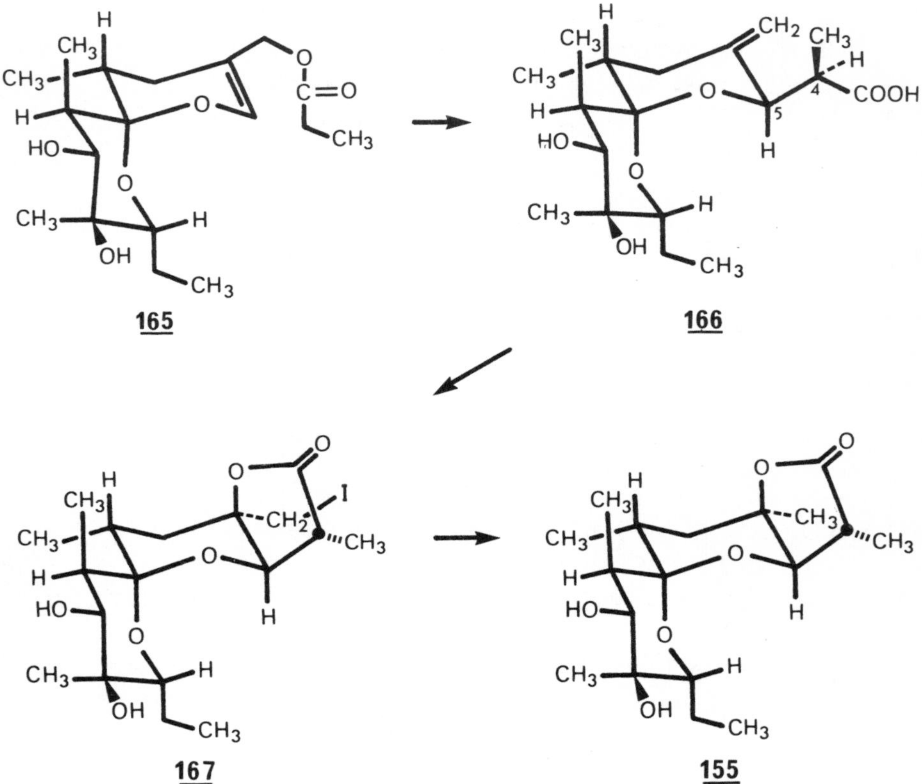

REFERENCES

(1) Eliel, E.L.; Nader, F.W. J. Am. Chem. Soc. 1969, 91, 536; 1970, 92, 584.

(2) Bailey, W.F.; Croteau, A.A. Tetrahedron Lett. 1981, 22, 545.

(3) Spencer, T.A.; Schmiegel, K.K.; Williamson, K.L. J. Am. Chem. Soc. 1963, 85, 3785.

(4) Spencer, T.A.; Neel, H.S.; Ward, D.C.; Williamson, K.L. J. Org. Chem. 1966, 31, 434.

(5) Marshall, J.A.; Fanta, W.I. J. Org. Chem. 1964, 29, 2501.

(6) Hajos, Z.G.; Parrish, D.R. J. Org. Chem. 1974, 39, 1612.

(7) Fujita, E.; Shibuya, M.; Nakamura, S.; Okada, Y.; Fujita, T. J. Chem.

Soc., Perkin Trans 1 1974, 165.

(8)   Schoemaker, H.E.; Speckamp, W.N. Tetrahedron Lett. 1978, 1515.

(9)   Schoemaker, H.E.; Speckamp, W.N. Tetrahedron Lett. 1978, 4841.

(10) Evans, D.A.; Thomas, E.W. Tetrahedron Lett. 1979, 411.

(11) Aratani, M.; Dunkerton, L.V.; Fukuyama, T.; Kishi, Y.; Kakoi, H.;
      Sugiura, S.; Inoue, S. J. Org. Chem. 1975, 40, 2009.

(12) Fukuyama, T.; Dunkerton, L.U.; Aratani, M.; Kishi, Y. J. Org. Chem.
      1975, 40, 2011.

(13) Harding, K.E.; Cooper, J.L.; Puckett, P.M. J. Org. Chem. 1979, 44,
      2834.

(14) Schoemaker, H.E.; Kruk, C.; Speckamp, W.N. Tetrahedron Lett. 1979,
      2437.

(15) Maurer, B.; Hauser, A.; Thommen, W.; Schulte-Elte, K.H.; Ohloff, G.
      Helv. Chim. Acta 1980, 63, 293.

(16) Kikuchi, O. Tetrahedron Lett. 1980, 1055.

(17) Wolfe, S.; Lee, S.L.; Ducep, J.B.; Kannengiesser, G.; Lee, W.S. Can.
      J. Chem. 1975, 53, 497.

(18) Baldwin, J.E.; Christie, M.A. J. Am. Chem. Soc. 1978, 100, 4597.

(19) Baldwin, J.E.; Christie, M.A.; Haber, S.B.; Kruse, L.I. J. Am. Chem.
      Soc. 1976, 98, 3045.

(20) Ohuchida, S.; Hamanaka, N.; Hayashi, M. J. Am. Chem. Soc. 1981, 103,
      4597.

(21) Noyori, R.; Kobayashi, H.; Sato, T. Tetrahedron Lett. 1980, 2573.

(22) Krow, G.R. Tetrahedron 1981, 37, 2697.

(23) Lattes, A.; Oliveros, E.; Rivière, M.; Belzecki, C.; Mostowicz, D.;
      Abramskj, W.; Piccini-Leopardi, C.; Germain, G.; van Meerssche, M.
      J. Am. Chem. Soc. 1982, 104, 3929.

(24) Baldwin, J.E.; Norris, R.K. J. Org. Chem. 1981, 46, 697.

(25) Norris, R.K.; Sternhell, S. Aust. J. Chem. 1966, 19, 841.

(26) Norris, R.K.; Sternhell, S. Aust. J. Chem. 1969, 22, 935.

(27) Norris, R.K.; Sternhell, S. Aust. J. Chem. 1971, 24, 1449.

(28) Norris, R.K.; Sternhell, S. Aust. J. Chem. 1972, 25, 1907.

(29) Vasella, A. Helv. Chim. Acta 1977, 60, 1273.

(30) Vasella, A.; Voeffray, R. J. Chem. Soc, Chem. Commun. 1981, 97.

(31) Vasella, A.; Voeffray, R. Helv. Chim. Acta 1982, 65, 1134.

(32) Dolson, M.G.; Swenton, J.S. J. Am. Chem. Soc. 1981, 103, 2361.

(33) Burton, G.W.; Le Page, Y.; Gabe, E.J.; Ingold, K.U. J. Am. Chem. Soc.
      1980, 102, 7792.

(34) Cram, D.J. "Fundamentals of Carbanion Chemistry"; Academic Press Inc.:

New York, N.Y., 1965.

(35) Hartmann, A.A.; Eliel, E.L. J. Am. Chem. Soc. 1971, 93, 2572.

(36) Eliel, E.L.; Hartmann, A.A.; Abatjoglou, A.G. J. Am. Chem. Soc. 1974, 96, 1807.

(37) Eliel, E.L.; Abatjoglou, A.; Hartmann, A.A. J. Am. Chem. Soc. 1972, 94, 4786.

(38) Eliel, E.L. Tetrahedron 1974, 30, 1503.

(39) Abatjoglou, A.G.; Eliel, E.L.; Kuyper, L.F. J. Am. Chem. Soc. 1977, 99, 8262.

(40) Eliel, E.L.; Koskimies, J.K.; Lohri, B. J. Am. Chem. Soc. 1978, 100, 1614.

(41) Meyers, A.I.; Campbell, A.L.; Abatjoglou, A.G.; Eliel, E.L. Tetrahedron Lett. 1979, 4159.

(42) Rauk, A.; Buncel, E.; Moir, R.Y.; Wolfe, S. J. Am. Chem. Soc. 1965, 87, 5498.

(43) Baldwin, J.E.; Hackler, R.E.; Scott, R.M. J. Chem. Soc., Chem. Commun. 1969, 1415.

(44) Durst, T.; Fraser, R.R.; McClory, M.R.; Swingle, R.B.; Viau, R.; Wigfield, Y.Y. Can. J. Chem. 1970, 48, 2148.

(45) Durst, T.; Viau, R.; McClory, M.R. J. Am. Chem. Soc. 1971, 93, 3077.

(46) Nishihata, K.; Nishio, M. J. Chem. Soc., Perkin Trans 2 1972, 1730.

(47) D'Amore, M.B.; Brauman, J.T. J. Chem. Soc., Chem. Commun. 1973, 398.

(48) Viau, R.; Durst, T. J. Am. Chem. Soc. 1973, 95, 1346.

(49) Nishihata, K.; Nishio, M. Tetrahedron Lett. 1972, 4839.

(50) Fraser, R.R.; Schuber, F.J.; Wigfield, Y.Y. J. Am. Chem. Soc. 1972, 94, 8795.

(51) Hofer, O.; Eliel, E.L. J. Am. Chem. Soc. 1973, 95, 8045.

(52) Barbarella, G.; Garbesi, A.; Boicelli, A.; Fava, A. J. Am. Chem. Soc. 1973, 95, 8051.

(53) Wolfe, S.; Chamberlain, P.; Garrard, T.F. Can. J. Chem. 1976, 54, 2847.

(54) Bernardi, F.; Csizmadia, I.G.; Mangini, A.; Schlegel, H.B.; Whangbo, M.H.; Wolfe, S. J. Am. Chem. Soc. 1975, 97, 2209.

(55) Streitwieser, Jr., A.; Ewing, S.P. J. Am. Chem. Soc. 1975, 97, 190.

(56) Streitwieser, Jr., A.; Williams, Jr., J.E. J. Am. Chem. Soc. 1975, 97, 191.

(57) Rauk, A.; Wolfe, S.; Csizmadia, I.G. Can. J. Chem. 1969, 47, 114.

(58) Wolfe, S.; Rauk, A.; Tel, L.M.; Csizmadia, I.G. J. Chem. Soc., Chem. Commun. 1970, 96.

(59) Wolfe, S.; Tel, L.M.; Csizmadia, I.G. Can. J. Chem. 1973, 51, 2423.

(60) Radom, L.; Hehre, W.J.; Pople, J.A. J. Am. Chem. Soc. 1972, 94, 2371.

(61) Lehn, J.-M.; Wipff, G. J. Am. Chem. Soc. 1976, 98, 7498.

(62) Oae, S.; Tagaki, W.; Ohno, A. Tetrahedron 1964, 20, 417, 427.

(63) Orchin, M.; Jaffe, H.H. "The Importance of Antibonding Orbitals"; Houghton Mifflin: Boston, Mass., 1967.

(64) Wolfe, S. Acc. Chem. Res. 1972, 5, 102.

(65) Lehn, J.-M.; Wipff, G.; Demuynck, J. Helv. Chim. Acta 1977, 60, 1239.

(66) Kishi, Y.; Fukuyama, T.; Nakatsuka, S. J. Am. Chem. Soc. 1973, 95, 6490, 6492.

(67) Kishi, Y.; Nakatsuka, S.; Fukuyama, T.; Havel, M. J. Am. Chem. Soc. 1973, 95, 6493.

(68) Nakatsuka, S.; Fukuyama, T.; Kishi, Y. Tetrahedron Lett. 1974, 1549.

(69) Sasaki, K.; Fukuyama, T.; Nakatsuka, S.; Kishi, Y. J. Chem. Soc., Chem. Commun. 1975, 542.

(70) Meyers, C.Y.; Ho, L.L.; Ohno, A.; Kagami, N. Tetrahedron Lett. 1974, 729, and references cited therein.

(71) Poje, M.; Sikirica, M.; Vicković, I.; Bruvo, M. Tetrahedron Lett. 1980, 3089.

(72) Lehn, J.-M.; Wipff, G. J. Chem. Soc., Chem. Commun. 1975, 800.

(73) Gorenstein, D.G.; Findlay, J.B.; Luxon, B.A.; Kar, D. J. Am. Chem. Soc 1977, 99, 3473.

(74) Gorenstein, D.G.; Luxon, B.A.; Findlay, J.B.; Momii, R. J. Am. Chem. Soc. 1977, 99, 4170.

(75) Gorenstein, D.G.; Luxon, B.A.; Findlay, J.B. J. Am. Chem. Soc. 1977, 99, 8048.

(76) Gorenstein, D.G.; Luxon, B.A.; Findlay, J.B. J. Am. Chem. Soc. 1979, 101, 5869.

(77) Gorenstein, D.G.; Luxon, B.A.; Goldfield, E.M. J. Am. Chem. Soc. 1980, 102, 1757.

(78) Gorenstein, D.G.; Rowell, R.; Findlay, J.B. J. Am. Chem. Soc. 1980, 102, 5077.

(79) Gorenstein, D.G.; Rowell, R. J. Am. Chem. Soc. 1979, 101, 4925.

(80) Rowell, R.; Gorenstein, D.G. J. Am. Chem. Soc. 1981, 103, 5894.

(81) Hall, C.R.; Inch, T.D. Tetrahedron 1980, 36, 2059.

(82) Erhardt, J.M.; Grover, E.R.; Wuest, J.D. J. Am. Chem. Soc. 1980, 102, 6365.

(83) Atkins, T.J. J. Am. Chem. Soc. 1980, 102, 6364.

(84) Erhardt, J.M.; Wuest, J.D. J. Am. Chem. Soc. 1980, 102, 6363.

(85) Bohlmann, F. Chem. Ber. 1958, 91, 2157.

(86) Crabb, T.A.; Newton, R.F.; Jackson, D. Chem. Rev. 1971, 71, 109.

(87) Woodward, R.B.; Logusch, E.; Nambiar, K.P.; Sakan, K.; Ward, D.E.;
     Au-Yeung, B.-W.; Balaram, P.; Browne, L.J.; Card, P.J.; Chen, C.H.;
     Chênevert, R.B.; Fliri, A.; Frobel, K.; Gais, H.-J.; Garratt, D.G.;
     Hayakawa, K.; Heggie, W.; Hesson, D.P.; Hoppe, D.; Hoppe, I.; Hyatt,
     J.A.; Ikeda, D.; Jacobi, P.A.; Kim, K.S.; Kobuke, Y.; Kojima, K.;
     Krowicki, K.; Lee, V.J.; Leutert, T.; Malchenko, S.; Martens, J.; Mat-
     thews, R.S.; Ong, B.S.; Press, J.B.; Rajan Babu, T.V.; Rousseau, G.;
     Sauter, H.M.; Suzuki, M.; Tatsuta, K.; Tolbert, L.M.; Truesdale, E.A.;
     Uchida, I.; Ueda, Y.; Uyehara, T.; Vasella, A.T.; Vladuchick, W.C.;
     Wade, P.A.; Williams, R.M.; Wong, H.N.-C. J. Am. Chem. Soc. 1981, 103,
     3213, 3215.

(88) Deslongchamps, P.; Bernet, B.; Bishop, P.; Caron, M.; Kawamata, T.;
     Roy, B.; Ruest, L.; Sauvé, G.; Schwartz, D.; Soucy, P. Unpublished
     results.

(89) Evans, D.A.; McGee, L.R. Tetrahedron Lett. 1980, 3975.

(90) Ireland, R.E.; Mueller, R.H.; Willard, A.K. J. Am. Chem. Soc. 1976,
     98, 2868.

CHAPTER 9

BIOLOGICAL PROCESSES

Most of the organic reactions catalyzed by enzymes take place on function-alities which contain heteroatoms. This is the case for the hydrolysis of functional groups such as acetals (e.g. ß-galactosidase and lysozyme), es-ters and amides (e.g. chymotrypsin, trypsin, papain, carboxypeptidase) and phosphate esters (e.g. ribonuclease). This is also true for enzymes which accelerate oxidation-reduction processes (e.g. alcohol dehydrogenase), elim-inations, isomerizations, rearrangements, or with reactions that make or break carbon-carbon bonds (e.g. carboxylation, aldol or Claisen condensa-tion, decarboxylation) (1-4). The preceding Chapters have shown that the counterparts to most enzyme catalyzed organic reactions are strongly influ-enced by stereoelectronic effects when these reactions are carried out with simple reagents. Consequently, there appears to be no doubt that stereo-electronic effects must play an important role in enzymatic catalysis. In fact, these effects must play an even more important role with enzymes as these "highly sophisticated" reagents allow only one among a multitude of chemical transformations.

Indeed, this Chapter will show that the knowledge of transition state stere-ochemistry for the formation of various enzyme-reaction intermediates is gained through the application of stereoelectronic principles. The driving force of various bond-making or breaking processes and the required confor-mational changes of some specific enzymic transformations are more clearly understood. The importance of stereochemistry in proton transfer, hydrogen-bonding or metal-coordination also becomes an important parameter. As a result, a much deeper understanding of the events (in three dimensions) which take place during an enzyme-catalyzed process is gained through the application of stereoelectronic principles.

Enzymatic reactions

It has been pointed out in Chapter 2 (p. 39) (see also 1-4) that there is good evidence which indicates that in the hydrolysis of β-glycosides by lysozyme, the substrate must take a boat conformation in order to produce the half-chair oxocarbonium ion ($\underline{1} \rightarrow \underline{2} \rightarrow \underline{3}$). Lysozyme is therefore a good example which provides evidence that stereoelectronic effects play a key role in enzymic hydrolysis.

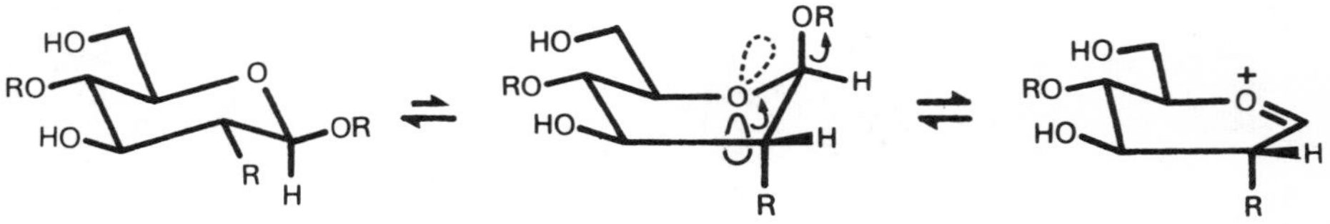

$$\underline{1} \qquad\qquad \underline{2} \qquad\qquad \underline{3}$$

The alcohol dehydrogenases are zinc metalloenzymes which can oxidize a wide variety of alcohols to their corresponding aldehydes or ketones using nicotinamide adenine dinucleotide (NAD^+) as coenzyme. These reactions are readily reversible so that carbonyl compounds may be reduced by NADH.

These reactions are essentially the result of hydride ion transfer from NADH (**4**) to the carbonyl compound. The hydrogen transfer is stereospecific and can occur <u>via</u> two different stereochemical pathways depending on the enzyme: one class catalyzes the transfer of the pro-R hydrogen while the other catalyzes the transfer of the pro-S hydrogen (5-7). Benner (8) has recently found that the stereochemical outcome (pro-R and pro-S) depends on the equilibrium constant (K_{eq}) of the reaction. If K_{eq} is less than 10^{-12} M^{-1}, the pro-R hydrogen is generally transferred and if K_{eq} is greater than 10^{-12} M^{-1}, the pro-S hydrogen is transferred. Benner (8) has suggested that

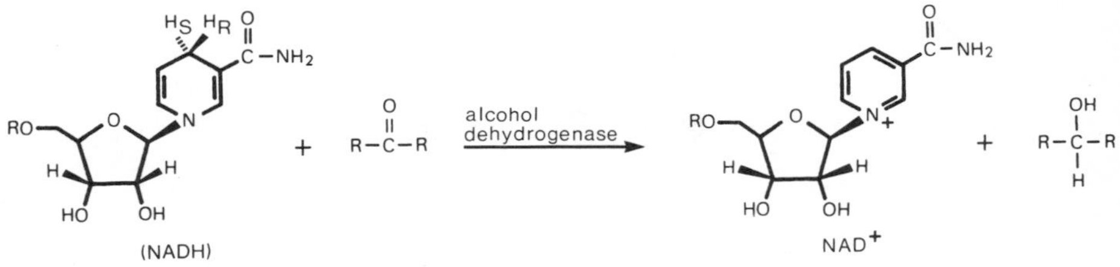

(NADH)

4

these results indicate that the more reactive carbonyl compounds are reduced by the pro-R hydrogen and the less reactive by the pro-S hydrogen of NADH. He has further suggested that the enzymic transfers of the pro-R and the pro-S hydrogens occur when the nicotinamide ring is in the <u>anti</u> and the <u>syn</u> conformation **5** and **6** respectively. Thus, <u>anti</u>-NADH (**5**) which delivers the pro-R hydrogen would be a weaker reducing agent than <u>syn</u>-NADH (**6**) which delivers the pro-S hydrogen. It was further concluded that this difference in reactivity can be explained by taking into account stereoelectronic effects due to the relative orientation of the ring nitrogen electron pair. Interestingly, this idea is supported by experimental data (8).

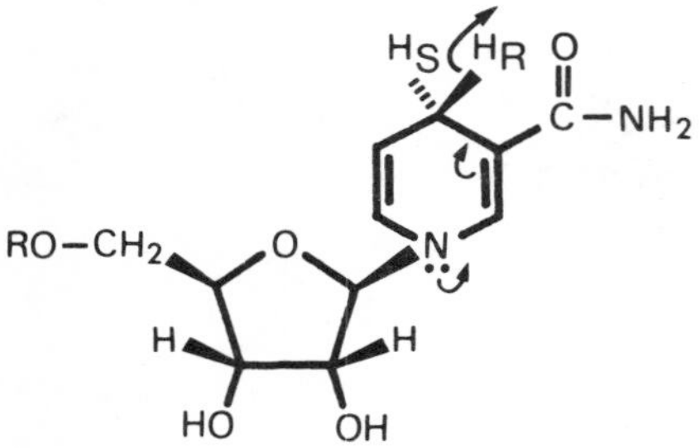

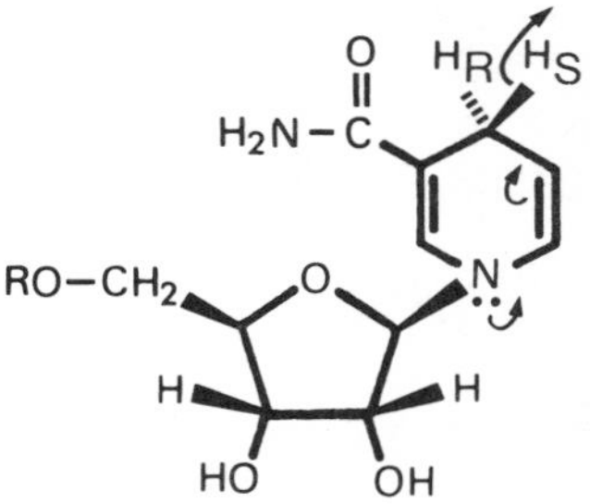

The stereochemical aspects of peptide hydrolysis catalyzed by chymotrypsin and related serine proteases has been recently analyzed with respect to requirements for stereoelectronic control of bond cleavage and this analysis has led to a much more complete understanding of the reaction mechanism (9-14).

Chymotrypsin, a serine endopeptidase, most readily reacts at the carboxyl group of the aromatic amino acid residue of proteins and polypeptides (or N-acyl aromatic amino acid esters) to form first a tetrahedral intermediate which then collapses into an acyl-enzyme ($\underline{7} \rightarrow \underline{8} \rightarrow \underline{9}$). The acyl-enzyme is then hydrolyzed by water to furnish the N-acylated aromatic amino acid again through the formation of a tetrahedral intermediate ($\underline{10} \rightarrow \underline{11} \rightarrow \underline{12}$) (1-4).

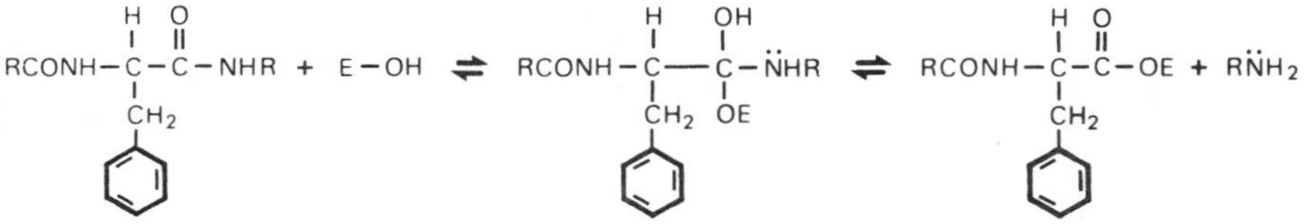

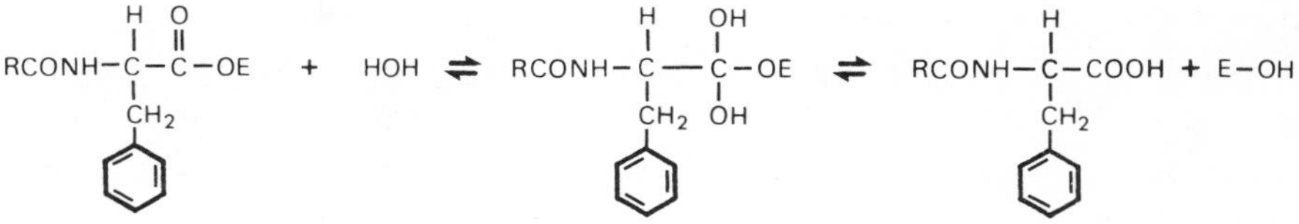

$$RCONH-\underset{\underset{\underset{\bigcirc}{\overset{|}{CH_2}}}{\overset{\overset{H}{|}}{C}}}{}-\overset{\overset{O}{\|}}{C}-OE \quad + \quad HOH \quad \rightleftharpoons \quad RCONH-\underset{\underset{\underset{\bigcirc}{\overset{|}{CH_2}}}{\overset{\overset{H}{|}}{C}}}{}-\underset{\overset{|}{OH}}{\overset{\overset{OH}{|}}{C}}-OE \quad \rightleftharpoons \quad RCONH-\underset{\underset{\underset{\bigcirc}{\overset{|}{CH_2}}}{\overset{\overset{H}{|}}{C}}}{}-COOH \quad + \quad E-OH$$

$$\underline{10} \qquad\qquad\qquad\qquad \underline{11} \qquad\qquad\qquad\qquad \underline{12}$$

In the above transformation, the hydroxyl group of Ser-195 plays the important role of the nucleophile. The enhanced nucleophilic activity of this
group is due to its interaction with Asp-102 and His-57. Indeed this proton
relay system (Asp-102 and His-57) through hydrogen bonding makes the Ser-195
hydroxyl group sufficiently nucleophilic to attack the carboxyl function
of the substrate with the resulting formation of a tetrahedral intermediate
(cf. 13 → 14). Note that this proton relay system is also used to deliver
a proton on the nitrogen of the tetrahedral intermediate, a necessary step
for the ejection of the amino group in the course of the formation of the
acyl-enzyme (cf. 14 → 15).

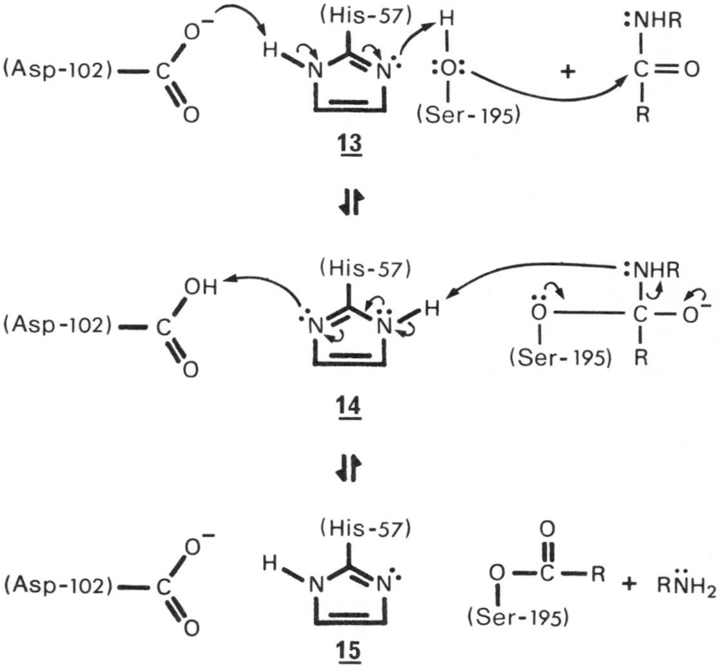

In order to confirm and to obtain a more detailed knowledge of the above
proposed mechanism, precise information is required on the structure and
stereochemistry of intermediates such as the enzyme-substrate complex, the
tetrahedral intermediate and the acyl-enzyme. Normally, this information
cannot be obtained directly because these intermediates are transient species. On the other hand, it can be obtained indirectly from the precise
structural information available from high resolution X-ray diffraction
studies of several enzyme derivatives or complexes with substrate analogs.

Using this approach, Bizzozero and Zweifel (9) and Bizzozero and Dutler
(10) have constructed molecular models of two intermediates (an enzyme-
substrate complex and a tetrahedral intermediate) by appropriate modification of the models of stable enzyme-species. The stable enzyme-species used
(15, 16) are trypsin-benzamidine complex (TR-B) (17), trypsin-pancreatic
trypsin inhibitor complex (TR-PTI) (18, 19) and tosyl-chymotrypsin (Tos-
CHT) (20) which are related to enzyme substrate complex, tetrahedral intermediate and acyl-enzyme respectively.

Comparison of the TR-B, TR-PTI and Tos-CHT models indicates no gross conformational changes in the main chain section carrying the catalytic residues
during catalysis. Also, the CH_2OH group of Ser-195 is maintained in one
specific conformation but the models show that the imidazole ring of His-
57 can take two distinct positions: the in- and the out-positions. More
interestingly, in the in-position, the imidazole nitrogen of His-57 can form
a good hydrogen bond with the oxygen atom of the hydroxyl group of Ser-195,
and in the out-position, the same imidazole nitrogen can form a hydrogen
bond with the nitrogen atom of the tetrahedral intermediate derived from
the amide nitrogen of the substrate. On that basis, it seems likely that
the in- and out-positions are related to the proposed two distinct functional roles of His-57, i.e. 13 → 14 and 14 → 15. It can further be assumed that
interconversion of the two positions takes place during catalysis. Interestingly, molecular models suggest that this interconversion takes place with
concomitant movement of the backbone part carrying the groups involved (Asp-
102 and His-57) in this network of hydrogen bonds.

Model studies suggest also that two CONH groups of the substrate adjacent to
the amide function which has reacted to give a tetrahedral intermediate, form
a hydrogen bond ($CONH\cdots O=C$) with the carboxyl oxygens of Phe-41 and Ser-214
(cf. 16 → 17). Thus, formation of these two hydrogen bonds guides the approach
of the substrate peptide chain toward a close contact with the enzyme surface.

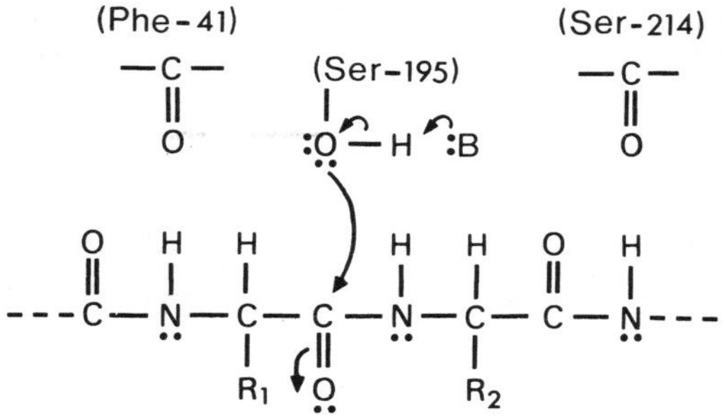

16

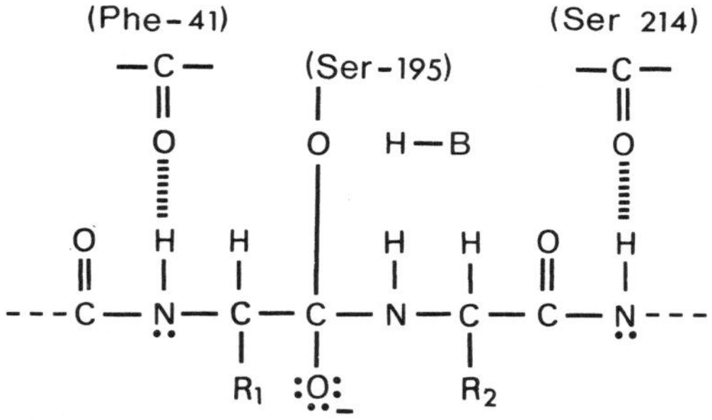

17

It is also interesting to point out that the direction of approach of the Ser-195 hydroxyl group is consistent with the Bürgi-Dunitz approach of nucleophilic attack on carbonyl (21). The only information which cannot be obtained from model building is the relative orientation at the nitrogen atom of the tetrahedral intermediate. This information can however be obtained by application of the principle of stereoelectronic control. Indeed under such conditions, the nucleophilic attack of the hydroxyl group of Ser-195 on the amide function of the substrate leads to a tetrahedral intermediate having the spatial arrangement **18**. Final adjustment of the model was then carried out to minimize the strain arising from the 1,3-synperiplanar arrangement between Ser-195 methylene group and the carbon C_A next to the nitrogen in the ideal fully staggered arrangement **18** (see arrow). In the final model (cf. **19**), the conformation about these two carbons becomes intermediate between eclipsed and staggered, but the relative orientation of bonds and electron pairs remains essentially the same as in the idealized model **18**.

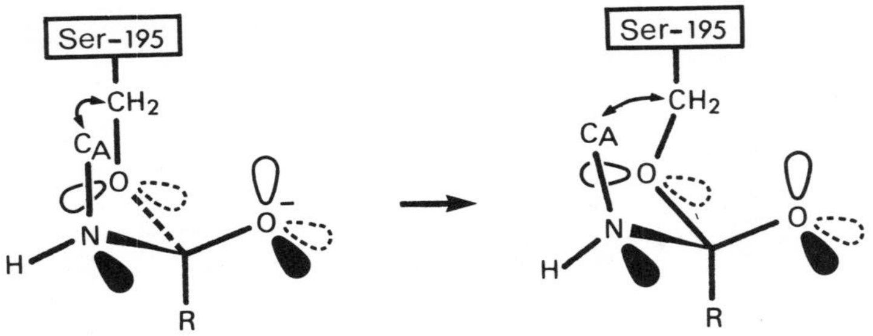

The complete stereochemical picture for the formation of the tetrahedral intermediate must therefore correspond to $\underline{20} \rightarrow \underline{21}$ (Fig. 1) where the imidazole ring occupies the in-position. The formation of the acyl-enzyme now requires two changes: (**a**) the leaving nitrogen of the tetrahedral intermediate must undergo a Walden type inversion ($\underline{21} \rightarrow \underline{22}$) and (**b**) the imidazole ring of His-57 must then move from the in- to the out-position ($\underline{22} \rightarrow \underline{23}$). Then,

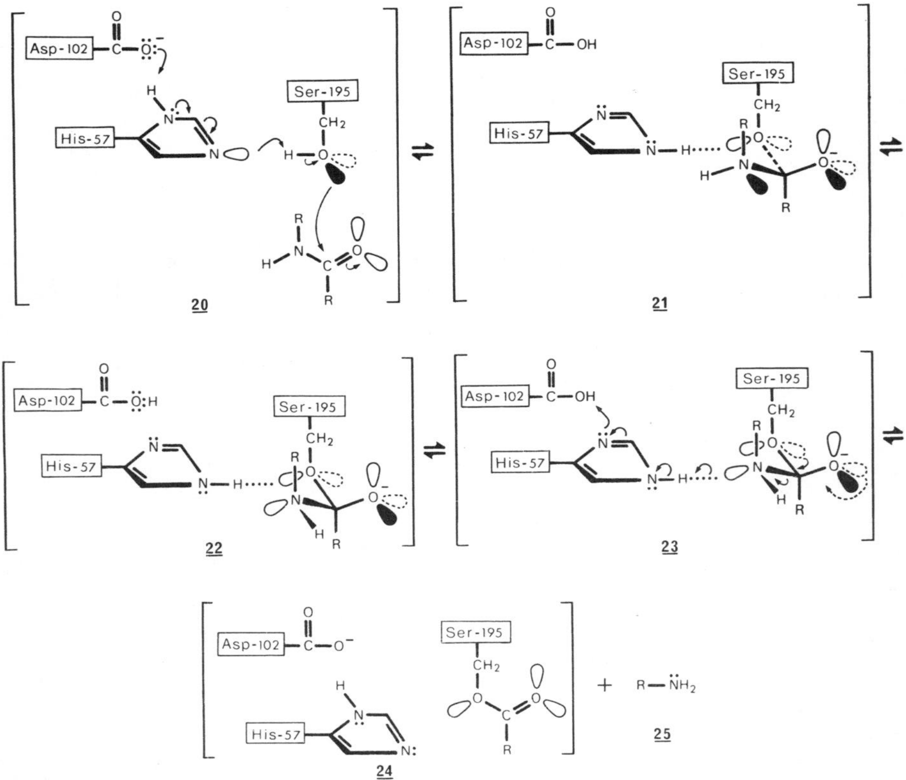

Fig. 1

proton transfer from the imidazole ring of His-57 to the nitrogen atom of
the tetrahedral intermediate can take place with concomitant break down to
give the acyl-enzyme **24** and the amine **25**. Note that this proton transfer
can only occur when the electron pair of the leaving nitrogen is oriented
toward His-57 (in **22** but not in **21**) and that the same proton which is first
abstracted from Ser-195 is subsequently donated to the leaving nitrogen.

The hydrolysis of the acyl-enzyme occurs through a similar process where
a proton would be abstracted from the entering nucleophile (H_2O) by His-
57 which is in the out-position (Fig. 2). This allows the formation of a
tetrahedral intermediate (**26** → **27**). His-57 then moves into the in-position
(**27** → **28**) and delivers a proton to Ser-195. This permits the breakdown of
the intermediate to give the free enzyme **29** and the hydrolyzed amino acid
residue **30**.

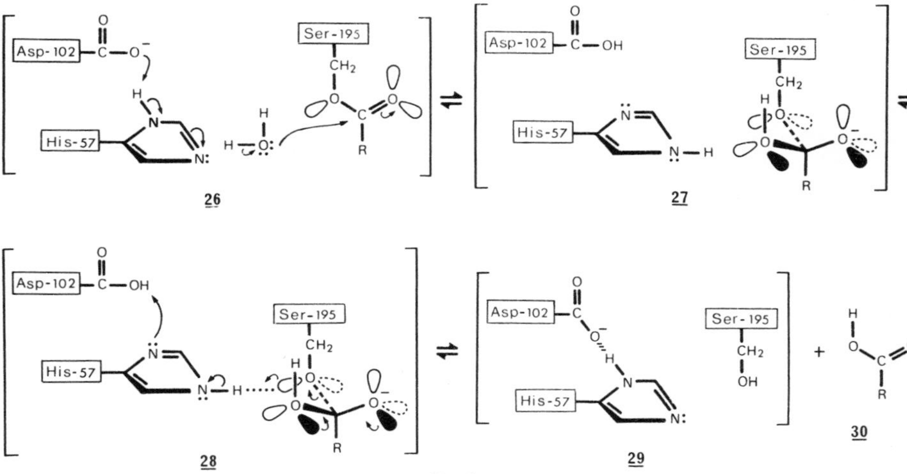

Fig. 2

Returning to the acyl-enzyme process **20** → **24**, it is interesting to point out
that if the equilibrium between the two tetrahedral intermediates **21** and **23**
is on the side of **23** due to a more favorable hydrogen bond (N—H⋯:N), the
tetrahedral intermediate becomes locked into a configuration where reversion
to the enzyme-substrate complex is no longer electronically allowed (the
nitrogen electron pair is no longer antiperiplanar to the leaving oxygen
of Ser-195). This implies that the leaving nitrogen via the N-inversion
step acts as a switch controlling the mode of cleavage of the intermediate.
The requirement for a conformational change involving His-57 fits well into

the first of six mechanistic schemes enumerated by Satterthwait and Jencks (22). According to this scheme, a rate-limiting conformational change of the enzyme during acylation had to be assumed in order to rationalize kinetic data. It is therefore quite reasonable to identify the assumed conformational change with the above mechanism proposed by Bizzozero, Zweifel, and Dutler (9, 10).

Experimental support for the above mechanistic interpretation comes from the work of Bizzozero and Zweifel (9) who have studied the behavior of N-acetyl-L-phenylalanyl-L-proline amide (31) and N-acetyl-L-phenylalanyl-sarcosine amide (32) toward enzymic hydrolysis with α-chymotrypsin. These two dipeptides were found to be good competitive inhibitors with a specific substrate (Ac-Phe-OCH$_3$ (33)) but no hydrolysis was observed. These two peptides thus form an enzyme-substrate complex and the reason for their non-reactivity has to be sought in the nature of the enzyme-substrate interactions occurring during the subsequent bond-change steps.

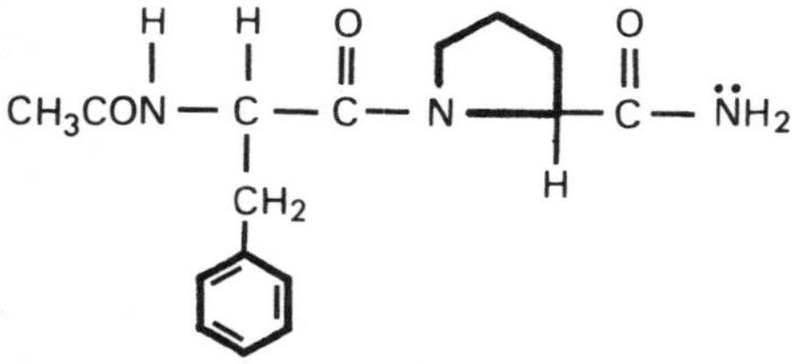

31

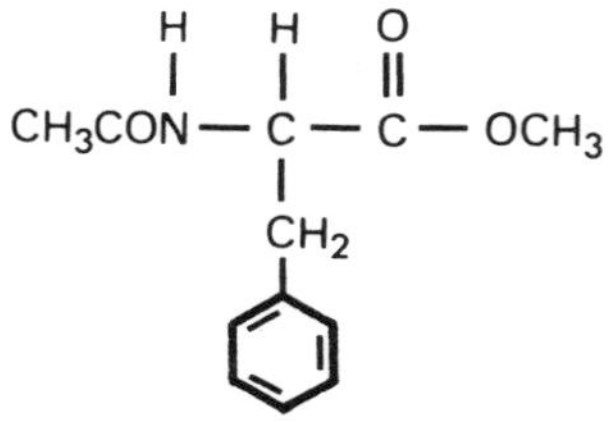

33

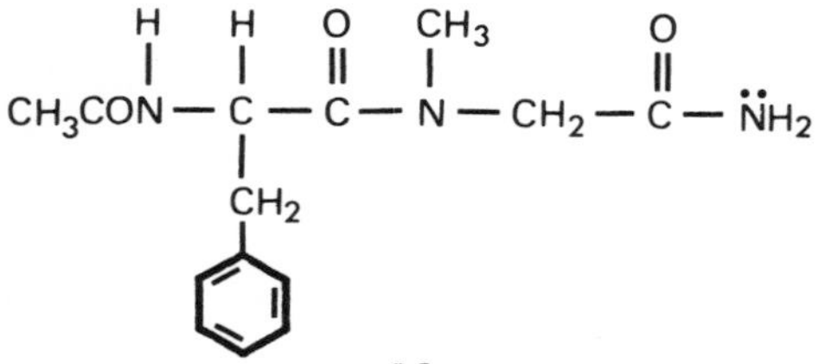

32

Application of the principle of stereoelectronic control predicts that the first tetrahedral intermediate must have conformation 21 rather than 23. As a result, the non-reactivity of the above two peptide substrates can be readily explained. In these two substrates, the amide nitrogen carries an extra alkyl substituent and molecular models show that this extra

alkyl substituent allows the formation of an enzyme-substrate complex (cf.
$\underline{20}$, $\begin{matrix} H \\ | \\ -N-R \end{matrix} \begin{matrix} R \\ | \\ = -N-R \end{matrix}$), but it does not permit the formation of a tetrahedral
intermediate in the required conformation $\underline{21}$ $\left(\begin{matrix} H \\ | \\ -N-R \end{matrix} \begin{matrix} R \\ | \\ = -N-R \end{matrix} \right)$. Indeed, in
$\underline{21}$ $\left(\begin{matrix} H \\ | \\ -N-R \end{matrix} \begin{matrix} R \\ | \\ = -N-R \end{matrix} \right)$, the extra N-alkyl substituent comes too close to the
imidazole ring of His-57. Dipeptides $\underline{31}$ and $\underline{32}$ are therefore inactive be-
cause steric hindrance prevents formation of the tetrahedral intermediate.
In normal substrates which have a hydrogen atom instead of an alkyl group
pointing towards His-57, there is no steric hindrance and formation of the
tetrahedral intermediate is possible ($\underline{20} \rightarrow \underline{21}$). Again, it is worth pointing
out that a conformation such as $\underline{22}$ $\left(\begin{matrix} H \\ | \\ -N-R \end{matrix} \begin{matrix} R \\ | \\ = -N-R \end{matrix} \right)$ (or $\underline{23}$), which is not
sterically hindered, cannot be obtained directly from the enzyme-substrate
complex unless the principle of stereoelectronic control is violated. The
experimental results indicate that the violation of this principle is not
allowed.

Petkov, Christova, and Stoineva (11) have reported a study on the hydrolysis
of N-acetyl-L-phenylalanine anilide derivatives with α-chymotrypsin: N-
methylated anilides $\underline{34}$ (R=CH$_3$) were found to be unreactive under the conditions used for the hydrolysis of N—H anilides $\underline{34}$ (R=H). These authors have
explained their results in a manner analogous to that described above, i.e.
no hydrolysis takes place because steric hindrance caused by the N-methyl
group prevents the formation of a tetrahedral intermediate in the N-methyl
anilide derivatives.

$$\begin{matrix} & H & H & O & R \\ & | & | & \| & | \\ CH_3CON & - & C & - & C & - & N & -\bigcirc & X \\ & & | & & & & & \\ & & CH_2 & & & & & \\ & & | & & & & & \\ & & \bigcirc & & & & & \end{matrix}$$

$\underline{34}$

Petkov and Stoineva (12) have more recently reported that the relative rate
of the α-chymotrypsin-catalyzed hydrolysis of p-alkoxycarbonyl anilide derivatives $\underline{35}$ of N-acetyl-L-phenylalanine is enhanced with an increase in
the size of the R alkyl group of the leaving group. This rate enhancement
specificity appears to be entropy controlled: the bulky alkyl groups increase both enthalpy and entropy of activation. These kinetic and thermodynamic data were interpreted in the following way: the bulky p-alkoxycarbonyl

aniline leaving group of the anilides does not fit sterically in the leaving
-group-binding pocket of the enzyme. As a result of this lack of leaving-
group-binding, rotation around the C—N bond is rendered easier in the transition state for the formation of the enzyme tetrahedral intermediate. It
was concluded that this study provides further experimental evidence in
favor of a conformational change at the nitrogen atom of the tetrahedral
intermediate prior to cleavage as predicted by the principle of stereoelectronic control.

$$CH_3CON - \overset{\overset{\displaystyle H}{|}}{\underset{\underset{\displaystyle CH_2}{|}}{C}} - \overset{\overset{\displaystyle O}{||}}{C} - \overset{\overset{\displaystyle H}{|}}{N} - \bigcirc - \overset{\overset{\displaystyle O}{||}}{C} - OR$$

35

The anilide substrates L-Ala-L-Ala-pNA (**36**) and L-Ala-L-Pro-pNA (**37**) are
hydrolyzed _via_ a different rate-determining step by the serine protease,
dipeptidyl-peptidase IV (12, 13). For **36**, the rate-determining-step is the
formation of the acyl-enzyme and for **37**, it is the cleavage of this intermediate. This different behavior has been rationalized on the basis of arguments derived from the principle of stereoelectronic control in the hydrolytic process.

$$NH_2 - \overset{\overset{\displaystyle H}{|}}{\underset{\underset{\displaystyle CH_3}{|}}{C}} - \overset{\overset{\displaystyle O}{||}}{C} - NH - \overset{\overset{\displaystyle H}{|}}{\underset{\underset{\displaystyle CH_3}{|}}{C}} - \overset{\overset{\displaystyle O}{||}}{C} - \overset{\overset{\displaystyle H}{|}}{N} - \bigcirc - R$$

36

$$NH_2 - \overset{\overset{\displaystyle H}{|}}{\underset{\underset{\displaystyle CH_3}{|}}{C}} - \overset{\overset{\displaystyle O}{||}}{C} - N - C - \overset{\overset{\displaystyle O}{||}}{C} - \overset{\overset{\displaystyle H}{|}}{N} - \bigcirc - R$$

37

Carboxypeptidase A is a metalloenzyme (containing Zn^{+2}) which hydrolyzes the C-terminal peptide bond in polypeptide chains (1-4). The hydrolysis occurs most readily when the terminal amino acid residue has an aromatic (or a large aliphatic) R group (cf. **38** → **39+40**).

The primary binding sites responsible for the catalytic activity of carboxypeptidase A are the following: (1) the metal ion (Zn^{+2}) coordinates the carbonyl oxygen of the amide function of the terminal amino acid. As a result, the carbonyl group reactivity is enhanced toward nucleophilic attack; (2) a hydrophobic pocket binds the aromatic side chain and the carboxylate ion of the terminal amino acid forms a salt linkage with the guanidinium cation of Arg-145. The hydrolytic process is illustrated in Fig. 3. The carboxylate ion of Glu-270 serves as a nucleophile which attacks the zinc-coordinated carbonyl group to form a tetrahedral intermediate (**41** → **42**). This intermediate then breaks down with concomitant proton transfer of the phenolic proton of Tyr-248 to the nitrogen atom yielding the acyl-enzyme, i.e. the mixed anhydride, and the terminal aromatic amino acid which is probably sterically prevented from leaving the active site until deacylation has occurred (23) (cf. **43**). The acyl-enzyme is then hydrolyzed by reaction with water (via the formation of another tetrahedral intermediate) regenerating carboxypeptidase A (**44**) and yielding the hydrolyzed polypeptide **45** (24-28) and liberating the terminal aromatic amino acid **46**.

Specific ester substrates are also hydrolyzed with carboxypeptidase A. For instance, Makinen, Fukuyama, and Kuo (27) have recently studied the enzymic hydrolysis of O-(trans-p-chlorocinnamoyl)-L-β-phenyllactate (CICPL) (**47**),and the spin labeled nitroxide ester substrate O-3-(2,2,5,5-tetramethylpyrrolinyl-1-oxyl)-propen-2-oyl-L-β-phenyllactate (TEPOPL) (**48**). They have shown that these reactions take place via the formation of a covalent intermediate (the mixed anhydride) which can be isolated under subzero temperature conditions. The hydrolysis of CICPL and TEPOPL catalyzed by carboxypeptidase A is consequently governed by the rate-limiting breaking of the acyl-enzyme.

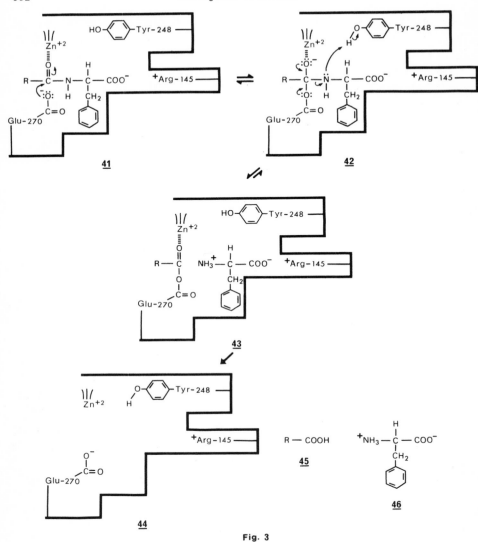

Fig. 3

The same authors have also pointed out that the experimental results of others suggest that, with peptide substrates, the kinetics are different as the formation of the mixed anhydride intermediate is rate determining.

Kuo, Fukuyama, and Makinen (28) have further observed using molecular graphics techniques that the binding of the spin label TEPOPL (**48**) in the active site of carboxypeptidase A takes place with geometric strain and torsional distortion of the substrate. This is so because in the extended trans configuration, the bulky substituents of the pyrrolinyl ring of the spin label **48** are hindered by residues Ser-197, Tyr-198, Ser-199, and Ph-279. These amino acid residues have been defined as the sites of secondary substrate recognition. There is evidence of similar steric hindrance with oligopeptide substrates. The process of bond cleavage is thus dependent both upon interaction with the hydrophobic pocket and Arg-145 which fixes the substrate in the active site, and upon strain and distortion of the substrate by the site of secondary substrate recognition.

Interestingly, according to these authors, torsional distortion through interaction of the substrate with the sites of secondary substrate recognition is a requirement of the stereoelectronic effects which govern the cleavage of tetrahedral intermediates in the hydrolytic process. Inspection of molecular models shows that only one tetrahedral conformer is compatible with the structural constraints imposed by the active site. Indeed, substrate binding leads to the stereoelectronically allowed conformer **50** (Fig. 4) because of (**a**) coordination of the carbonyl group oxygen to the metal ion, (**b**) the binding (with Arg-145) of the COOH-terminus residue in the hydrophobic pocket and (**c**) the direction of approach of the nucleophilic γ-carboxylate group of Glu-270. (For the carboxylate ion preferred direction of approach towards the electrophile, see ref. 29).

Conformer **50** has the proper electron pair orientation to break down to give either the free ester substrate and the enzyme (cf. **49**) or the mixed anhydride intermediate **52**. The former process is however highly favored over the latter because (**a**) a carboxylate group is a much better leaving group than an alkoxy group and (**b**) the electron pair of an alkoxy group ($R\ddot{O}-$) is a better electron donor than that of an acyloxy group $\left(\begin{smallmatrix} & O & \\ & \| & \\ R-&C-&\ddot{O}- \end{smallmatrix}\right)$. On the other hand, breakdown of conformer **51** obtained by rotation around the C−O alkyl bond in **50** can only give the mixed anhydride intermediate. Indeed, breakdown of conformer **51** to the free ester is not permitted electronically.

Thus, the collapse of the tetrahedral adduct to the mixed anhydride intermediate is determined purely by rotation around a single C − O bond of the
tetrahedral species formed after substrate binding. The above authors have
suggested that the torsional distortion of the substrate by the sites of
secondary recognition provides the mechanical driving force that causes the
required bond rotation to convert **50** into **51**. This interpretation is supported by inspection of molecular models.

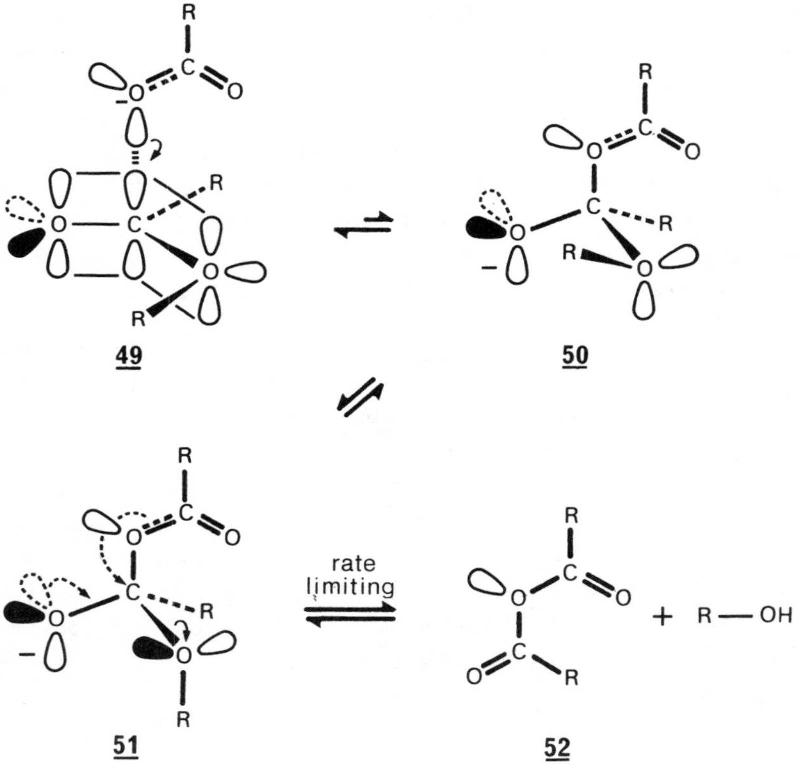

Fig. 4

The pathway by which the tetrahedral adduct formed in peptide hydrolysis is
cleaved can be predicted by the same principle. Nucleophilic attack by the
γ-carboxylate of Glu-270 on an oligopeptide substrate (**53**) in the active
site of carboxypeptidase A must give the conformer **54** (Fig. 5). Conformer **54**
has the proper electron pair orientation to break down to give either the
peptide substrate and the enzyme (cf. **53**) or the mixed anhydride intermediate **56**. The former process is however highly favored over the latter because
(**a**) a carboxylate group is a much better leaving group than an amino group

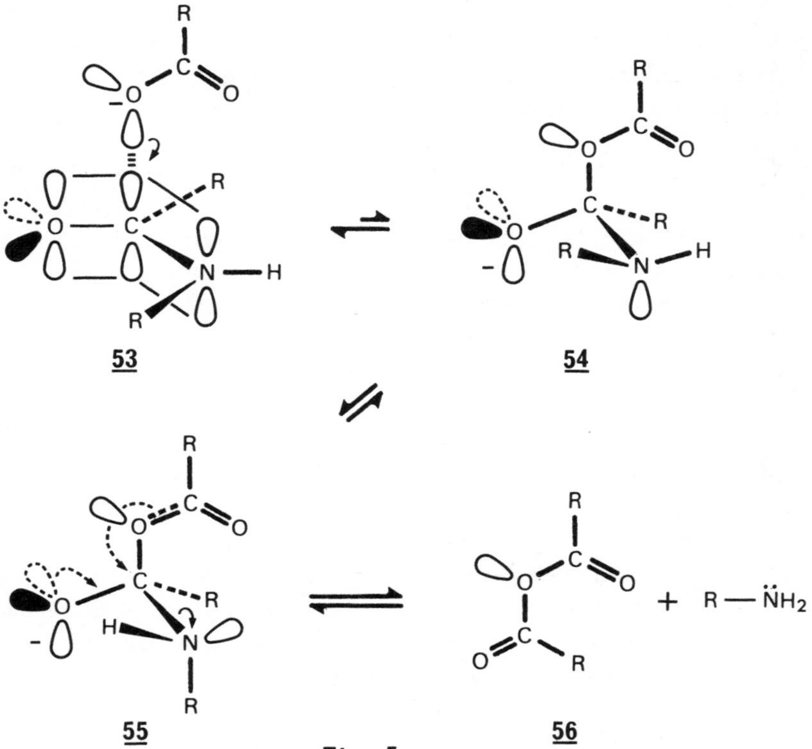

Fig. 5

and (b) the electron pair of an amino group is a better electron donor than that of an acyloxy group. As a result, the adduct **54** can be cleaved only non productively to reactants (→**53**). On the other hand, rotation around the C − N bond provides conformer **55** which can give the mixed anhydride **56** while the reversal to **53** is stereoelectronically forbidden. This rotation process is again favored by the secondary steric interactions. Thus, conformer **55** permits a productive cleavage of the C − N bond (→**56**) which takes place after or with concomitant proton transfer from Tyr-248 to the nitrogen atom (cf. **42** → **43**).

It has been mentioned that the formation and the hydrolysis of the acyl-enzyme are respectively the rate-determining steps for oligopeptide and ester substrates. This different behavior can be readily explained by taking into account the different reactivity of the amide and the ester function. The amide function is electronically more stable than the ester function due to greater π-bonding. The amide carbonyl group is a relatively poor electrophile which is less vulnerable to nucleophilic attack than the ester carbonyl group. Consequently, the formation of a tetrahedral intermediate may require in the case of the amide an energy barrier sufficiently high to change the rate-limiting step of the process.

It can therefore be concluded that the sites of secondary substrate recognition of the enzyme which impose a strain on the substrate and cause a geometrical distortion of the tetrahedral intermediate, are an obligatory part of the catalytic action of carboxypeptidase A which takes place under stereoelectronically controlled conditions.

Structure-activity relationship

Belleau and collaborators (30-32) have recently pointed out that one aspect of conformation-activity relationships which has escaped attention concerns the importance of stereoelectronic effects. More specifically, they have proposed that stereoelectronic effects about the basic nitrogen of morphinans as opposed to stereoisomerism about chiral carbons play an important role at the analgesic receptor level. They have presented concrete evidence that the relative spatial orientation of the nitrogen lone pair in morphinans is of critical importance for productive interaction with the opiate receptors.

The opiate receptor displays absolute chiral specificity towards morphinans but relatively little towards structurally non-rigid narcotics of the piperidine and methadone classes. This is not surprising since flexible molecules are conformationally adaptable and thus more responsive to the geometric needs of the binding sites. The contrasting absolute chiral specificity of the receptor towards morphinans is very much like that of many enzymes towards their substrates. Interestingly, complete loss of receptor affinity was observed upon contraction of the six-membered piperidine ring found in morphinan (57) to a five-membered ring (58, D-normorphinan). Since it is known (30, 33) that in 58 the N-methyl group projects in a direction opposite to the phenyl ring while in 57, the same methyl group projects toward that ring, it was concluded that D-normorphinan (58) is inactive because the nitrogen lone pair responsible for the binding is not properly oriented. It is on this basis that Belleau and collaborators (30) suggested that the free N-lone pair of morphinans exerts an important stereoelectronic effect at the receptor level.

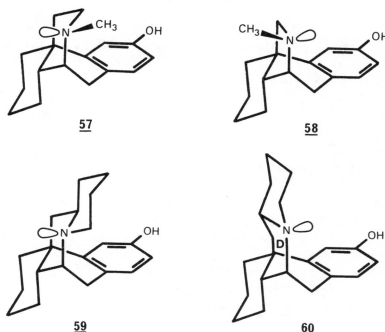

57   58

59   60

Further strong evidence that the lone pair orientation effect is indeed a major parameter controlling productive binding comes from the results obtained with the isomeric 16-α- and 16-β-butanomorphinans (31). The isomer 16-α-butanomorphinan is known to exist in conformation **60** where the piperidine ring D is locked in the boat conformation (34). Therefore, the nitrogen electron pair in **60** is rigidly oriented towards the phenyl ring (as in **58**) and inability of this morphinan to bind on the opiate receptor should be achieved. Indeed, **60** is inactive. Contrary to this result, the 16-β-isomer displayed analgesic activity and it exists in conformation **59** where the nitrogen lone pair is in the opposite orientation (as in **57**). Thus, the nitrogen lone pair orientation effect first noted with D-normorphinan **58** is of equal importance in genuine morphinans.

It was also suggested that the nitrogen lone pair is involved in a proton transfer when it interacts with the receptor. Thus, proton transfer would be required for a strong analgesics. This is in agreement with the effect of quaternization of the tertiary nitrogen (32) and deuterium isotope effect studies (31) which further support the conclusion that the N-lone pair of rigid morphinans should be properly oriented for facile proton transfer and induction of analgesia.

REFERENCES

(1) Walsh, C. "Enzymatic Reaction Mechanisms"; W.H. Freeman and Co.: San Francisco, 1979.

(2) Fersht, A. "Enzyme Structure and Mechanism"; W.H. Freeman and Co.: San Francisco, 1977.

(3) Stryer, L. "Biochemistry"; W.H. Freeman and Co.: San Francisco, 1975.

(4) Dugas, H.; Penney, C. "Bioorganic Chemistry"; Springer-Verlag: New York, N.Y., 1981.

(5) Fisher, H.F.; Ofner, P.; Conn, E.E.; Vennesland, B.; Westheimer, F.H. J. Biol. Chem. 1953, 202, 687.

(6) Simon, H.; Kraus, A. "Isotopes in Organic Chemistry"; Buncell, E.; Lee, C.C., Eds; Elsevier: Amsterdam, 1976; pp. 153-229.

(7) You, K.; Arnold, Jr., L.J.; Allison, W.S.; Kaplan, N.O. Trends in Biochem. Sciences 1978, 3, 265.

(8) Benner, S.A. Experientia 1982, 38, 633.

(9) Bizzozero, S.A.; Zweifel, B.O. Febs Lett. 1975, 59, 105.

(10) Bizzozero, S.A.; Dutler, H. Bioorg. Chem. 1981, 10, 46.

(11) Petkov, D.; Christova, E.; Stoineva, I. Biochem. Biophys. Acta 1978, 527, 131.

(12) Petkov, D.; Stoineva, I. Bioorg. Chem. 1980, 9, 318.

(13) Fischer, G.; Küllertz, G.; Barth, A. Tetrahedron 1978, 34, 2123.

(14) Barth, A.; Fischer, G.; Neubert, K.; Heins, J.; Mager, H. Wiss. Z. TH Leuna-Merseburg 1980, 22, 352.

(15) Blow, D.M. Acc. Chem. Res. 1976, 9, 145.

(16) Huber, R.; Bode, W. Acc. Chem. Res. 1978, 11, 114.

(17) Bode, W.; Schwager, P. J. Mol. Biol. 1975, 98, 693.

(18) Rühlmann, A.; Kukla, D.; Schwager, P.; Bartels, K.; Huber, R. J. Mol. Biol. 1973, 77, 417.

(19) Huber, R.; Kukla, D.; Bode, W.; Schwager, P.; Bartels, K.; Dreisenhofer, J.; Steigemann, W. J. Mol. Biol. 1974, 89, 73.

(20) Birktoft, J.J.; Blow, D.M. J. Mol. Biol. 1972, 68, 187.

(21) Bürgi, H.B.; Dunitz, J.D.; Shefter, E. J. Am. Chem. Soc. 1973, 95, 5065.

(22) Satterthwait, A.C.; Jencks, W.P. J. Am. Chem. Soc. 1974, 96, 7018.

(23) Hartsuck, J.A.; Lipscomb, W.N. "The Enzymes"; Vol. III; Boyer, P.D., Ed.; 3rd Ed.; Academic Press: New York, N.Y., 1971; pp. 1-46.

(24) Makinen, M.W.; Kuo, L.C.; Dymowski, J.J.; Jaffer, S. J. Biol. Chem. 1979, 254, 356.

(25) Makinen, M.W.; Yamamura, K.; Kaiser, E.T. Proc. Natl. Acad. Sci. U.S.A.
     1976, 73, 3882.

(26) Kuo, L.C.; Makinen, M.W. J. Biol. Chem. 1982, 257, 24.

(27) Makinen, M.W.; Fukuyama, J.M.; Kuo, L.C. J. Am. Chem. Soc. 1982, 104,
     2667.

(28) Kuo, L.C.; Fukuyama, J.M.; Makinen, M.W. J. Mol. Biol. In Press.

(29) Gandour, R.D. Bioorg. Chem. 1981, 10, 169.

(30) Belleau, B.; Conway, T.; Ahmed, F.R.; Hardy, A.D. J. Med. Chem. 1974,
     17, 907.

(31) DiMaio, J.; Ahmed, F.R.; Schiller, P.; Belleau, B. "Recent Advances
     in Receptor Chemistry"; Gualtieri, F.; Giannella, M.; Melchiorre, C.,
     Eds.; Elsevier: North Holland Publishers, Amsterdam, 1979, p. 221.

(32) Belleau, B. "Stereoelectronic Regulation of the Opiate Receptor Some
     Conceptual Problems, in "Chemical Regulation of Biological Mechanism","
     Cambridge, 1982, Burlington House, London W1V OBN.

(33) Hardy, A.D.; Ahmed, F.R. Acta Crystallogr., Sect. B 1975, 31, 2919.

(34) Ahmed, F.R. Acta Crystallogr., Sect. B 1981, 37, 188.

AUTHOR INDEX

SUBJECT INDEX